Forum for Interdisciplinary Mathematics

Volume 1

W0225843

The Forum for Interdisciplinary Mathematics (FIM) series publishes high-quality monographs and lecture notes in mathematics and interdisciplinary areas where mathematics has a fundamental role, such as statistics, operations research, computer science, financial mathematics, industrial mathematics, and bio-mathematics. It reflects the increasing demand of researchers working at the interface between mathematics and other scientific disciplines.

More information about this series at http://www.springer.com/series/13386

Michel-Marie Deza · Mathieu Dutour Sikirić
Mikhail Ivanovitch Shtogrin

Geometric Structure of Chemistry-Relevant Graphs

Zigzags and Central Circuits

 Springer

Michel-Marie Deza
École Normale Supérieure, Paris
and University of Campinas
Campinas
Brazil

Mikhail Ivanovitch Shtogrin
Russian Academy of Sciences
Steklov Mathematical Institute
Moscow
Russia

Mathieu Dutour Sikirić
Institute Rudjer Boúsković
Zagreb, Hrvatska
Croatia

ISSN 2364-6748   ISSN 2364-6756   (electronic)
Forum for Interdisciplinary Mathematics
ISBN 978-81-322-3419-7   ISBN 978-81-322-2449-5   (eBook)
DOI 10.1007/978-81-322-2449-5

Springer New Delhi Heidelberg New York Dordrecht London
© Springer India 2015
Softcover reprint of the hardcover 1st edition 2015

Printed on acid-free paper

Springer (India) Pvt. Ltd. is part of Springer Science+Business Media (www.springer.com)

Preface

The central actors in the present book are the *zigzags* and *central circuits* of three- or four-regular plane graphs, which allow to obtain a double covering or covering of the edgeset. This book, a companion of the authors' book [DeDu08], mainly focuses on specific classes of bifaced plane graphs, that is, those without faces of negative curvature. It contains, as a particular case, the *fullerenes*, that is, three-regular plane graphs with faces of size five or six, which are prominent chemistry-relevant graphs. The class also contains the *octahedrites*, that is, four-regular plane graphs with faces of size three or four. We also consider three classes of graphs, which are self-dual. For all those graphs, we consider how to enumerate them, their possible symmetry groups, their connectivity, and other structural properties. We also study the *icosahedrites*, that is, five-regular plane graphs with faces of size three or four; these have faces of negative curvature and so, their number grows exponentially. Finally, we consider *disk-fullerenes*, that is, three-regular partitions of a disk by five- and six-gons.

For all these classes of graphs, we treat the notion of zigzags and central-circuits, sometimes, at the same time. We consider simplicity of circuits, possible configuration, tightness, and enumeration of the tight graphs with simple circuits. We also address extremal questions, such as the maximum number of circuits of tight graphs.

For the classes of graphs with maximal symmetry, such as the fullerenes of icosahedral symmetry, a special construction called *Goldberg-Coxeter construction* allows to describe them explicitly in terms of two integer parameters k and l. This construction is studied systematically for three-, four-, and six-valent graphs and allow us to describe many classes in a simple way. We study the zigzags and central-circuits of the obtained graphs and build a new (k, l)-*product* algebraic formalism that allows us to describe the zigzags and central-circuits of the obtained graphs explicitly.

For classes of graphs with non-maximal symmetry, more complex description is needed. We explain how this can be done in practice by presenting the formalism of hyperbolic complex geometry derived by William Thurston in [Thur98].

For dimensions higher than two, the possible similar structures are more complicated. In that case, we limit ourselves to zigzags and compute them for several infinite families of complexes and the regular, semiregular, and regular-faced polytopes.

Michel-Marie Deza
Mathieu Dutour Sikirić
Mikhail Ivanovitch Shtogrin

References

[DeDu08] Deza, M., Dutour Sikirić, M.: Geometry of Chemical Graphs: polycycles and two faced maps, Encyclopedia of Mathematics and its Applications, vol. 119. Cambridge University Press, Cambridge (2008)

[Thur98] Thurston, W.P.: Shapes of polyhedra and triangulations of the sphere. In: Rivin, J., Rourke, C., Series, C. (eds.) Geometry and Topology Monographs 1, The Epstein Birthday Schrift, pp. 511–549. Geometry and Topology Publishing, Coventry (1998)

Contents

About the Authors

Michel-Marie Deza is former research director at the French National Research Centre and École Normale Supérieure, Paris. Later, he was research professor at the Japan Advanced Institute of Science and Technology. He is author of about 300 research publications on discrete geometry, combinatorics, and their applications to chemistry and crystallography. He is also the author of seven books on mathematics, including popular *Geometry of Cuts and Metrics* and *Encyclopedia of Distances*, both published by Springer.

Mathieu Dutour Sikirić a senior associate researcher of Institute Rudjer Boúsković in Zagreb, Croatia. He is the author of 70 articles and two books in the fields of science and engineering: mathematics, computer science, physics, chemistry, crystallography, oceanography, and meteorology. His mathematical work concerns graph theory, polytope theory, tiling and their applications, geometry of numbers, and computational topology.

Mikhail Ivanovitch Shtogrin is a leading research associate of Steklov Institute of Mathematics in the Russian Academy of Sciences, Moscow, Russia. He is the author of over 120 research publications and two books on discrete geometry and geometric crystallography.

Chapter 1
Introduction: Main ZC-Notions

In this chapter we summarize the main notions considered in this book and, briefly, the results that we obtain. Specifically, we define the pure graph theoretic and plane graph theoretic notions needed for this work with emphasis on symmetries. Then we define the main object of this book, i.e., zigzags and central circuits for 3- and 4-valent graphs and the corresponding notions of intersections. We explain how perfect matchings can be obtained from zigzags and knots from central circuits.

The notion of curvature of faces is also considered and the 10 classes of plane bifaced graphs without faces of negative curvature, that we consider, are described. We summarize in the tables the main obtained results on growth rates, symmetry groups, connectivity, zigzags/central circuits, etc. Finally, we give the smallest examples of such graphs occurring on projective plane, torus, and Klein bottle.

1.1 Graphs

A *graph* $G = (V, E)$ consists of a vertex-set $V = V(G)$ and an edge-set $E = E(G)$, such that either one or two vertices are assigned to each edge as its ends. A graph is said to be *simple* as it has no loops or multiple edges.

A *k-connected graph* is the one that remains connected after removing any set of $k - 1$ vertices. So, a plane graph is not 3-connected if it has a 2-gon (doubled edge) and not 2-connected if it has a 1-gon (loop). We mainly deal with 3- or 2-connected plane graphs, whose vertex- and edge-sets are finite.

A *plane graph* is a particular embedding (i.e., drawing) of a graph in the Euclidean plane $\mathbb{E}^2$ using smooth curves that cross each other only at the vertices of the graph. A graph, which has at least one such drawing, is called *planar*. In general, given a surface $\mathbb{F}^2$, an $\mathbb{F}^2$-*graph* is a graph embedded in $\mathbb{F}^2$.

The main structural vectors of an $\mathbb{F}^2$-graph G are the *v-vector* $\mathbf{v}(G) = (\ldots, v_i, \ldots)$ and *P-vector* $\mathbf{p}(G) = (\ldots, p_i, \ldots)$, which enumerates the numbers v_i of vertices of degree i and, respectively, the numbers p_i of faces of *gonality* i, i.e., having i sides. The problem of existence of plane graphs with a fixed p-vector is an active subject of research; see, [YHZQ10]. A *k-regular graph* is the one with $v_i = 0$ for $i \neq k$.

© Springer India 2015
M. Deza et al., *Geometric Structure of Chemistry-Relevant Graphs*,
Forum for Interdisciplinary Mathematics 1, DOI 10.1007/978-81-322-2449-5_1

For a *connected* (i.e., 1-connected) $\mathbb{F}^2$-graph G, its *dual graph* is the graph G^* on the set of faces of G with two faces being adjacent if they share an edge. Clearly, $\mathbf{v}(G^*) = \mathbf{p}(G)$ and $\mathbf{p}(G^*) = \mathbf{v}(G)$. A graph G is *self-dual* if $G = G^*$.

A *truncation* (or $\frac{1}{3}$-*truncation*) $Tr(G)$ of a plane graph G is an operation replacing, for any i with $v_i > 0$, its i-valent vertices by "small" i-gons. The *leapfrog graph* of G is $Tr(G^*)$. The *medial* (or $\frac{1}{2}$-*truncation*) will be defined in Sect. 1.3.

A *bipartite graph* is the one with $V = V_1 \cup V_2$, $V_1 \cap V_2 = \emptyset$, such that every edge connects a vertex in V_1 to one in V_2. A plane graph is bipartite if and only if its faces have even gonality, i.e., its dual is a *Eulerian graph*. A connected graph is *Eulerian* if its vertices have even degrees or, equivalently, it has a *Eulerian circuit*, i.e., a circuit that passes through every vertex and uses every edge exactly once.

The *skeleton* of a polytope P is the graph $G(P)$ formed by its vertices, with two vertices being adjacent if they generate a face. The famous *Steinitz Theorem* [Ste16] is that a graph is the skeleton of a *polyhedron* (3-dimensional polytope) if and only if it is planar and 3-connected. A polyhedron is usually represented by the *Schlegel diagram* (projection from $\mathbb{R}^3$ into $\mathbb{R}^2$ through a point beyond one of its facets) of its skeleton; the program used is *CaGe*. Such drawing is unique up to diffeomorphisms $\mathbb{R}^2$, but there exist many distinct polyhedra having the same graph as their skeleton; all such polyhedra are of the same *combinatorial type*. A *simple polyhedron* is one with 3-regular skeleton.

1.2 Symmetries

A *point group* is a finite subgroup of the group $O(3)$ of isometries of $\mathbb{R}^3$, fixing the origin (0). The connection with graphs come from representing them on the sphere centered at (0). The point group of isometries of a polyhedron is a subgroup of the algebraic *symmetry group* $Aut(G)$ of its skeleton G, consisting of automorphisms of G. The *rotation group* $Rot(G)$ consists of all rotations preserving G; it is a subgroup of index 1 or 2 of $Aut(G)$. We use Schoenflies nomenclature for point groups.

By the *Mani's Theorem* ([Ma71], a refinement of Steinitz's theorem) valid for 3-connected plane graphs, there is, for each combinatorial type, at least one polyhedron, for which this inclusion becomes equality, i.e., $Aut(G)$ can be realized as the point group of a convex polyhedron with the skeleton G. So, one can identify the polyhedron and its graph, as well as $Aut(G)$ and the point group; the maximal symmetry groups of plane graphs are identified with the corresponding point groups.

Theorem 1.1 *For any plane graph G, the group $Aut(G)$ of symmetries preserving the set of vertices, edges, and faces and their incidence can be identified with a group of isometries of 3-space.*

Proof For a plane graph G, we define $order(G)$ to be the planar graph of the *Hasse diagram* of G, i.e., the vertex-set of $order(G)$ is formed by the vertices, edges, and faces of G. Two vertices of $order(G)$ are adjacent if their corresponding vertex, edge, or face in G are incident. Clearly, $order(G)$ is a plane graph as well. The faces of $order(G)$ correspond to the triples (v, e, f) with v, e and f being vertex, edge, and face of G satisfying $v \in e \subset f$.

If G is a general plane graph, i.e., possibly, with loops, vertices of degree 1 and disconnecting vertices, then $order(G)$ is 2-connected. If G is 2-connected, then $order(G)$ is 3-connected. Any symmetry of G preserving incidence between vertices, edges, and faces induces a symmetry of $order(G)$ as well. So, if G a plane graph, then $G_2 = order(order(G))$ is a 3-connected plane graph. So, Mani's Theorem implies that the symmetry group of G_2 is isomorphic to a group of symmetries of 3-space. Since $Aut(G)$ is a subgroup of $Aut(G_2)$, it can also be realized in 3-space. □

In the presence of 2-gons, i.e., multiple edges, one cannot speak of convex polyhedra and so, Mani's Theorem has no equivalent in that setting.

The list of point groups is split into two classes: the infinite families and the sporadic cases. Every point group has a normal subgroup formed by the rotations.

The group, denoted C_m, is the cyclic group of rotations by angle $\frac{2\pi}{m}k$ with $0 \le k \le m - 1$ around a fixed axis Δ. Both groups C_{mv} and C_{mh} contain C_m as normal subgroups of rotations. The group C_{mh} (respectively, C_{mv}) is the group, generated by C_m and a symmetry of plane P, with P being orthogonal to Δ (respectively, containing Δ). The group D_m is the group, generated by C_m and a rotation by angle π, whose axis is orthogonal to Δ. The point group D_{mh} (respectively, D_{md}) is generated by C_{mv} and a rotation by angle π, whose axis is orthogonal to Δ and contained in a plane of symmetry (respectively, going between two planes of symmetry). Both D_{mh} and D_{md} contain D_m as a normal subgroup. If N is even, one defines the point group S_N as the cyclic group generated by the composition of a rotation by angle $\frac{2\pi}{N}$ with axis Δ and a symmetry of plane P with P being orthogonal to Δ. The particular cases $C_1, C_s = C_{1v} = C_{1h}, C_i = S_2$ correspond to the trivial group, plane symmetry group, and central symmetry inversion group.

The point groups T_d, O_h, I_h are the symmetry groups of Tetrahedron, Cube, Icosahedron; the point groups T, O, I are their normal subgroups of rotations. The point group T_h is formed by all f and $-f$ with $f \in T$.

More detailed descriptions of point groups are available, for example, from [FoMa95, Dut04]. Since those groups have been classified long ago and are much used in Chemistry, we will use the chemical nomenclature for them.

On the pictures of this book, in order to express better the (maximal) symmetry of an i-hedrite, we put a double arrow, in order to represent an edge passing at infinity, and a quadruple arrow, in order to represent a vertex at infinity.

1.3 Zigzag and Central Circuits

Let G be a 3-regular plane graph. Then all edges, incident to a vertex x, can be labeled in counterclockwise order as $e_1, e_2, \ldots, e_k$, where k is the degree of x in G. For any edge e_i, $1 \le i \le k$, the edges e_{i+1}, and e_{i-1} (with $i + 1$ and $i - 1$ being

Fig. 1.1 Two types of zigzag
intersections in an edge

Type I   Type II

addition modulo k) are called, respectively, the *left* and the *right*. A circuit of edges
of G is called a *zigzag* (or a *Petrie path* [Cox73], *geodesic* [GrünMo63], *left-right
path* [Sh75]), if, in tracing the circuit, we alternately select, as the next edge, the left
neighbour and the right neighbor; see Fig. 1.1. In a 3-regular plane graph, any pair
of edges sharing a vertex define a zigzag.

Given an edge, there are two possible directions for extending it to a zigzag.
Hence, each edge is covered exactly twice by zigzags.

Let Z and Z' be (possibly, $Z = Z'$) zigzags of a plane graph G and let an
orientation be selected on them. An edge e of intersection $Z \cap Z'$ is called of *type I* or
type II, if Z and Z' traverse e in opposite or same direction, respectively; see Fig. 1.1.
The types of edges depend on orientation chosen on zigzags except if $Z = Z'$. So,
the *edge of self-intersection* of a zigzag Z is called of *type I* or *type II* [GrünMo63],
if Z traverses it twice in the opposite or in the same direction, respectively. Edges
of type I and type II correspond to edges of *cocycle* and *cycle character* in [Sh75].
The *signature of a zigzag* is the pair (α_1, α_2), where α_1 and α_2 are the numbers of
its edges of type I and type II, respectively.

A zigzag on a non-orientable surface can have either even or odd number of edges.
However, on an orientable surface (so, including a sphere or a disk), the length of
each zigzag is even, because we consecutively take left and right turns. The number
of intersections of two simple zigzags in a graph on a surface of genus $g \geq 1$ can be
odd; see (in Table 1.1 and in Fig. 1.7) Klein and Dyck maps of the hyperbolic plane
$\mathbb{H}^2$. In our Euclidean plane $\mathbb{E}^2$ case, the number of intersections can be only even.

While for dual graphs v- and p-vectors are interchanged, the z-vector **z** remains
the same, except that type I and type II in the signature are interchanged. Other kinds
of dualities—interchanging z- and p-vector **p** (or z- and v-vector), but preserving
v-vector **v** (or, respectively, p-vector) are considered in Sect. 8.4.

Let us give an example of application of zigzags. Given a plane graph G with
vertices, $V_1, \ldots V_v$, to every zigzag Z of G corresponds [Sh75] an element $x(Z) \in
\{0, 1\}^v$ which is a basic element of the *kernel* (equation $Lx = 0$ over $\{0, 1\}$) of the
Laplacian $L(G) = D(G) - A(G)$ of G. Here $D(G)$ is a diagonal matrix with d_{ii}
being the degree of vertex V_i and $A(G)$ is the adjacency matrix $A(G)$ of G. So, the
number of zigzags $Z(G)$ is equal to corank of $L(G)$ over $\mathbb{Z}^2$ and corank of any minor
of $L(G)$ is $Z(G) - 1$. Using this and the *Kirchhoff's theorem* (number of spanning
trees of G is equal to the determinant of a minor of $L(G)$), it was proved in [GoRo01]
that the number of spanning trees of G is odd if and only if G is z-knotted, i.e., has
unique zigzag.

Table 1.1   Zigzag structure of Platonic and semiregular polyhedra; also, of Klein and Dyck maps

Nr. edges	Polyhedron	z-vector $\mathbf{z}$	Int. vector $\mathbf{Int}$
6	*Tetrahedron* {3; 3}	4^3	2^2
12	*Cube* {4; 3}	6^4	2^3
12	*Octahedron* {3; 4}	6^4	2^3
30	*Dodecahedron* {5; 3}	10^6	2^5
30	*Icosahedron* {3; 5}	10^6	2^5
24	*Cuboctahedron*	8^6	$2^4, 0$
60	*Icosidodecahedron*	10^{12}	$2^5, 0^6$
48	*Rhombicuboctahedron*	12^8	$2^6, 0$
120	*Rhombicosidodecahedron*	20^{12}	$2^{10}, 0$
72	*Truncated Cuboctahedron*	18^8	$6, 2^6$
180	*Truncated Icosidodecahedron*	30^{12}	$10, 2^{10}$
18	*Truncated Tetrahedron*	12^3	6^2
36	*Truncated Octahedron*	12^6	$4, 2^4$
36	*Truncated Cube*	18^4	6^3
90	*Truncated Icosahedron*	18^{10}	2^9
90	*Truncated Dodecahedron*	30^6	6^5
60	*Snub Cube*	$30^4_{3,0}$	8^3
150	*Snub Dodecahedron*	$50^6_{5,0}$	8^5
3m	*Prism$_m$*, $m \equiv 0$ (mod 4)	$(\frac{3m}{2})^4$	$(\frac{m}{2})^3$
3m	*Prism$_m$*, $m \equiv 2$ (mod 4)	$(3m_{\frac{m}{2},0})^2$	$2m$
3m	*Prism$_m$*, $m \equiv 1, 3$ (mod 4)	$6m_{m,2m}$	
4m	*APrism$_m$*, $m \equiv 0$ (mod 3)	$(2m)^4$	$(\frac{2m}{3})^3$
4m	*APrism$_m$*, $m \equiv 1, 2$ (mod 3)	$2m; 6m_{0,2m}$	
84	*Klein map* {3; 7}	8^{21}	$1^8, 0^{12}$
48	*Dyck map* {3; 8}	6^{16}	$1^6, 0^9$

A *central circuit* of a plane Eulerian graph (i.e., see Sect. 1.1, having only vertices of even degree) is a circuit, which is obtained by starting with an edge and continuing at each vertex by the edge opposite to the entering one. Such circuit is also called *traverse* [GaKe94], *straight ahead* [Harb97], *straight Eulerian* (Chap. 17 of [GoRo01]), *cut-through* [Jeo95], *intersecting*, etc.

Clearly, the edge-set of G is partitioned by all its central circuits. The length of a central circuit is even, since it is twice the number of its points of self-intersection plus the sum of its intersections with other circuits. Two central circuits intersect in an even number of vertices. However, it is harder to define the intersection type of two central circuits. Like in the zigzag case, the intersection type of two central circuits C and C' at a vertex V will depend on the orientation of the central circuits except if $C = C'$. For a 4-regular plane graph, the dual graph is bipartite, therefore we can choose a *chess coloring* of the faces in white or black such that for any two face of a given color, the adjacent faces are of opposite color. If an orientation on C, C', and a chess-like coloring are chosen, then the vertex can be called of *type I* or *type II* according to Fig. 1.2. The *signature of a central circuit* is the pair (α_1, α_2), where α_1 and α_2 are the numbers of its vertices of type I and type II, respectively.

Some examples of applications of plane 4-regular graphs are *projections of links*, *rectilinear embedding* in VLSI and *Gauss crossing problem* for plane graphs (see, for example, [Liu98]). A *link* is one or more circles (components of a link) embedded into the space $\mathbb{R}^3$; a link with one component is called *Knot*. A *projection of a link* is a drawing of it on the plane with gaps representing underpass and solid line representing overpass. Knot Theory is concerned with characterizing different plane presentations of links (see [Lic97] for a pleasant introduction). An *alternating link* is the one admitting a plane representation in which overlappings and underlappings alternate. (Apropos, there is no known topological characterization of such links.)

So, such projection can be seen just as a 4-regular plane graph. We will consider only *minimal projections*, i.e., those without 1-gons (loops). Clearly, each edge belongs to exactly one central circuit and any 4-regular graph without 1-gons can be seen as a minimal projection of an alternating link with components corresponding to its central circuits going alternately under and over at consecutive intersections; see an example in Fig. 1.3.

Proposition 1.1 *(i) For any plane bipartite graph G there exists an orientation of zigzags, with respect to each edge that has type I.*

(ii) For any 6-regular plane graph, there exist an orientation on zigzags and central circuits such that any edge, respectively, vertex of intersection is of type II.

Proof We represent the graph on the sphere. The list of vertices, adjacent to a given vertex, can be organized in counterclockwise order. Let the vertex-set V be partitioned into the two subsets V_1 and V_2 of the bipartition.

(i) Fix a zigzag Z; it will turn left at vertices, say, $v \in V_1$ and right at vertices $v \in V_2$. It is easy to see that the edges of self-intersection of Z can be only of

Fig. 1.2 Two types of central circuits intersection in a vertex

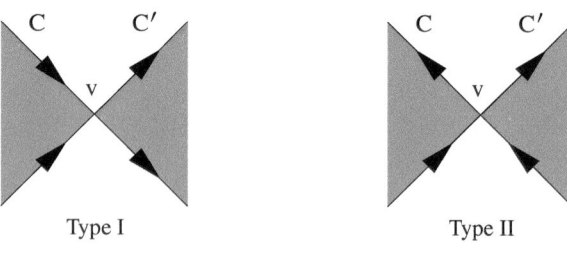

Type I   Type II

Fig. 1.3 The link 6_2^3 (*Borromean rings*) corresponding to Octahedron

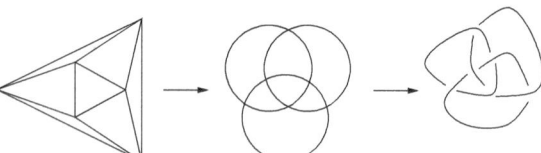

Fig. 1.4 The orientation of the edges of the face

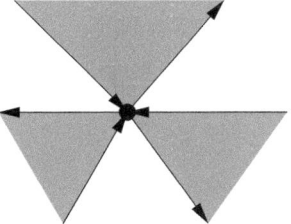

type I. Take another zigzag Z', having a common edge e with Z. We choose an orientation on Z', such that e is an edge of type I. Then Z' will turn left at vertices $v \in V_1$ and right at vertices $v \in V_2$. Iterating this construction, all zigzags will be oriented and all edges will have type I with respect to this orientation of zigzags; (i) is proved.

(ii) For a 6-regular plane graph G, the dual graph G^* is bipartite. Let us take one color c of the faces of G and orient the edges of the face of color c in such a way that they turn clockwise around the face (see Fig. 1.4). It is apparent that such orientation carries over to the zigzags and central circuits of G and that with this orientation all the intersection are of type II. $\qquad \square$

In the case of one zigzag, this proposition was already known [Sh75]. It is also valid for any bipartite graph, which is embedded in an oriented surface, in view of the well-known topological fact that any two-dimensional orientable manifold admits coherent orientation of its faces.

If all edges of plane graph are of the same type for some orientation of its zigzags, then this orientation is unique. In fact, the orientation of one zigzag defines orientation of any zigzag, with which it have a common edge, and so, by a connectivity argument, orientation of any other zigzag. In a graph G, a *perfect matching* is a family PM of edges such that any vertex belongs to exactly one edge in PM.

Proposition 1.2 *[DDF04] Let G be a plane graph; for any orientation of all zigzags of G, we have:*

(i) The number of edges of type II, which are incident to any fixed vertex, *is even.*

(ii) The number of edges of type I, which are incident to any fixed face, *is even.*

(iii) If all v vertices of G have odd degree, then the number of edges of type I is at least $\frac{v}{2}$, and in the case of equality, these edges form a perfect matching of G.

Proof (i) For each vertex the number of times that a zigzag enters it should be equal to the number of times that a zigzag leaves it. This forces the number of edges of type II leaving to be equal to the number of edge II entering to be equal. So, the number of edge of type II incident to a vertex is even. See an example in Fig. 1.5.

(ii) Passing to the dual graph, the type of edges are interchanged.

(iii) follows from (i), since in this case any vertex is incident to odd (and so, positive) number of edges of type I. If the number is exactly $\frac{v}{2}$ then this number is 1 and we have a perfect matching. $\qquad \square$

(a)   **(b)**

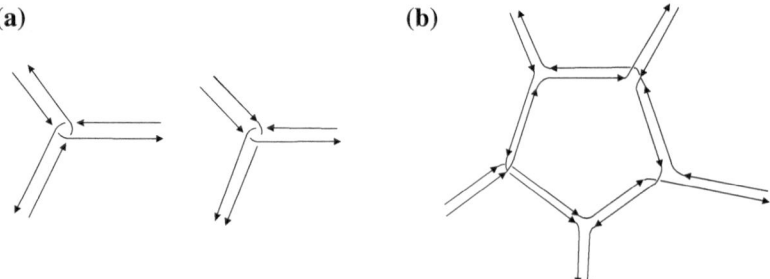

Fig. 1.5 Pictorial proof of (i) and (ii) in Proposition 1.2

In Table 1.1, zigzag notions are illustrated by the skeletons of Platonic and *semi-regular polyhedra*, i.e., such that their symmetry group is transitive on vertices.

Proposition 1.3 *(i) If G is a k-valent, k odd, plane v-vertex graph having only simple zigzags, then v is divisible by* 4.
(ii) If G is a 4-valent, k even, v-vertex plane graph having only simple central circuits, then vt is even.

Proof (i) Any edge of $G = (V, E)$ is the intersection of two zigzags Z and Z'. However, Z and Z' intersect in an even number of edges. Hence, the edge-set E of G is a union of parts of even size and so, has even cardinality itself. By double counting we get easily $kv = 2|E|$, which proves that kv is divisible by 4 and so, v as well.

(ii) Any vertex of G is the intersection of two central circuits C and C'. The circuits C and C' intersect in an even number of vertices. Hence, the vertex-set of G is a union of parts of even size and so, has even cardinality itself. □

A *road* in a 3- or 4-regular plane graph is a nonextensible sequence (possibly, with self-intersections) of either 6-gonal faces or of 4-gonal faces, such that any non-end face is adjacent to its neighbors on opposite edges. If the sequence stops on a nonhexagon or, respectively, a nonquadrilateral face, then it is called a *pseudo-road*; otherwise, it is called a *Railroad* and it is a circuit by finiteness of the graph.

We associate to each railroad a *representing plane curve* in the following way: in each of its faces one connects, by an arc, the midpoints of opposite edges, on which it is adjacent to its two neighbors. The sequence of those arcs can be seen as a closed curve in the plane and self-intersections of railroad correspond to self-intersections of the curve. Those self-intersections can be only double or, when a railroad consists of hexagons, triple. If the railroad has no triple self-intersection points, then this curve can be seen as a projection (not minimal, if there are 1-gons) of an alternating knot (see, for example, [Kaw96, Rol76]) with v crossings, where v is the number of self-intersection points. If there are several railroads without triple intersections and triple self-intersections, then the set of plane curves representing them can be seen as a projection of an alternating link.

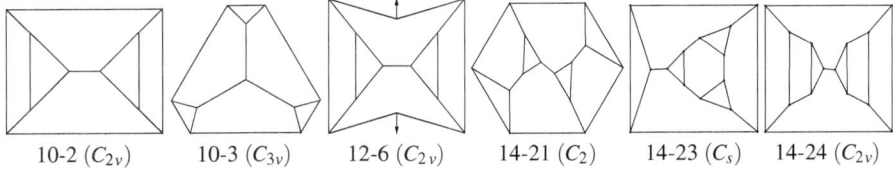

| 10-2 (C_{2v}) | 10-3 (C_{3v}) | 12-6 (C_{2v}) | 14-21 (C_2) | 14-23 (C_s) | 14-24 (C_{2v}) |

Fig. 1.6 Some z-uniform 3-regular graphs with their symmetry group; the numbers of graphs are those given by the program *plantri*

Many results for 3- and 4-regular graphs will be similar; in such case we will use general notion of "either zigzag, or central circuit" and call it *ZC-circuit*. A zigzag is called *simple* if no edge occurs twice and a central circuit is called *simple* if no vertex occurs twice; so, we defined a *simple ZC-circuit*. A simple railroad is called a *belt*; it is bounded by two simple ZC-circuits.

A simple ZC-circuit can be seen as a *Jordan curve*, i.e., a simple and closed plane curve, which is a homeomorphic image of the unit circle. A *(k,t)-arrangements of pseudocircles* (or (k, t)-*AP*) is a set of k Jordan curves where any two intersect (triple or tangent points excluded) exactly in t points; so, there are $t(k\text{-}1)$ points. It is a tight pure cc-uniform 4-regular graph with k central circuits of length $t(k\text{-}1)$, pairwise intersecting in t points, i.e., with (defined below) $\mathbf{cc} = (tk - t)^k$, $\mathbf{Int} = t^{k-1}$. An *Grünbaum arrangements* of plane curves is an $(k, 2)$-AP.

For example, the central circuits of the *medials* (defined in this section) of four truncated polyhedra—Tetrahedron, Cube, Icosahedron, Dodecahedron—form $(3, 6)$-, $(4, 6)$-, $(10, 2)$-, $(6, 6)$-APs; see Table 1.1. The medials of seven z-uniform fullerenes from Fig. 2.8 give $(k, 2)$-APs with $k = 6, 7, 9, 10, 10, 12, 15$. The central circuits of five cc-uniform 8-hedrites from Fig. 4.11 form $(k, 2)$-APs with $k = 3, 4, 4, 5, 6$. See Fig. 1.6 for some z-uniform 3-regular graphs.

The *z-vector* (or *cc-vector*) of a graph G is the vector $\mathbf{z}$ (or $\mathbf{cc}$) enumerating *lengths*, i.e., the numbers of edges, of all its zigzags (or, respectively, central circuits) with their signature as subscript. In general, we will use the term *ZC-vector*. The simple ZC-circuits are put in the beginning, in nondecreasing order of length, without their signature $(0, 0)$, and separated by a semicolon from others. The self-intersecting ones are also ordered by nondecreasing lengths. If there are $m > 1$ ZC-circuits of the same length l and the same signature (α_1, α_2), then we write l^m if $\alpha_1 = \alpha_2 = 0$ and $l^m_{\alpha_1,\alpha_2}$, otherwise. For a ZC-circuit ZC, its *intersection vector* $(\alpha_1, \alpha_2); \ldots, c_k^{m_k}, \ldots$ is such that $\ldots, c_k, \ldots$ is an increasing sequence of sizes of its intersection with all other ZC-circuits, while m_k denote respective multiplicity.

A 3- or 4-regular graph G is called:

- *pure graph* if all its ZC-circuits are simple;
- *tight graph* if it has no railroads;
- *ZC-knotted graph* if it has only one ZC-circuit.

If G has more than one ZC-circuit, it is called:

- *ZC-transitive graph* if $Aut(G)$ acts transitively on ZC-circuits;

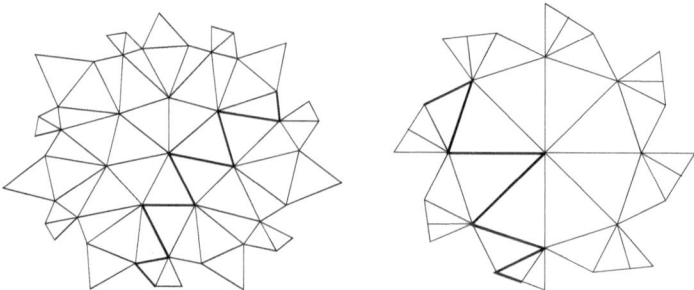

Fig. 1.7 A zigzag of length 8 in Klein map {3; 7} and of length 6 in Dyck map {3; 8}

- *ZC-uniform graph* if all its ZC-circuits have the same length and signature;
- *ZC-balanced graph* if all its ZC-circuits of the same length and signature have identical intersection vectors **Int**.

Clearly, ZC-transitivity implies ZC-uniformity and ZC-balanceness. In ZC-uniform case, the length of each of the r central circuits (respectively, zigzags) is $\frac{2v}{r}$ (respectively, $\frac{3v}{r}$). We do not know an example of a ZC-uniform, but not ZC-balanced, graph.

Zigzags and central circuits, being local notions, are defined for maps on any surface, even on non-orientable one; see, for example, in Fig. 1.7 a zigzag for the *Klein map* {3; 7} and the *Dyck map* {3; 8}, which are dual triangulations for such 3-regular maps. The notion of zigzag (respectively, central circuit) is used here (except 6- and 5-regular case in Chap. 5) in 3-regular (respectively, 4-regular) case, but they can be defined on any plane graph (respectively, Eulerian plane graph). [Harb97] considers central circuits for any drawing on the plane of any Eulerian graph, so that edges are mapped into simple curves with at most one crossing point.

The notion of a zigzag was generalized also to locally-finite infinite plane 3-connected graphs. All such *edge-transitive* graphs without self-intersecting zigzags (i.e., simple circuits and doubly infinite rays are the only zigzags) were classified in [GrSh87]: these are the three regular partitions {6; 3}, {3; 6}, {4; 4} of the Euclidean plane $\mathbb{E}^2$, the Archimedean partition (3.6.3.6), its dual, and several infinities of partitions of the hyperbolic plane $\mathbb{H}^2$. Also the notion of zigzag (Petrie polygon) was extended in [Cox73], p. 223 for n-dimensional polytopes and honeycombs.

The *medial graph* of a plane graph G, denoted by $Med(G)$, is defined by taking, as vertex-set, the set of edges of G with two edges being adjacent if they share a vertex and belong to the same face of G. $Med(G)$ is 4-regular and its central circuits $C_1, \ldots, C_r$ correspond to zigzags $Z_1, \ldots, Z_r$ of G (see Fig. 1.8) and the same, up to change of types I and II in the signatures, zigzags of G^*. It holds $Med(G) = Med(G^*)$. Moreover, an orientation of a zigzag Z_i induces an orientation of a central circuit C_i. The set of faces of $Med(G)$ corresponds to the set of vertices and faces of G. If we keep the same orientation, the intersection numbers of C_i and C_j are the same as the intersection numbers of Z_i and Z_j.

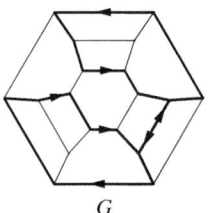 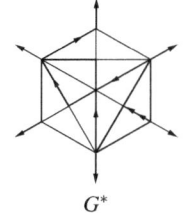 

G G^* $Med(G) = Med(G^*)$

Fig. 1.8 Example of a zigzag $18_{1,0}$ in a plane graph G, the corresponding zigzag $18_{0,1}$ in G^* and the corresponding to them central circuit in $Med(G)$

 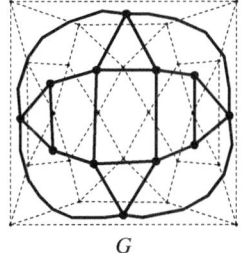 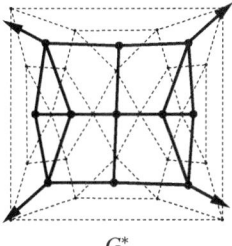

$H = Med(G) = Med(G^*)$ G G^*

Fig. 1.9 Inverse medial graphs G and G^* of a chess-colored bipartite graph H are $G = H_{\text{black}}$ and $G^* = H_{\text{white}}$ of the black and white faces of H

The graph $(Med(G))^*$ is bipartite, that is, the face-set of $Med(G)$ is split into two sets, which correspond to the vertices and faces of the graph G. So, any 4-regular plane graph H is the medial graph for a pair of following mutually dual plane graphs: one can assign two colors to the faces of H in the "chess way," such that no two adjacent faces of H have the same color. Two faces of the same color are said to be *adjacent* if they share a vertex. See example in Fig. 1.9.

Removing of a central circuit is explained in Fig. 1.10. Removing a zigzag in a graph G consists of removing corresponding central circuit in $Med(G)$ and taking one of two inverse medial graphs. A central circuit is called *reducible* if on one of its sides there are only 4-gons. This sequence of 4-gons can be completely eliminated to get a *reduced graph*. For a 3-regular plane graph, a zigzag is called *reducible* if on one side there are only 6-gons. We can reduce the graph by eliminating those 6-gons only if the zigzag is simple. Moreover, there are several possibilities for this reduced graph, while in the 4-regular case, the reduction is uniquely defined. By removing one of two boundary central-circuits for each railroad, any 4-regular graph can be reduced to a tight one. There is no analogous reduction for the 3-regular case.

Proposition 1.4 *Given a tight k-regular map G, $k \in \{3, 4\}$ with p-vector $\mathbf{p} = (\ldots, p_i, \ldots)$; denote $\frac{2k}{k-2}$ by k'. The number of ZC-circuits (zigzags or central circuits for 3- or 4-regular case, respectively) of G is at most*

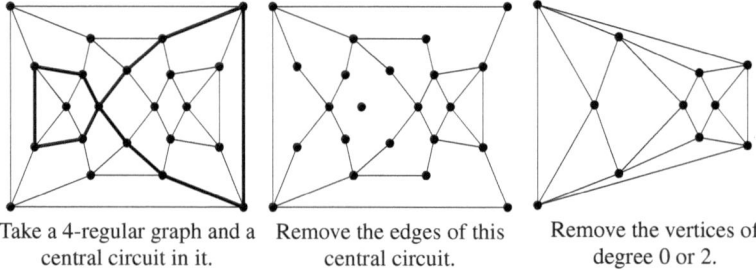

| Take a 4-regular graph and a central circuit in it. | Remove the edges of this central circuit. | Remove the vertices of degree 0 or 2. |

Fig. 1.10 Removing a central circuit

$$\frac{1}{2}\sum_{i\neq k'} i p_i.$$

Proof In fact, each ZC-$circuit$ has a non-k'-gonal face on each of its sides, since, otherwise, it would have a railroad on this side; so, the number of incidences between ZC-circuits and non-k'-gons is at least twice the number of ZC-circuits. On the other hand, the number of those incidences is exactly $\sum_{i\neq k'} i p_i$. □

1.4 Curvature of Faces

Given $R \subset \mathbb{N}$ and a connected closed (compact and without boundary) surface $\mathbb{F}^2$, let R-$\mathbb{F}^2$ denote a map M on $\mathbb{F}^2$ whose faces have gonalities $i \in R$ and let (R, k)-$\mathbb{F}^2$ denote such a k-regular map. We suppose also $\max\{i \in R\} \geq 3 \leq k$.

The *Euler characteristic* $\chi = \chi(M) = \chi(\mathbb{F}^2)$ is $v - e + f$, where $v, e = \frac{1}{2}\sum_{i\geq 1} i p_i$ and $f = \sum_i p_i$ are the numbers of vertices, edges, and faces of M. So, $v = \chi + \frac{1}{2}\sum_{i\geq 1} p_i(i - 2)$ and in the k-regular case, $v = \frac{1}{k}\sum_{i\geq 1} i p_i$.

Since $kv = 2e = \sum_i i p_i$, the Euler formula becomes also $\chi = \sum_{i\geq 1} p_i \kappa_i$, where

$$\kappa_i = 1 + \frac{i}{k} - \frac{i}{2}$$

denote the *curvature of a face* of gonality i. Euler formula is a discrete analog of the *Gauss–Bonnet formula*, $2\pi \chi(\mathbb{F}) = \int_{\mathbb{F}} K(x)dx$, for the Gaussian curvature K.

A map (R, k)-$\mathbb{F}^2$ with nonnegatively curved faces has $\chi(\mathbb{F}^2) \geq 0$. So, such irreducible surface $\mathbb{F}^2$ is only the sphere $\mathbb{S}^2$, torus $\mathbb{T}^2$ (two orientable ones), real projective plane $\mathbb{P}^2$ or Klein bottle $\mathbb{K}^2$ with $\chi = 2, 0, 1, 0$, respectively. Clearly, the set of all maps (R, k)-$\mathbb{F}^2$ with only positively curved faces is finite and consists of

(i) $(\{1, 2, 3, 4, 5\}, 3)$-, $(\{1, 2, 3\}, 4)$-, $(\{1, 2, 3\}, 5)$-maps on $\mathbb{S}^2$ and
(ii) their antipodal quotients, when they are centrally symmetric, on $\mathbb{P}^2$.

For example, all maps (i) with $\min\{i \in R\} \geq 3$, are only Octahedron, Icosahedron, and nine spheres ($\{3, 4, 5\}$, 3)-$\mathbb{S}^2$: all eight dual *convex deltahedra* (all faces are equilateral triangles) and *Dürer's octahedron* (truncation of Cube on two opposite vertices). All such non-Platonic spheres are given in Fig. 1.11.

All maps (ii) with $\min\{i \in R\} \geq 3$, are the antipodal quotients of Octahedron, Icosahedron, Cube, Dodecahedron, and Dürer's octahedron.

Also, all maps (i) with $\min\{i \in R\} \geq 2$ and $|R| = 2$, are 8, 2, 4 maps ($\{a, b\}$, 3)-, ($\{2, 3\}$, 4)-, ($\{2, 3\}$, 5)-$\mathbb{S}^2$, given on Fig. 2.1 [DeDu08].

For fixed R and k, the set of maps (R, k)-$\mathbb{F}^2$ with nonnegatively curved faces is infinite if and only if the faces of maximal gonality (say, r) are *flat faces* (i.e., $\kappa_r = 0$), implying $r = \frac{2k}{k-2}$ and so, $(r, k) = (6, 3), (4, 4)$ or $(3, 6)$.

Hence, all such infinite families are:

(i) ($\{1, 2, 3, 4, 5, 6\}$, 3)-, ($\{1, 2, 3, 4\}$, 4)-, ($\{1, 2, 3\}$, 6)-maps on $\mathbb{S}^2$,
(ii) their antipodal quotients, when they are centrally symmetric, on $\mathbb{P}^2$,
(iii) ($\{6\}$, 3)-, ($\{4\}$, 4)-, ($\{3\}$, 6)-maps on the torus $\mathbb{T}^2$ and their antipodal quotients, when they are centrally symmetric, on $\mathbb{K}^2$.

Exclusion of negatively curved faces simplifies enumeration, while number p_r of flat faces not being restricted, there is an infinity of such (R, k)-spheres. The number of such v-vertex (R, k)-spheres with $|R| = 2$ increases polynomially with v. Such spheres admit parametrization and description in terms of rings of (*Gaussian* if $k = 4$ and *Eisenstein* if $k = 3, 6$) *integers*.

There are 16 (see Sect. 6.4) infinite families of *non-bifaced*, i.e., $|\{3 \leq i \leq 5 : p_i > 0\}| \geq 2$, ($\{3, 4, 5, 6\}$, 3)-spheres. Their symmetries are listed explicitly in [DDF09]. It is also done by Proposition 1.5; "iff" there means "if and only if".

Proposition 1.5 *Let G be a non-bifaced ($\{3, 4, 5, 6\}$, 3)-sphere. Then its possible symmetries are given by the criterion below.*

(i) *G admit groups C_1, C_s.*
(ii) *G admits C_i iff $\gcd(p_3, p_4, p_5)$ is even, and then it admits C_2, C_{2v}, C_{2h} also. Moreover, G admits S_4, D_2, D_{2h}, D_{2d} iff $\frac{p_5}{2}$ is even and D_m, D_{mh}, D_{md} iff $m = \frac{p_5}{2} > 2$.*
(iii) *If $\gcd(p_3, p_4, p_5)$ is odd, then G admits C_2, C_{2v} iff p_5 is even and it admits C_3, C_{3v} iff p_5 is divided by 3. Moreover, G admits C_{3d} iff $\gcd(p_3, p_4) = 3$ and it admits D_m, D_{mh} iff $m = p_4 > 2$ and p_3 is even.*

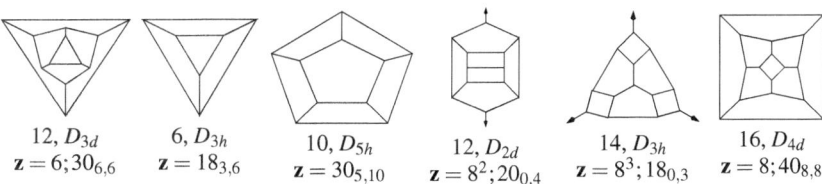

| 12, D_{3d} | 6, D_{3h} | 10, D_{5h} | 12, D_{2d} | 14, D_{3h} | 16, D_{4d} |
| $\mathbf{z} = 6; 30_{6,6}$ | $\mathbf{z} = 18_{3,6}$ | $\mathbf{z} = 30_{5,10}$ | $\mathbf{z} = 8^2; 20_{0,4}$ | $\mathbf{z} = 8^3; 18_{0,3}$ | $\mathbf{z} = 8; 40_{8,8}$ |

Fig. 1.11 Dürer's octahedron and all non-Platonic dual deltahedra

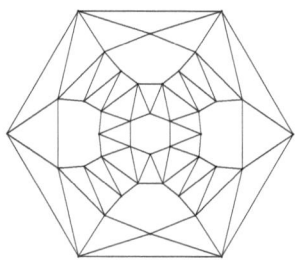

Fig. 1.12 Borosphere has
symmetry D_{2d},
$\mathbf{v} = (v_4 = 16, v_5 = 24)$,
$\mathbf{p} = (p_3 = 48, p_6 = 2,$
$p_7 = 4)$, $\mathbf{z} = (16^2, 36;$
$116_{8,24})$ and 286, 224 perfect
matchings in 36, 090 orbits

There are three families of non-bifaced ($\{2, 3, 4\}$, 4)-spheres (with $(p_2, p_3) =$ (3, 2), (2, 4), (1, 6)) and one family of non-bifaced self-dual $\{2, 3, 4\}$-spheres (with $(p_2, p_3) = (1, 2)$). All of them admit groups C_1, C_s, C_2, C_{2v}. Moreover, first family admits D_3, D_{3h} and second one admits $C_i, C_{2h}, D_2, D_{2h}, D_{2d}$.

1.5 Bifaced Maps

In particular, any *bifaced map* (i.e., ($\{a, b\}, k$)-$\mathbb{F}^2$, $a < b$) have, by Euler formula,

$$p_a = \frac{b(k-2) - 2k}{2k - a(k-2)} + \chi(\mathbb{F}^2)2k.$$

Besides, 16 maps with only positively curved faces (11 on $\mathbb{S}^2$ and 5 on $\mathbb{P}^2$, see Sect. 1.4), each such map has $a < \frac{2k}{k-2} \leq b$ and negatively curved faces are excluded exactly if $b = \frac{2k}{k-2}$, i.e., if b-gons are flat. In this case, $(b, k) = (6, 3), (4, 4)$ or $(3, 6)$, Euler formula $\chi = \kappa_a p_a$ becomes $\chi = (1 - \frac{a}{b})p_a$, the number p_a is a constant $\frac{\chi b}{b-a}$ and all possible (a, p_a) for a ($\{a, b\}, k$)-$\mathbb{S}^2$ are:

- (5, 12), (4, 6), (3, 4), (2, 3) for $(b, k) = (6, 3)$,
- (3, 8), (2, 4) for $(b, k) = (4, 4)$ and
- (2, 6), (1, 3) for $(b, k) = (3, 6)$.

These eight families can be seen as spherical analogs of the regular plane partitions $\{6; 3\}, \{4; 4\}, \{3; 6\}$ with p_a *disclinations* ("defects") of the curvature κ_a, added to get the curvature 2 of the sphere.

Denote by a_v any v-vertex map ($\{a, 6\}, 3$)-$\mathbb{S}^2$. The smallest (with $p_6 = 0$) 3_v, 4_v and 5_v are Tetrahedron, Cube, and Dodecahedron. In fact, all maps ($\{a, b\}, 3$)-$\mathbb{S}^2$ with $3 \leq a < b < 6$, besides them, are those given in Fig. 1.11.

The 5_v are, actually, the (geometric) *fullerenes*, well known in Organic Chemistry (they will be considered in Chap. 2), while 4_v represent *boron nitrides*. See Fig. 1.12 for a recently discovered by Wang et al. (see https://news.brown.edu/articles/2014/ 07/buckyball and [Boro2014]) *Borosphere*.

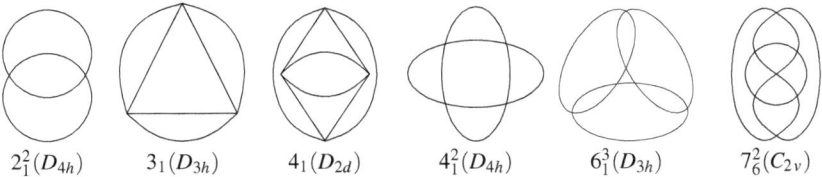

$2_1^2(D_{4h})$ $3_1(D_{3h})$ $4_1(D_{2d})$ $4_1^2(D_{4h})$ $6_1^3(D_{3h})$ $7_6^2(C_{2v})$

Fig. 1.13 Minimal plane projections of some alternating links

Chapter 4 treats the maps $(\{2, 3, 4\}, 4)\text{-}\mathbb{S}^2$ and self-dual maps $(\{2, 3, 4\})\text{-}\mathbb{S}^2$. Chapter 5 treats the maps $(\{1, 2, 3\}, 6)\text{-}\mathbb{S}^2$ and $(\{3, 4\}, 5)\text{-}\mathbb{S}^2$.

Note that all $(\{3, 6\}, 3)$-, $(\{4, 6\}, 3)$-, $(\{2, 4\}, 4)$- and $(\{1, 3\}, 6)\text{-}\mathbb{S}^2$ are Hamiltonian, but $(\{2, 6\}, 3)\text{-}\mathbb{S}^2$ with $v \equiv 0 \pmod 4$ are not; see [Good77, GrZa74].

Denote by $i \times G$ the graph G with each edge replaced by i edges. The $Bundle_m$ is defined as m-regular graph $m \times K_2$; its z-vector is $2m_{m,0}$ for odd m and $2m_{0,m}$ for even m. The $Foil_m$ is defined as 4-regular graph $2 \times C_m$; its cc-vector is $2m$ for odd m and m^2 for even m. Clearly, $Foil_m$ with $m = 2, 4$ and 3 are (projections of) links 2_1^2, 4_1^2 and knot Trefoil 3_1 (see Fig. 1.13). The m-rose is the $2m$-regular 1-vertex plane graph $m \times K_1$ with one m-gonal and m 1-gonal faces; $Trifolium$ is the 3-rose.

The maps $(R, k)\text{-}\mathbb{S}^2$ with negatively curved faces are much more complicate to study. We will consider in detail only the simplest cases: $icosahedrites$ $((\{3, 4\}, 5)$-spheres), c-$disk$-$fullerenes$ $((\{5, 6, c\}, 3)$-spheres with $c > 6$ and $p_c = 1$) and any $(\{a, b\}, k)\text{-}\mathbb{S}^2$ with $p_b \le 3$. See Table 1.2. The last two lines there give the families of bifaced self-dual plane maps with nonnegatively curved faces.

The criteria of existence there were obtained in [GrünMo63] for the $(\{a, 6\}, 3)$-spheres with $3 \le a \le 5$, in [Grün67] for $(\{3, 4\}, 4)$- and in [GrZa74] for $(\{2, 6\}, 3)$- and $(\{1, 3\}, 6)$-spheres. For an infinite family $\mathcal{M}$ of spheres, let $Nr_v(\mathcal{M})$ be the number of such v-vertex maps. In the column "Order" we give the order of magnitude

Table 1.2 Main families of considered maps $(\{a, b\}, k)\text{-}\mathbb{S}^2$ and two self-dual $\{a, b\}\text{-}\mathbb{S}^2$

k	(a, b)	Smallest one	Exists if and only if	Connect	p_a	v	Order	$NrGr$
3	$(5, 6)$	Dodecahedron	$p_6 \ne 1$	3-conn.	12	$20 + 2p_6$	v^9	28
3	$(4, 6)$	Cube	$p_6 \ne 1$	3-conn.	6	$8 + 2p_6$	v^3	16
4	$(3, 4)$	Octahedron	$p_4 \ne 1$	3-conn.	8	$6 + p_4$	v^5	18
6	$(2, 3)$	$Bundle_6 = 6 \times K_2$	p_3 is even	2-conn.	6	$2 + \frac{p_3}{2}$	v^4	22
3	$(3, 6)$	Tetrahedron	p_6 is even	Proposition 3.1	4	$4 + 2p_6$	v	5
4	$(2, 4)$	$Bundle_4 = 4 \times K_2$	p_4 is even	2-conn.	4	$2 + p_4$	v	5
3	$(2, 6)$	$Bundle_3 = 3 \times K_2$	$p_6 = (k^2 + kl + l^2) - 1$	2-conn.	3	$2 + 2p_6$	v	2
6	$(1, 3)$	Trifolium	$p_3 = 2(k^2 + kl + l^2) - 1$	1-conn.	3	$\frac{1 + p_3}{2}$	v	3
5	$(3, 4)$	Icosahedron	$p_4 \ne 1$	3-conn.?	$20 + 2p_4$	$12 + 2p_4$	Exp.	38
Self-dual	$(3, 4)$	Tetrahedron	$p_4 \ge 0$	2-conn.	4	$4 + p_4$	v^3	16
Self-dual	$(2, 4)$	$Bundle_2 = 2 \times K_2$	$p_4 \ge 0$	1-conn.	2	$2 + p_4$	v	5

Table 1.3 Number of complex parameters needed for describing spheres from Table 1.2 (see Chap. 6); when this number is 1, a description with Goldberg–Coxeter construction is given; for the self-dual case, we restrict to odd $k + l$, see Theorem 4.10

k	(a, b)	Groups of symmetry and number of complex parameters	
3	$(5, 6)$	(C_1, C_s, C_i), 10	$(C_2, C_{2v}, C_{2h}, S_4)$, 6
		$(C_3, C_{3v}, C_{3h}, S_6)$, 4	(D_2, D_{2h}, D_{2d}), 4
		(D_3, D_{3h}, D_{3d}), 3	(D_5, D_{5h}, D_{5d}), 2
		(D_6, D_{6h}, D_{6d}), 2	(T, T_h, T_d), 2
		(I, I_h), 1, $GC_{k,l}(Dodecahedron)$	
3	$(4, 6)$	(C_1, C_s, C_i), 4	(C_2, C_{2v}, C_{2h}), 3
		(D_2, D_{2h}, D_{2d}), 2	(D_3, D_{3h}, D_{3d}), 2
		(D_6, D_{6h}), 1, $GC_{k,l}(Prism_6)$	(O, O_h), 1, $GC_{k,l}(Cube)$
4	$(3, 4)$	(C_1, C_s, C_i), 6	$(C_2, C_{2v}, C_{2h}, S_4)$, 4
		(D_2, D_{2h}, D_{2d}), 3	(D_3, D_{3h}, D_{3d}), 2
		(D_4, D_{4h}, D_{4d}), 2	(O, O_h), 1, $GC_{k,l}(Octahedron)$
6	$(2, 3)$	(C_1, C_s, C_i), 4	$(C_2, C_{2v}, C_{2h}, S_4)$, 3
		$(C_3, C_{3v}, C_{3h}, S_6)$, 2	(D_2, D_{2h}, D_{2d}), 2
		(D_3, D_{3h}, D_{3d}), 2	(D_6, D_{6h}), 1, $GC_{k,l}(6 \times K_2)$
		(T, T_d, T_h), 1, $GC_{k,l}(K_2 \times Tetrahedron)$	
3	$(3, 6)$	(D_2, D_{2h}, D_{2d}), 2	(T, T_d), 1, $GC_{k,l}(Tetrahedron)$
4	$(2, 4)$	(D_2, D_{2h}, D_{2d}), 2	(D_4, D_{4h}), 1, $GC_{k,l}(Bundle_4)$
3	$(2, 6)$	(D_3, D_{3h}), 1, $GC_{k,l}(Bundle_3)$	
6	$(1, 3)$	(C_3, C_{3h}, C_{3v}), 1, $GC_{k,l}(Trifolium)$	
5	$(3, 4)$	(C_1, C_i, C_s), ∞	$(C_2, C_{2h}, C_{2v}, S_4)$, ∞
		$(C_3, C_{3h}, C_{3v}, S_6)$, ∞	$(C_4, C_{4h}, C_{4v}, S_8)$, ∞
		$(C_5, C_{5h}, C_{5v}, S_{10})$, ∞	(D_2, D_{2h}, D_{2d}), ∞
		(D_3, D_{3h}, D_{3d}), ∞	(D_4, D_{4h}, D_{4d}), ∞
		(D_5, D_{5h}, D_{5d}), ∞	(O, O_h), ∞
		(T, T_d, T_h), ∞	(I, I_h), ∞
Self-dual	$(3, 4)$	(C_1, C_i, C_s)	$(C_2, C_{2h}, C_{2v}, S_4)$
		(C_3, C_{3v})	(C_4, C_{4v})
		(D_2, D_{2d}, D_{2h})	(T, T_d), 1, $GC_{k,l}(Tetrahedron)$
Self-dual	$(2, 4)$	(C_2, C_{2v}, C_{2h})	(D_2, D_{2h}), 1, $GC_{k,l}(Bundle_2)$

of $Nr_v(\mathcal{M})$. In the column "NrGr", the number of symmetry groups of maps from $\mathcal{M}$ is given. In Tables 1.2 and 1.4, '?' means "conjectured" or "unknown". In Table 1.3 we give the number of complex parameters needed to describe the considered classes of graphs and the possible groups.

In Table 1.4, we give information on the ZC-structure of considered graphs. Among 11 families, given in Table, the ($\{2, 6\}$, 3)- and ($\{1, 3\}$, 6)-spheres come by the Goldberg–Coxeter construction from Bundle$_3$ and Trifolium, i.e., admit a descrip-

Table 1.4 ZC-structure of maps from Table 1.2; $zcMax$ is the maximum number of zc-circuits of tight graphs; First zcMax is the first graph attaining the bound; zcPT is the number of z- or cc-tight pure graphs

k	(a, b)	Circuits	zcMax	First zcMax	zcPT
3	(5, 6)	Zigzags	15	$GC_{2,1}(Dodecahedron)$	9?
3	(4, 6)	Zigzags	8?	$4_{88}(D_{2h})$?	2
4	(3, 4)	Central circuits	6	$GC_{2,1}(Cube)$	8
6	(2, 3)	Zigzags	6	$GC_{2,1}(Bundle_6)$	3
		Central circuits	6	$GC_{2,1}(Bundle_6)$	1
3	(3, 6)	Zigzags	3	$Tetrahedron$	∞
4	(2, 4)	Central circuits	2	Bundle$_4$	∞
3	(2, 6)	Zigzags	3	$GC_{1,1}(Bundle_3)$	0
6	(1, 3)	Zigzags	0	0	0
		Central circuits	0	0	0
5	(3, 4)	Zigzags	∞?	N/A?	3?
		Weak zigzags	∞?	N/A?	∞?
Self-dual	(3, 4)	Zigzags	6	$GC_{2,1}(Tetrahedron)$	6
Self-dual	(2, 4)	Zigzags	2	Bundle$_2$	∞

For the case $k = 5$, or 6, distinct notions of tightness are used, see Chap. 5 for details.

tion by 1 complex parameter. All their ZC-circuits self-intersect. See Fig. 1.14 for smallest known $(\{a, b\}, k)$-spheres with maximal number of ZC-circuits.

The $(\{3, 6\}, 3)$- [GrünMo63] and $(\{2, 4\}, 4)$-spheres [DeSt03] admit a simple description by 2 complex parameters, or, moreover, by 3 natural ones: pseudo-road length, number of circumscribing railroads, and shift. All their ZC-circuits are simple. Such pure sphere is tight if and only if it has the minimal number (3 or, respectively, 2) ZC-circuits. There is a tight v-vertex $(\{2, 4\}, 4)$-sphere for any possible (i.e., even) v. A tight v-vertex $(\{3, 6\}, 3)$-sphere exists if and only if $\frac{v}{4}$ is odd; all such spheres are tight if and only if $\frac{v}{4}$ is prime > 2. See Chaps. 3 and 4 for details.

Aggregating groups as $\mathbf{C_m} = \{C_m, C_{mv}, C_{mh}, S_{2m}\}$, $\mathbf{D_m} = \{D_m, D_{mh}, D_{md}\}$, $\mathbf{C'_m} = \mathbf{C_m} \setminus \{S_{2m}\}$, $\mathbf{D'_m} = \mathbf{D_m} \setminus \{D_{md}\}$ and $\mathbf{T} = \{T, T_d, T_h\}$, $\mathbf{O} = \{O, O_h\}$, $\mathbf{I} = \{I, I_h\}$, $\mathbf{T'} = \mathbf{T} \setminus \{T_h\}$, the symmetries of families of bifaced plane maps become simpler to see:

1. 28 for $(\{5, 6\}, 3)$- [FoMa95]: $\mathbf{C_1}, \mathbf{C_2}, \mathbf{C_3}, \mathbf{D_2}, \mathbf{D_3}, \mathbf{D_5}, \mathbf{D_6}, \mathbf{T}, \mathbf{I}$
2. 22 for $(\{2, 3\}, 6)$- [DeDu12]: $\mathbf{C_1}, \mathbf{C_2}, \mathbf{C_3}, \mathbf{D_2}, \mathbf{D_3}, \mathbf{D'_6}, \mathbf{T}$
3. 16 for $(\{4, 6\}, 3)$- [DeDu05]: $\mathbf{C_1}, \mathbf{C'_2}, \mathbf{D_2}, \mathbf{D_3}, \mathbf{D'_6}, \mathbf{O}$
4. 18 for $(\{3, 4\}, 4)$- [DeDuSh03]: $\mathbf{C_1}, \mathbf{C_2}, \mathbf{D_2}, \mathbf{D_3}, \mathbf{D_4}, \mathbf{O}$
5. 5 for $(\{3, 6\}, 3)$- [FoCr97]: $\mathbf{D_2}, \mathbf{T'}$
6. 5 for $(\{2, 4\}, 4)$- [DeDuSh03]: $\mathbf{D_2}, \mathbf{D'_4}$
7. 2 for $(\{2, 6\}, 3)$- [GrZa74]: $\mathbf{D'_3}$
8. 3 for $(\{1, 3\}, 6)$- [DeDu12]: $\mathbf{C'_3}$
9. 38 for $(\{3, 4\}, 5)$- [DDS13a]: $\mathbf{C_1}, \mathbf{C_2}, \mathbf{C_3}, \mathbf{C_4}, \mathbf{C_5}, \mathbf{D_2}, \mathbf{D_3}, \mathbf{D_4}, \mathbf{D_5}, \mathbf{T}, \mathbf{O}, \mathbf{I}$.
10. 16 for self-dual $(\{3, 4\})$- [DuDe11]: $\mathbf{C_1}, \mathbf{C_2}, \{C_3, C_{3v}\}, \{C_4, C_{4v}\} \mathbf{D_2}, \mathbf{T'}$.
11. 5 for self-dual $(\{2, 4\})$- [DuDe11]: $\mathbf{C'_2}, \mathbf{D'_2}$.

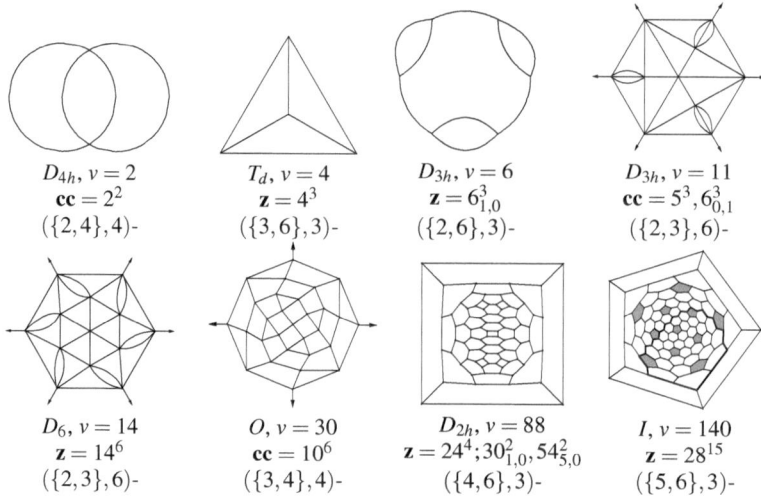

$D_{4h}, v = 2$
$\mathbf{cc} = 2^2$
$(\{2,4\},4)$-

$T_d, v = 4$
$\mathbf{z} = 4^3$
$(\{3,6\},3)$-

$D_{3h}, v = 6$
$\mathbf{z} = 6^3_{1,0}$
$(\{2,6\},3)$-

$D_{3h}, v = 11$
$\mathbf{cc} = 5^3, 6^3_{0,1}$
$(\{2,3\},6)$-

$D_6, v = 14$
$\mathbf{z} = 14^6$
$(\{2,3\},6)$-

$O, v = 30$
$\mathbf{cc} = 10^6$
$(\{3,4\},4)$-

$D_{2h}, v = 88$
$\mathbf{z} = 24^4; 30^2_{1,0}, 54^2_{5,0}$
$(\{4,6\},3)$-

$I, v = 140$
$\mathbf{z} = 28^{15}$
$(\{5,6\},3)$-

Fig. 1.14 Smallest tight $(\{a, b\}, k)$-spheres with maximal number of ZC-circuits

The (R, k)-maps on the projective plane $\mathbb{P}^2$ are the antipodal quotients of centrosymmetric maps (R, k)-$\mathbb{S}^2$; so, with halving of their v- and p-vector. The point groups with inversion are: $T_h, O_h, I_h, C_{mv}, D_{mh}$ with even m and D_{md}, S_{2m} with odd m. So, the symmetries of the above 11 families on $\mathbb{P}^2$ are:

1. 9 for $(\{5, 6\}, 3)$-: $C_i, C_{2h}, S_6, D_{2h}, D_{3d}, D_{5d}, D_{6h}, T_h, \mathbf{I_h}$
2. 7 for $(\{2, 3\}, 6)$-: $C_i, C_{2h}, S_6, D_{2h}, D_{3d}, \mathbf{D_{6h}}, \mathbf{T_h}$
3. 6 for $(\{4, 6\}, 3)$-: $C_i, C_{2h}, D_{2h}, D_{3d}, \mathbf{D_{6h}}, \mathbf{O_h}$
4. 6 for $(\{3, 4\}, 4)$-: $C_i, C_{2h}, D_{2h}, D_{3d}, D_{4h}, \mathbf{O_h}$
5. 2 for $(\{2, 4\}, 4)$-: $D_{2h}, \mathbf{D_{4h}}$
6. 1 for $(\{3, 6\}, 3)$-: D_{2h}
7. 0 for $(\{2, 6\}, 3)$- and $(\{1, 3\}, 6)$-
8. 12 for $(\{3, 4\}, 5)$-: $C_i, C_{2h}, C_{4h}, S_6, S_{10}, D_{2h}, D_{3d}, D_{4h}, D_{5d}, T_h, O_h, I_h$
9. 2 for self-dual $(\{3, 4\})$- and for self-dual $(\{2, 4\})$-: C_{2h}, D_{2h}.

Each of the 7 above families with bold-faced symmetry is described by one natural parameter and contains $O(\sqrt{v})$ spheres with at most v vertices.

The smallest $(\{a, 6\}, 3)$-$\mathbb{P}^2$ for $a = 5, 4, 3$ and $(\{3, 4\}, 5)$-$\mathbb{P}^2$ are K_6^* (*Petersen graph*), K_4 (smallest $\mathbb{P}^2$-quadrangulation), K_4 truncated on 2 vertices and K_6, respectively. The smallest $(\{a, 4\}, 4)$-$\mathbb{P}^2$ for $a = 3, 2$ and $(\{2, 3\}, 6)$-$\mathbb{P}^2$ are K_5 and points with 2, 3 loops; smallest ones without loops are Bundle$_4$, Bundle$_6$ but on $\mathbb{P}^2$. The smallest self-hedrite maps on $\mathbb{P}^2$ without 1- and 2-gons are given in Table 1.5; here $K_{2,2,2} - P_2$ is the antipodal quotient of the 12-th one in Fig. 4.21. See Figs. 1.15, 1.16, 1.17 and 1.18 for some drawings of some of the minimal maps.

Clearly, for an $(\{a, b\}, k)$-sphere with $a, b, k \geq 3$, all possible (a, k) are $(3, 3)$, $(4, 3), (3, 4), (5, 3)$ and $(3, 5)$. Permitting $a = 2$, one get a b-vertex $(\{2, b\}, k)$-sphere with $p_b = 2$ for any $b, k \geq 3$, not both odd: repeat every of disjoint $\lfloor \frac{b}{2} \rfloor$ edges of the

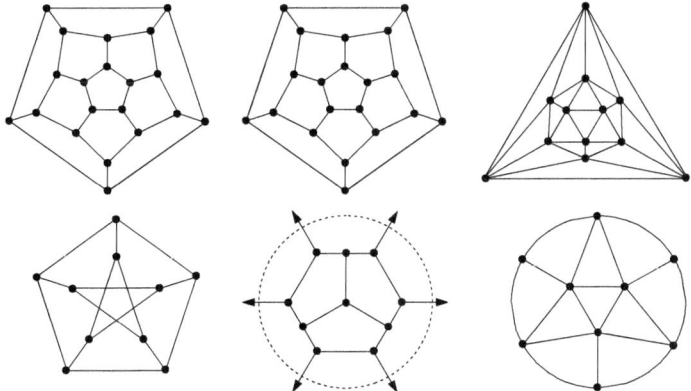

Fig. 1.15 The Petersen graph is the smallest $\mathbb{P}^2$-fullerene. Its $\mathbb{P}^2$-dual, K_6, is the smallest $\mathbb{P}^2$-icosahedrite (*half-Icosahedron*) and smallest $\mathbb{P}^2$-triangulation.

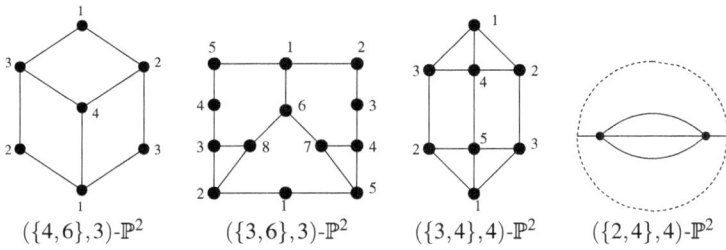

$(\{4,6\},3)$-$\mathbb{P}^2$ $(\{3,6\},3)$-$\mathbb{P}^2$ $(\{3,4\},4)$-$\mathbb{P}^2$ $(\{2,4\},4)$-$\mathbb{P}^2$

Fig. 1.16 Some bifaced maps on projective plane

Table 1.5 Smallest bifaced maps without loops on irreducible surfaces

k	(a,b)	Smallest on $\mathbb{S}^2$	on $\mathbb{P}^2$	on $\mathbb{T}^2$	on $\mathbb{K}^2$
3	$(5,6)$	Dodecahedron	K_6^*	K_7^*	$K_{3,3,3}^*$
3	$(4,6)$	Cube $= K_2^3$	K_4	K_7^*	$K_{3,3,3}^*$
3	$(3,6)$	Tetrahedron $= K_4$	2-trunc. K_4	K_7^*	$K_{3,3,3}^*$
3	$(2,6)$	Bundle$_3 = 3 \times K_2$	–	K_7^*	$K_{3,3,3}^*$
4	$(3,4)$	Octahedron $= K_{2,2,2}$	K_5	K_5	$K_{2,2,2}$
4	$(2,4)$	Bundle$_4 = 4 \times K_2$	$4 \times K_2$	K_5	$K_{2,2,2}$
6	$(2,3)$	Bundle$_6 = 6 \times K_2$	$6 \times K_2$	K_7	$K_{3,3,3}$
6	$(1,3)$	Trifolium	–	K_7	$K_{3,3,3}$
5	$(3,4)$	Icosahedron	K_6	K_6	?
Self-dual	$(3,4)$	Tetrahedron $= K_4$	$K_{2,2,2} - P_2$	K_5	$K_{2,2,2}$
Self-dual	$(2,4)$	Bundle$_2 = 2 \times K_2$	$K_5 - P_3$	K_5	$K_{2,2,2}$

cycle C_b i times, $1 \leq i \leq k-1$, and repeat every other edge $k-i$ times (it should be $i = k - i$ for odd b). Permitting $a = 1$, one get, say, a $\frac{2b}{3}$-vertex ($\{1, b\}$, 3)-sphere with $p_b = 2$ if $b \equiv 0$ (mod 3): put 1-gons, $\frac{b}{3}$ inside and $\frac{b}{3}$ outside, on the $\frac{2b}{3}$-cycle.

Now we give full (except only conjectured completeness for the case $(a, k) = (3, 5)$, $p_b = 3$) listing of ($\{a, b\}$, k)-spheres with $p_b \leq 3 \leq a, b, k$, obtained explicitly in [DDS13a]. The ($\{a, b\}$, k)-spheres with $p_b \leq 1$ are five Platonic solids (a; k): Tetrahedron, Cube ($Prism_4$), Octahedron ($APrism_3$), Dodecahedron (snub $Prism_5$), Icosahedron (snub $APrism_3$). Given two circuits $u_1, \ldots, u_m$ and $v_1, \ldots, v_m$, an m-sided prism $Prism_m$ is formed when every u_i is joined to v_i by an edge. An m-sided antiprism $Prism_m$ is formed by adding the cycle $u_1, v_2, u_2, v_3, \ldots, v_m, u_m, v_1, u_1$.

There is one *trivial* 3-connected ($\{a, b\}$, k)-sphere with $p_b = 2$ for each of $(a, k) = (4, 3), (3, 4), (5, 3)$ and $(3, 5)$ – $Prism_b$, $APrism_b$, snub $Prism_b$ and – defined as two b-gons separated by b-ring of 4-gons, $2b$-ring of 3-gons, two b-rings of 5-gons and two $3b$-rings of 3-gons, respectively.

All the remaining such spheres are 10, for any $t \geq 2$, *non-trivial* ($\{a, at\}$, k)-spheres with $p_{at} = 2$: 1, 1, 2, 2 and 4 for $(a, k) = (3, 3), (4, 3), (5, 3), (3, 4)$ and $(3, 5)$, respectively. All have symmetry D_{th}, except two ($\{3, at\}$, 5)-spheres, one of C_{th} and one of D_t symmetry. All nontrivial ($\{a, ta\}$, k)-spheres with $p_{ta} = 2$ and $k = 3, 4, 5$, are given, for the case $t = 2$, in Fig. 1.19.

Let $k \geq 3 \leq a < b$. An ($\{a, b\}$, k)-sphere with $p_b = 3$ existing if and only if $b \equiv 2, a, 2(a - 1)$ (mod $2a$) and, in addition for $a = 5$, $b \equiv 4, 6$ (mod 10). Such sphere are unique if b is not $\equiv a$ (mod $2a$) and then their symmetry is D_{3h}, except when $(a, k) = (3, 5)$ when the symmetry is D_3. (There is also unique such ($\{a, a - 1\}$, k)-sphere for $(a, k) = (5, 3), (3, 4), (3, 5)$.) For $b \equiv a$ (mod $2a$), there are 1, 2, 5, 3 and ≥ 15 of such spheres for $(a, k) = (3, 3), (4, 3), (5, 3), (3, 4)$ and $(3, 5)$, respectively.

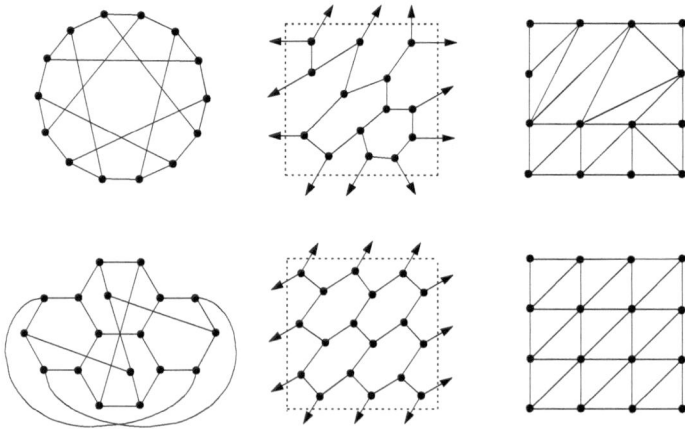

Fig. 1.17 The *Heawood graph* is the smallest ($\{6\}$, 3)-$\mathbb{T}^2$; its $\mathbb{T}^2$-dual K_7, is the smallest ($\{3\}$, 6)-$\mathbb{T}^2$. The $K_{3,3,3}$ is the smallest ($\{3\}$, 6)-$\mathbb{K}^2$; its $\mathbb{K}^2$-dual is the smallest ($\{6\}$, 3)-$\mathbb{K}^2$. The K_5 and $K_{2,2,2}$ are the smallest ($\{4\}$, 4)-$\mathbb{T}^2$ and ($\{4\}$, 4)-$\mathbb{K}^2$

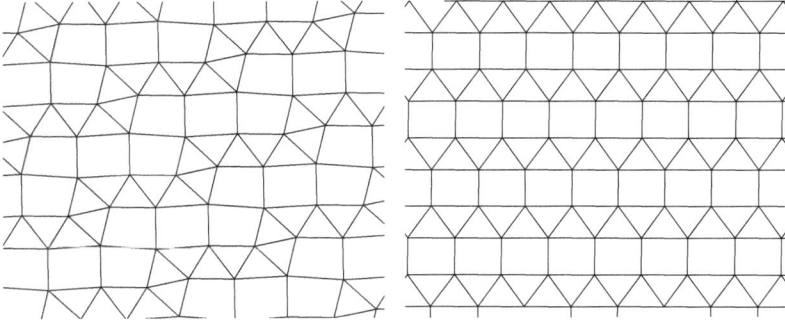

Fig. 1.18 The plane icosahedrites $(\{3, 4\}, 5)$-$\mathbb{E}^2$, the quotient of which are the smallest $(\{3, 4\}, 5)$-$\mathbb{T}^2$; their graphs are isomorphic to K_6

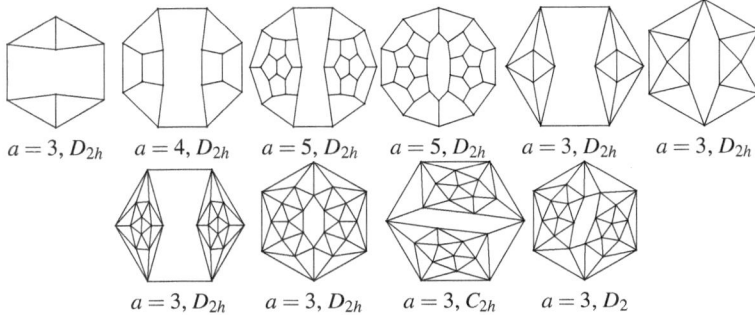

$a = 3, D_{2h}$ $a = 4, D_{2h}$ $a = 5, D_{2h}$ $a = 5, D_{2h}$ $a = 3, D_{2h}$ $a = 3, D_{2h}$

$a = 3, D_{2h}$ $a = 3, D_{2h}$ $a = 3, C_{2h}$ $a = 3, D_2$

Fig. 1.19 All non-trivial $(\{a, 2a\}, k)$-spheres with $p_{2a} = 2$ and $k = 3, 4, 5$

1.6 Computer Generation of the Families

Many results below are based on extensive computer computations. Main technique: exhaustive search. Sometimes, speedup by proving that a group of faces cannot be completed to the desired graph. The GAP computer algebra system [GAP], Plantri [BrMK01], for general graphs, and CaGe [BFDH97], for plane graph drawings, were used. The program CPF [BrHaHe03] generates 3-regular plane graphs with specified p-vector. ENU [Heid98, BrHaHe03] does the same for 4-regular plane graphs. CGF [Har00] generates 3-regular orientable maps with specified genus and p-vector. The second author adapted ENU to deal with 2-gonal faces also. His package PlanGraph [Dut02] was used for handling planar graphs in general.

References

[BrMK01] Brinkmann, G., McKay, B.: The Plantri program. http://cs.anu.edu.au/people/ bdm/plantri/ (2001)

[BFDH97] Brinkmann, G., Delgado Friedrichs, O., Dress, A., Harmuth, T.: CaGe— a virtual environment for studying some special classes of large molecules. MATCH: Commun. Math. Comput. Chem. **36**, 233–237 (1997). (http://www. mathematik.uni-bielefeld.de/CaGe/Archive/)

[BrHaHe03] Brinkmann, G., Harmuth, T., Heidemeier, O.: The construction of cubic and quartic planar maps with prescribed face degrees. Discrete Appl. Math. **128– 2**(3), 541–554 (2003)

[Cox73] Coxeter, H.S.M.: Regular Polytopes. Dover Publications, New York (1973)

[Dut04] Dutour, M.: Point Groups. http://www.liga.ens.fr/dutour/PointGroups/ (2004)

[Dut02] Dutour, M.: PlanGraph, a GAP package for Planar Graph. http://www.liga. ens.fr/dutour/PlanGraph

[DeSt03] Deza M., Shtogrin, M.: Octahedrites, Special Issue "Polyhedra in Science and Art", Symmetry: Culture and Science. The Quarterly of the International Society for the Interdisciplinary Study of Symmetry, **11/1-4**, 27–64 (2003)

[DeDu05] Deza, M., Dutour, M.: Zigzag structure of simple two-faced polyhedra. Comb Probab Comput **14**, 31–57 (2005)

[DeDu08] Deza, M., Dutour Sikirić, M.: Geometry of Chemical Graphs: Polycycles and Two-faced Maps, Encyclopedia of Mathematics and its Applications, vol. 119. Cambridge University Press, Cambridge (2008)

[DuDe11] Dutour Sikirić, M., Deza, M.: 4-regular and self-dual analogs of fullerenes, Mathematics and topology of fullerenes. In: Ori, O., Graovac, A., Cataldo, F.: Carbon Materials, Chemistry and Physics, vol. 4, pp. 103–116, Springer, New York (2011)

[DeDu12] Deza, M., Dutour Sikirić, M.: Zigzag and central circuit structure of (1, 2, 3, 6)-spheres. Taiwanese J. Math. **16–3**, 913–940 (2012)

[DeDuSh03] Deza, M., Dutour, M. Shtogrin, M.: 4-valent plane graphs with 2-, 3- and 4-gonal faces. In: Advances in Algebra and Related Topics (in memory of B.H.Neumann; Proceedings of ICM Satellite Conference on Algebra and Combinatorics, Hong Kong 2002), pp. 73–97, World Scientific Publishing Co. (2003)

[DDF04] Deza, M., Dutour, M., Fowler, P.W.: Zigzags railroads and knots in fullerenes. J. Chem. Inf. Comput. Sci. **44**, 1282–1293 (2004)

[DDF09] Deza, M., Dutour Sikirić, M., Fowler, P.W.: The symmetries of cubic poly- hedral graphs with face size no larger than 6, MATCH-Commun. Math. Co. **61**, 589–602 (2009)

[DDS13a] Deza, M., Dutour Sikirić, M., Shtogrin, M.: Fullerene-like spheres with faces of negative curvature. In: Diudea, M.V., Nagy, C.L. (eds.) Diamond D5 and Related Nanostructures Carbon Materials: Chemistry and Physics, vol. 6, pp. 251–274. Springer, New York (2013)

[FoMa95] Fowler, P.W., Manolopoulos, D.E.: An Atlas of Fullerenes. Clarendon Press, Oxford (1995)

[FoCr97] Fowler, P.W., Cremona, J.E.: Fullerenes containing fused triples of pentagonal rings. J. Chem. Soc., Faraday Trans. **93–13**, 2255–2262 (1997)

[Good77] Goodey, P.R.: A class of Hamiltonian polytopes. J. Graph Theory **1**, 181–185 (1977)

[Grün67] Grünbaum, B.: Convex Polytopes. Wiley-Interscience, New York (1967); second edition, Graduate Texts in Mathematics, vol. 221. Springer, New York (2003)

[GrünMo63] Grünbaum, B., Motzkin, T.S.: The number of hexagons and the simplicity of geodesics on certain polyhedra. Can. J. Math. **15**, 744–751 (1963)

[GrZa74] Grünbaum, B., Zaks, J.: The existence of certain planar maps. Discrete Math. **10**, 93–115 (1974)

[GrSh87] Grünbaum, B., Shepard, G.C.: Edge-transitive planar graphs. J. Graph Theory **11–2**, 141–155 (1987)

[GaKe94] Gargano, M.L., Kennedy, J.W.: Gaussian graphs and digraphs. Congressus Numerantium **101**, 161–170 (1994)

[GoRo01] Godsil, C., Royle, G.: Algebraic Graph Theory, Graduate Texts in Mathematics, vol. 207. Springer, Berlin (2001)

[Boro2014] http://en.wikipedia.org/wiki/Borospherene

[Harb97] Harborth, H.: Eulerian straight ahead cycles in drawings of complete bipartite graphs, Bericht 97/23, Institute für Mathematik. Universität Braunschweg, Tech (1997)

[Har00] Harmuth, T.: The construction of cubic maps on orientable surfaces, PhD thesis Bielefeld (2000)

[Heid98] Heidemeier, O.: Die Erzeugung von 4-regulären, planaren, simplen, zusammenhängenden Graphen mit vorgegebenen Flächentypen. Diplomarbeit, Universität Bielefeld, Fakultät für Wirtschaft und Mathematik (1998)

[Jeo95] Jeong, D.: Realizations with a cut-through Eulerian circuit. Discrete Math. **137**, 265–275 (1995)

[Kaw96] Kawauchi, A.: A Survey of Knot Theory. Birkhäuser, Basel (1996)

[Lic97] Lickorish, W.B.R.: An Introduction to Knot Theory, Springer, New York (1997)

[Liu98] Liu, Y.: Embedding in Graphs. Kluwer, Dodrecht (1998)

[Ma71] Mani, P.: Automorphismen von polyedrischen Graphen. Math. Ann. **192**, 279–303 (1971)

[Rol76] Rolfsen, D.: Knots and Links, Mathematics Lecture Series 7, Publish or Perish, Berkeley, (1976); second corrected printing: Publish or Perish, Houston (1990)

[Ste16] Steinitz, E.: Polyeder und raumeinteilungen. In: Meyer, W.F., Mohrmann. H., Teubner, B.G. (eds.) Encyklopädie der mathematischen Wissenschaften, Dritter Band: Geometrie, III.1.2., Heft 9, Kapitel III A B 12, pp. 1–139. Leipzig (1922)

[Sh75] Shank, H.: The theory of left-right paths. In: Combinatorial Mathematics III, Proceedings of 3-rd Australian Conference, St. Lucia. Lecture Notes in Mathematics, vol. 452, pp. 42–54. Springer, New York, 1975 (1974)

[GAP] The GAP Group, GAP—Groups, Algorithms, and Programming, Version 4.3 (2002). http://www.gap-system.org

[YHZQ10] Yao, H., Hu, G., Zhang, H., Qiu, W.Y.: The construction of 4-regular polyhedra containing triangles, quadrilaterals and pentagons. MATCH **64**, 345–358 (2010)

Chapter 2
Zigzags of Fullerenes and c-Disk-Fullerenes

The *fullerenes*, i.e., the maps $(\{5, 6\}, 3)$-$\mathbb{S}^2$, are of particular interest in Carbon Chemistry. Denote by $F_v(G)$ any v-vertex fullerene of symmetry G. Denote by C_v and call *IP fullerene* any F_v with *isolated* (i.e., no two of them are adjacent) 5-gons. A number of C_v's with $60 \leq v < 100$, including $C_{60}(I_h)$, $C_{70}(D_{5h})$, $C_{76}(D_2)$, and some with $v = 78, 82, 84$ have been characterized as all-carbon molecular cages.

When v is of moderate size, a useful notation taken up in IUPAC nomenclature, is to label $F_v(G)$ as $v : m$ or $v : m(G)$ where m is the place of the fullerene in the *spiral lexicographical order* of general, or IP fullerenes [FoMa95].

In fact, Goldberg introduced the notion of fullerenes (as putative best approximations of $\mathbb{S}^2$) already in [Gold34]; he mentioned there that Kirkman, in 1882, found over 80 out of the 89 44-vertex fullerenes. Goldberg defined a *medial polyhedron* as a simple v-vertex ($4 \leq v \neq 18, 22$) polyhedron with only $\lfloor 6 - \frac{24}{v+4} \rfloor$- and $\lfloor 7 - \frac{24}{v+4} \rfloor$-gonal faces. Clearly, the first eight, i.e., ones with $v \leq 20$, medial polyhedra are the 8 dual convex deltahedra (3 simple Platonic solids and all, but the 1-st one, on Fig. 1.12) and the medial polyhedra with $v \geq 20$ are exactly the fullerenes.

We first give statistical information on the zigzags of fullerenes. Then we consider the Kekulé structure, that one can obtain from zigzags, and the z-knotted fullerenes. Later we consider railroads structure, that one can define on fullerenes, and prove that a fullerene without railroads has at most 15 zigzags. Our emphasis then shifts to c-disk-fullerenes, i.e., $(\{5, 6, c\}, 3)$-spheres with $p_c = 1$ for which we consider existence, symmetries, and zigzags structure.

2.1 Zigzags Statistics for Small Fullerenes

Table 2.1 lists all z-vectors for the set of fullerenes with $v \leq 38$ vertices. Table 2.2 lists z-vectors for the IP fullerenes with $v \leq 82$ vertices and 4 of 24 C_{84} including the experimentally observed ones. So, $C_{78} : 1$ and $C_{84} : 23$ are all experimental fullerenes sharing their zigzag structure with a less stable IP *isomer*, i.e., C_v with the same v.

© Springer India 2015
M. Deza et al., *Geometric Structure of Chemistry-Relevant Graphs*,
Forum for Interdisciplinary Mathematics 1, DOI 10.1007/978-81-322-2449-5_2

Table 2.1 Zigzag structure of the fullerenes F_v with $v \leq 38$ vertices

v	z-vectors					
20	10^6					
24	$12; 60_{12,12}$					
26	$12^3; 42_{0,9}$					
28	$12; 32_{0,4}, 40_{0,8}$	12^7				
30	$10^2; 70_{15,10}$	$22_{0,1}, 68_{6,18}$	$12^3; 54_{2,13}$			
32	$14; 82_{10,24}$	$14^2; 34_{0,4}^2$	$14^3; 54_{0,12}$	$12^3; 30_{0,3}, 30_{3,0}$		$12; 84_{18,18}$
34	$102_{17,34}$	$12, 14^3; 48_{1,8}$	$12; 30_{0,3}, 60_{7,9}$	$14^2; 74_{7,18}$		$30_{0,3}, 72_{12,12}$
36	$46_{0,7}, 62_{4,11}$	$16; 22_{0,1}^2, 48_{4,4}$	$12, 14; 36_{0,4}, 46_{1,7}$	$12, 14; 82_{19,11}$	$12, 16; 80_{20,8}$	$12; 30_{0,3}^2, 36_{0,4}$
	$14; 94_{12,28}$	$14^2; 30_{0,3}, 50_{0,9}$	$14^4; 26_{0,1}^2$	$46_{0,8}, 62_{4,12}$	$14^2; 30_{0,3}, 50_{0,9}$	$14^5; 38_{0,4}$
	$26_{0,1}^3, 30_{0,3}$		$12^2; 14^6$			
38	$16; 98_{12,29}$	$16^3; 22_{0,1}^3$	$12; 102_{20,25}$	$16; 36_{0,4}, 62_{7,8}$	$114_{21,36}$	$114_{29,28}$
	$14, 16; 84_{11,18}$	$14; 26_{0,1}, 74_{5,17}$	$114_{27,30}$	$114_{19,38}$	$14; 38_{0,4}, 62_{2,13}$	$16; 48_{2,6}, 50_{1,8}$
	$14^2; 86_{9,22}$	$14^3; 26_{0,1}, 46_{0,8}$	$12^2, 14^2; 62_{1,14}$	$14^6; 30_{0,3}$	$14^4; 58_{1,12}$	

For any v, z-vectors are listed by fullerene, in spiral lexicographic order

Table 2.2 Zigzag structure of the IP fullerenes C_v with $v \leq 82$ and 4 (among 24) C_{84}: 84 : 19(D_{3d}), 84 : 20(T_d), 84 : 23(D_{2d}) and 84 : 24(D_{6h})

v	z-vectors				
60	18^{10}				
70	$20^5; 110_{0,25}$				
72	$108_{0,24}, 108_{12,12}$				
74	$20^9; 42_{0,3}$				
76	$20; 56_{0,4}, 152_{8,40}$	$20^3; 42_{0,3}^4$			
78	$18^2; 198_{39,42}$	$38_{0,1}, 196_{22,58}$	$20^2; 64_{0,8}, 130_{4,31}$	$18^2; 198_{39,42}$	$42_{0,3}, 64_{0,8}^3$
80	$22^5; 130_{0,30}$	$22^2; 98_{0,16}^2$	$20, 22^2; 86_{2,13}, 90_{5,10}$	$20^3; 22^3; 114_{12,12}$	$42_{0,1}, 90_{5,10}, 108_{6,18}$
	$20^2; 90_{5,10}, 110_{15,10}$	20^{12}			
82	$20^2; 22^4; 118_{7,18}$	$20, 22^3; 160_{15,34}$	$22^2; 202_{23,58}$	$22^2; 202_{25,56}$	$42_{0,1}^2, 162_{17,32}$
	$20^2; 42_{0,1}^2, 122_{9,16}$	$42_{0,1}^3, 120_{12,12}$	$20^6; 42_{0,1}^3$	$20^4; 42_{0,1}^2, 82_{1,8}$	
84	$20^6; 22^6$	$42_{0,1}^6$	$42_{0,1}^6$	$20^6; 22^6$	

Table 2.3 z-uniform not z-knotted fullerenes F_v with $v \leq 60$

Fullerene $v{:}m$	Group	Orbits	Zigzag	Int. vector
20:1	I_h	6	10	2^5
28:2	T_d	4, 3	12	2^6
40:40	T_d	4	$30_{0,3}$	8^3
44:73	T	3	$44_{0,4}$	18^2
44:83	D_2	2	$66_{5,10}$	36
48:84	C_2	2	$72_{7,9}$	40
48:188	D_3	3, 3, 3	16	2^8
52:237	C_3	3	$52_{2,4}$	20^2
52:437	T	3	$52_{0,8}$	18^2
56:293	C_2	2	$84_{7,13}$	44
56:349	C_2	2	$84_{5,13}$	48
56:393	C_3	3	$56_{3,5}$	20^2
60:1193	C_2	2	$90_{7,13}$	50
60:1197	D_2	2	$90_{13,8}$	48
60:1803	D_3	6, 3, 1	18	2^9
60:1812	I_h	10	18	2^9

Table 2.4 z-uniform not z-knotted IP fullerenes C_v with $v \leq 100$

Fullerene $v{:}m$	Group	Orbits	Zigzag	Int. vector
60:1812	I_h	10	18	2^9
80:7	I_h	12	20	2^{10}
84:20	T_d	6	$42_{0,1}$	8^5
84:23	D_{2d}	4, 2	$42_{0,1}$	8^5
86:19	D_3	3	$86_{1,10}$	32^2
88:34	T	12	22	2^{11}
92:86	T	6	$46_{0,3}$	8^5
94:110	C_3	3	$94_{2,13}$	32^2
100:387	C_2	2	$150_{13,22}$	80
100:438	D_2	2	$150_{15,20}$	80
100:432	D_2	2	$150_{17,16}$	84
100:445	D_2	2	$150_{17,16}$	84

Table 2.3 lists z-uniform fullerenes F_v with $v \leq 60$ vertices, and Table 2.4 lists the z-uniform IP fullerenes C_v with $v \leq 100$ vertices. Note that in fullerenes $44 : 37(T), 52 : 437(T), 60 : 1812(I_h)$ each 6-gon is adjacent to exactly three 6-gons, while each 5-gon is adjacent to exactly 2, 1, 0 5-gons, respectively.

The smallest not z-balanced fullerene is a F_{52} (D_{2d}) with $\mathbf{z} = (16^4; 92_{12,12})$; not all of its zigzags of length 16 have the same intersection vector. The smallest *pure*

(a)   **(b)**

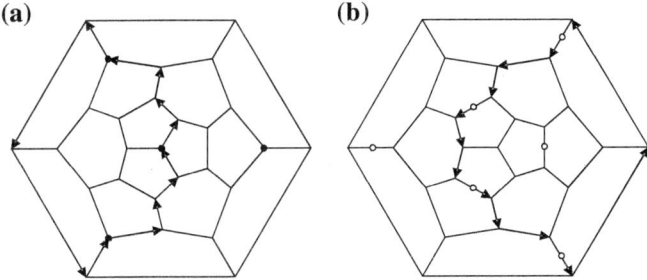

Fig. 2.1 The smallest fullerene, $F_{28}(T_d)$, that is z-uniform, $\mathbf{z} = 12^7$, but not z-transitive. The zigzag in (**a**) belongs to an orbit of size 4; the zigzag in (**b**) belongs to an orbit of size 3

not z-balanced fullerenes are $F_{108}(D_{2d})$ with $\mathbf{z} = (24^8, 26^4, 28)$ and $F_{144}(D_3)$ with $\mathbf{z} = (28^{12}, 32^3)$. Any z-uniform or tight pure fullerene F_v with $v \leq 200$ is z-balanced.

Icosahedral (i.e., of the symmetry I_h or I) fullerenes are called *Goldberg polyhedra*, since they are all available by the Goldberg–Coxeter construction (see [Gold37] and Chaps. 6 and 7) as GC_{kl} (Dodecahedron) with integers $0 \leq l \leq k$ specifying a net on the triangulation of the plane. Such fullerenes have $v = 20(k^2 + kl + l^2)$ vertices. The fullerene has I_h symmetry whenever $l = 0$ or $l = k$ and has I symmetry otherwise. The z-vectors of the icosahedral fullerenes C_v with $v < 2,000$ are listed in Table 2.5. All icosahedral fullerenes are IP and z-uniform, but they are z-transitive only for $gcd(k, l) = 1, 2$. It is sufficient to study only the case where k and l are coprime. In general, if the coprime parent (k, l) has $\mathbf{z} = s^t$, then the derived (ik, il) has $\mathbf{z} = is^{it}$ with $\lfloor \frac{i}{2} \rfloor$ orbits of size $2t$ and one orbit of size t, when i is odd. In fact [DuDe04], t is 6, 10 or 15; see Theorem 7.10. The icosahedral fullerenes may admit multiple solutions (k, l). There are fullerenes $C_{980}(I)$ with $(k, l) = (5, 3)$ (see Table 2.5) and $C_{980}(I_h)$ with $(k, l) = (7, 0)$ having $\mathbf{z} = 70^{42}$ and intersection vector 2^{35}.

2.2 Kekulé Graphs

A useful, though not infallible, indicator of stability of a π*framework* of molecule is that the corresponding plane graph should have a large number of perfect matchings, i.e., in chemical terms, *Kekulé structures*. Every fullerene has at least three [KlLi92] and typically very many more; $C_{60}(I_h)$ has 12,500 [KHCS85] and Borosphere (Fig. 1.13) has 286,224 of them. Call a *Kekulé graph* any 3-regular v-vertex plane graph admitting orientations of its zigzags, in which the number of edges of type I reaches the minimum $\frac{v}{2}$, according to the Proposition 1.2 (iii). A particularly simple perfect matching is formed by edges of type I of this graph.

Proposition 2.1 *Let G be a Kekulé graph with such orientation of its zigzags that the number e_I of edges of type I is $\frac{v}{2}$. If $p_i = 0$ for $i \geq 8$, then it holds*

Table 2.5 Icosahedral fullerenes C_v for $v \leq 2,000$. Only those cases defined by coprime pairs (k, l) are listed. N_z is the number of zigzags

v	k, l	N_z	Zigzag	Int. vector
20	1, 0	6	10	2^5
60	1, 1	10	18	2^9
140	2, 1	15	28	2^{14}
260	3, 1	10	$78_{0,3}$	8^9
380	3, 2	6	$190_{0,15}$	32^5
420	4, 1	6	$210_{5,10}$	36^5
620	5, 1	6	$310_{15,10}$	52^5
740	4, 3	6	$370_{0,25}$	64^5
780	5, 2	6	$390_{0,25}$	68^5
860	6, 1	10	$258_{0,9}$	$24^3, 28^6$
980	5, 3	10	$294_{0,9}$	$28^3, 32^6$
1140	7, 1	15	$228_{0,4}$	$14^8, 18^6$
1220	5, 4	15	$244_{0,4}$	$14^4, 18^{10}$
1340	7, 2	15	$268_{0,8}$	18^{14}
1460	8, 1	10	$438_{0,18}$	$42^3, 46^6$
1580	7, 3	10	$474_{0,18}$	$46^3, 50^6$
1820	6, 5	6	$910_{0,70}$	154^5
1820	9, 1	6	$910_{30,40}$	154^5
1860	7, 4	6	$930_{0,70}$	158^5
1920	8, 3	6	$970_{10,70}$	162^5

 (i) G has $f = \frac{v}{2} + 2$ faces, and two of them are empty, i.e., have no edges of type I, while others have exactly two such edges;
(ii) the two empty faces define uniquely the set of edges of type I over G; they are antipodal, in the sense that they span a diameter in the dual graph.

Proof (i). Any 3-regular v-vertex plane graph has $2 + \frac{v}{2}$ faces by Euler's formula. By Proposition 1.2 (ii), and since $p_i = 0$ for $i \geq 8$, each face have either 0 or 2 edges of type I. By Proposition 1.2 (i), equality $e_I = \frac{v}{2}$ implies that the edges of type I form a perfect matching of G. So, the number of faces with two edges of type I is e_I.

(ii). Let $F_1, \ldots, F_p$. be the set of faces adjacent to one empty face F_0. Every pair F_i, F_{i+1} is incident to one edge, say e_i, that must be of type I since e_i is incident to one vertex of F_0. The face F_i is incident to edges e_i and e_{i-1} and so, all other edges of F_i are of type II. By induction, one is able to assign a type to every edge.   □

2.3 *z*-knot Fullerenes

As a result of computations—all *z*-knotted simple polyhedra ($v \leq 24$), all *z*-knotted fullerenes F_n ($v \leq 74$) and all IP *z*-knotted fullerenes C_n ($v \leq 120$)—we conjectured:

Conjecture 2.1 [DDF04] *For any z-knotted 3-regular plane graph, e_1 is odd.*

The condition of 3-regularity is necessary, as, for example, the dual of the odd-gonal *Prism$_m$* has $e_1 = 2m$; see Table 1.1.

There are only two *z*-knotted F_v with $v \leq 40$ vertices; they are shown in Fig. 2.2. Tables 2.6, 2.7 and 2.8 give the statistics of *z*-knotted ones among 3-regular polyhedra, general fullerenes, and IP fullerenes. Their proportion among fullerenes should be small, because [ScZJ04] implies: for any m, the proportion, among 3-regular *v*-vertex plane graphs of those having at most m zigzags goes to 0 with $v \rightarrow \infty$. In fact, *z*-knot fullerenes account for only 19 out of a total of 2,706 IP fullerenes with $v \leq 100$, and of these 19 only 7 are Kekulé *z*-knotted ones (Table 2.8).

The unique zigzag of a *z*-knot fullerene must transform into itself (up to reversal of all arrows) under all symmetry operations belonging its group, G. Thus each sets of e_1 edges of type I and of e_2 edges of type II must comprise an integer number of orbits of G. Recall that, in a polyhedron, the site symmetry of an edge is C_{2v} or one of its subgroups C_2, C_s, or C_1. The orbits that may occur in the edge-sets are therefore limited to sizes $\frac{|G|}{4}$, $\frac{|G|}{2}$, or $|G|$.

The Conjecture 2.1 implies that the set of edges of type I must be spanned by an odd number of orbits of odd size. So, centrosymmetric groups are not possible for *z*-knot fullerenes with odd e_1. Application of the orbit-parity argument to the list of all 28 fullerene point groups reduces it to 11 candidates for *z*-knot fullerene groups: $D_{5h}, D_5, D_{3h}, D_3, C_{3h}, C_{3v}, C_3, C_{2v}, C_2, C_s, C_1$.

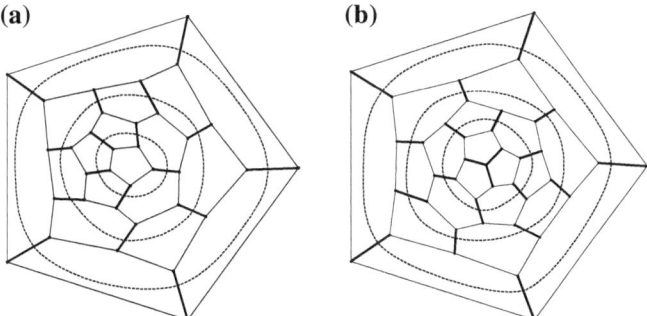

(a)   **(b)**

Fig. 2.2 a The smallest *z*-knot fullerene, 34:1 (C_2), with $\mathbf{z} = 102_{17,34}$. The $\frac{v}{2} = 17$ edges of self-intersection of type I are marked by *thick lines*. They span a perfect matching (Kekulé structure). *Dotted curves* indicate lines of latitude of the Föppl structure [Fö77] of the vertices. **b** The smallest *near-Kekulé* (i.e., $e_1 = \frac{v}{2} + 1$) *z*-knot fullerene 40:22 (C_1) with $\mathbf{z} = 120_{21,39}$

Table 2.6 Statistics of occurrence of knots in small 3-regular polyhedra

v	N_P	N_{knot}	N_{min}	Symmetries
4	1	0	0	–
6	1	1	1	$D_{3h}(1)$
8	2	0	0	–
10	5	3	1	$C_{2v}(1),C_{3v}(1),D_{5h}(1)$
12	14	4	0	$C_1(2),C_s(2)$
14	50	22	4	$C_1(8),C_2(3),C_{2v}(4),C_s(6),D_{7h}(1)$
16	233	70	0	$C_1(53),C_s(15),C_{3v}(2)$
18	1249	482	13	$C_1(398),C_s(45),C_2(27),C_{2v}(10),D_{3h}(1),D_{9h}(1)$
20	7595	2955	0	$C_1(2816),C_s(138),C_3(1)$
22	49566	17901	168	$C_1(17306),C_s(366),C_2(196),C_3(5),C_{2v}(33),D_{11h}(1)$
24	339722	114642	0	$C_1(113604),C_s(1026),C_{3v}(8),C_3(4)$

At each number v of vertices, N_P is the number of non-isomorphic polyhedra, N_{knot} and N_{min} are the number of z-knotted and Kekulé z-knotted ones among them. The final column gives their breakdown by symmetry

Table 2.7 Statistics on z-knot fullerenes with $v \leq 74$ vertices

v	N_{full}	N_{knot}	N_{min}	$N(C_1)$	$N(C_2)$	$N(C_3)$	$N(D_3)$
34	6	1	1	0	1	0	0
36	15	0	0	0	0	0	0
38	17	4	1	1	2	0	1
40	40	1	1	1	0	0	0
42	49	6	2	2	3	0	1
44	89	9	6	9	0	0	0
46	116	15	2	6	9	0	0
48	199	23	13	23	0	0	0
50	271	30	6	21	8	0	1
52	437	42	13	42	0	0	0
54	580	93	16	69	23	0	1
56	924	87	26	87	0	0	0
58	1205	186	11	155	30	1	0
60	1812	206	63	206	0	0	0
62	2385	341	20	297	41	2	1
64	3465	437	148	436	0	1	0
66	4478	567	64	507	59	0	1
68	6332	894	203	892	0	2	0
70	8149	1048	139	967	80	1	0
72	11190	1613	255	1612	0	1	0
74	14246	1970	200	1865	104	0	1

At each v, N_{full} is the number of such fullerenes; N_{knot}, N_{min}, and $N(G)$ are the number of z-knotted, Kekulé z-knotted, and those with symmetry G among them. The group D_5 appears 1-st for z-knot fullerene F_{90}

Table 2.8 IP z-knot fullerenes C_v with $v \leq 98$ vertices

v	Signature		Kekulé?
86	43, 86	C_2:2	Yes
90	47, 88	C_1:7	No
	53, 82	C_2:19	No
	71, 64	C_2:6	No
94	47, 94	C_1:60; C_2:26, C_2:126	Yes
	65, 76	C_2:121	No
	69, 72	C_2:7	No
96	49, 95	C_1:65	Near
	53, 91	C_1:7, C_1:37, C_1:63	No
98	49, 98	C_2:191, C_2:194, C_2:196	Yes
	63, 84	C_1:49	No
	75, 72	C_1:29	No
	77, 70	C_1:5; C_2:221	No

For each v and signature e_1, e_2, they are listed by symmetry and position in the spiral lexicographic order

All five pure rotation groups in above list (and only they) have been found as groups of some z-knot fullerene. So, in [DDF04] we conjectured that all z-knot fullerenes are chiral and that the set D_5, D_3, C_3, C_2, C_1 is the complete list of their point groups. The smallest D_5 z-knot fullerene is the unique $F_{90}(D_5)$ and smallest cases other 4 groups are given in Table 2.7. Trivial symmetry appears to dominate.

2.4 Railroads in Fullerenes

We conjectured in [DDF04] on the basis of calculations on fullerenes with $v \leq 74$ (general) and $v \leq 120$ (IP fullerenes) that there is at least one tight fullerene for each $v \geq 20$, $v \neq 22$. First and 7-th fullerenes on Fig. 2.3 are not tight.

The smallest fullerene with a belt is the 1-st on Fig. 2.3. See on Fig. 2.4 the smallest icosahedral fullerene with a belt. Figure 2.5a shows the smallest fullerene with a double self-intersection of a railroad, and Fig. 2.6a the smallest such IP fullerene (Figs. 2.7 and 2.8). A fullerene with a triple self-intersection of a railroad (conjectured to be such smallest) is shown in Fig. 2.9a.

Railroads without triple self-intersection can be seen as projections of alternating knots, and every fullerene with simple or doubly intersecting railroads has therefore an associated list of knots (Tables 2.9 and 2.10).

The *Conway graph* $(k \times m)^*$ (see, for example, [Kaw96]) is, for $k = 2$, $APrism_m$; for $k > 2$, it comes from $((k - 1) \times m)^*$ by inscribing an m-gon in the 1-st of its two m-gons. The group of $(k \times m)^*$ is D_{mh} for k even, and D_{md} for k odd, apart from

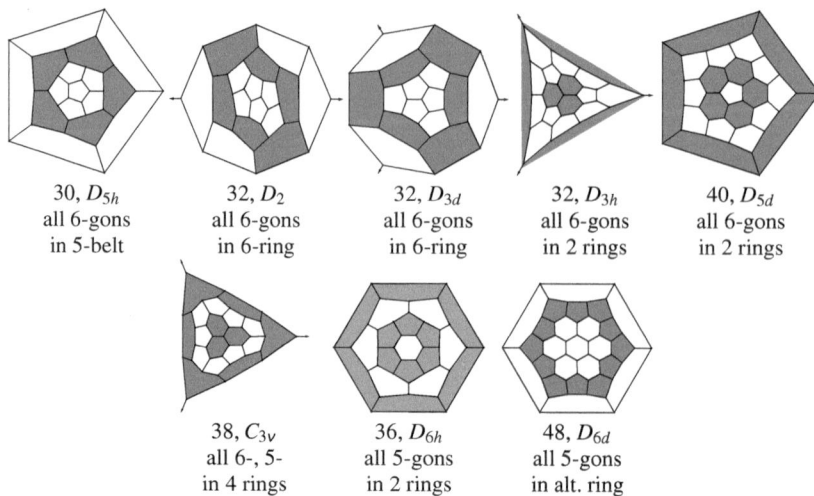

30, D_{5h}	32, D_2	32, D_{3d}	32, D_{3h}	40, D_{5d}
all 6-gons	all 6-gons	all 6-gons	all 6-gons	all 6-gons
in 5-belt	in 6-ring	in 6-ring	in 2 rings	in 2 rings

38, C_{3v}	36, D_{6h}	48, D_{6d}
all 6-, 5-	all 5-gons	all 5-gons
in 4 rings	in 2 rings	in alt. ring

Fig. 2.3 Some fullerenes with 6- or 5-gons, organized in rings. First four and 6-th, 7-th are given as 1-st, 2-nd, 3-rd, 5-th and 16-th, 5-th for $v = 30, 32, 32, 32$ and $38, 36$, respectively, in Table 2.1

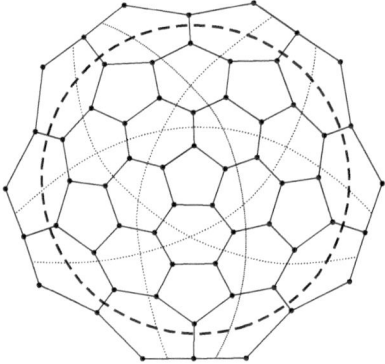

Fig. 2.4 $C_{80}(I_h)$, the smallest icosahedral fullerene with a belt, has $\mathbf{z} = 20^{12}$ and 6 belts

$(2 \times 3)^{\star}$ (i.e., Octahedron) and $(3 \times 4)^{\star}$ (i.e., Cuboctahedron) where it is O_h. In terms of links, $(2 \times 2)^{*} = 4_1$, $(2 \times 3)^{\star} = 6_2^3$, $(2 \times 4)^{\star} = 8_{18}$.

In the open nanotubes and the *graphite sheet* {6; 3}, both zigzags and railroads can become doubly infinite rays (which can be seen as circuits including the point at infinity). The graphite sheet has three parallel classes of infinite railroads, each of them simple; accordingly, it has three parallel classes of infinite simple zigzags (Fig. 2.7) that form a single orbit. *Nanotubes* are formally constructed by rolling up the graphite sheet, with each tube identified by a two-parameter lattice vector, and their zigzag structure derives from that of graphite itself. The three types of nanotube, *zigzag, armchair* and *chiral* [DDE96] have parameter signatures $(n, 0)$, (n, n), (n, m) $(n \neq m)$. Zigzag structures of these three types are:

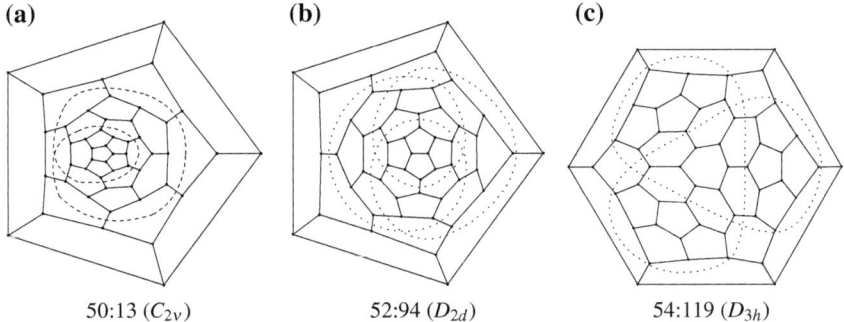

50:13 (C_{2v})   52:94 (D_{2d})   54:119 (D_{3h})

Fig. 2.5 Smallest fullerenes with self-intersecting railroads (indicated by *dashed lines*): **a** The smallest fullerene with a railroad that is a non-minimal projection of the trivial knot 0_1; **b** the smallest fullerene with a railroad that is a projection of a nontrivial knot (in this case the figure-of-eight knot 4_1); it is also the smallest not z-balanced fullerene; **c** the smallest fullerene with a railroad that is a projection of Trefoil knot 3_1

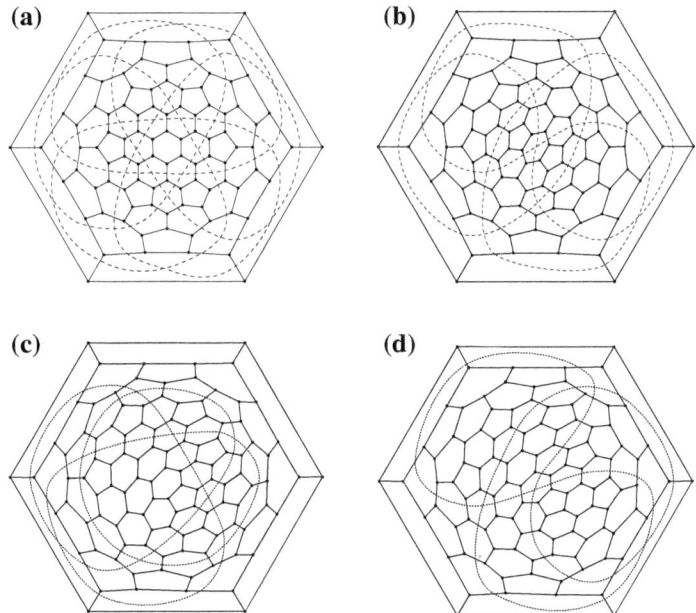

Fig. 2.6 IP fullerenes with self-intersecting railroads (indicated by *dashed lines*): **a** The smallest such fullerene, 96:187 (D_{6d}) = $GC_{2,0}(F_{24})$, has a railroad that is a projection of the knot $(4 \times 6)^*$. **b** 104:823 (D_{3h}) is the smallest fullerene with a railroad that is a projection of the knot 9_{40}. **c, d** 112:3341 has railroads corresponding to projections of two knots: the 4_1 (**c**) and the knot 8_{18} (**d**)

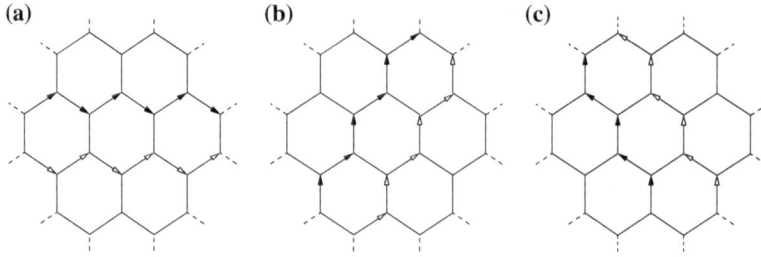

Fig. 2.7 The three parallel classes of zigzags in the graphite sheet $\{6; 3\}$

(a) for $(n, 0)$: two orbits of simple zigzags, one composed of concentric circuits of length $2n$ around the body of the tube, the other of two parallel classes of doubly infinite rays, related by reflection.

(b) for (n, n): two orbits of simple zigzags, both consisting of doubly infinite rays, one 'vertical' and one 'oblique'.

(c) for (n, m), $n \neq m$: 3 orbits of doubly infinite rays, each consisting of a parallel class of helically wound zigzags wrapping around the long axis of the tube.

In all 3 cases, the stack within each parallel class of zigzags corresponds to a stack of doubly infinite railroads.

2.5 Fullerenes 5_v with Simple Zigzags

Figure 2.8 gives all tight pure 5_v, $v \leq 200$; we conjecture that this list is complete. (Among them only $5_{60}(I_h)$, $5_{88}(T)$, $5_{140}(I)$ are IP and only $5_{76}(D_{2d})$ has 6 isolated pairs of adjacent pentagons.) The largest one, $5_{140}(I)$, has $\mathbf{z} = 28^{15}$; by Theorem 2.5, any tight pure 5_v has at most 15 zigzags. We expect that any tight 5_v has at most 15 zigzags. Also, we conjecture that a tight 5_v exists for any even $v \geq 20$, $v \neq 22$.

The central circuits of seven z-uniform pure fullerenes from Fig. 2.8 produce seven new *Grünbaum arrangements* of Jordan curves, defined in Sect. 1.3.

Theorem 2.1 *For any even number $h \geq 2$, there exists a fullerene 5_v, with $v = 18h - 8$, having two simple zigzags intersecting in exactly h edges. For $h = 2$, it is $5_{28}(T_d)$; for $h \geq 4$, it has symmetry D_{2h}, D_{2d} if $\frac{h}{2}$ is even, odd respectively.*

Proof For any even $h \geq 2$, there exists a unique h-vertex 4-regular plane graph H, whose faces are four 2-goons (in two pairs of adjacent ones) and $\frac{h}{2} - 2$ 4-gons only, and having two simple central circuits (see [DeSt03]). This graph has symmetry D_{4h} for $h = 2, 4$ and, for larger values, symmetry D_{2h}, D_{2d} if $\frac{h}{2}$ is even, odd respectively.

We identify the two central circuits of H with zigzags, Z_1 and Z_2, and each vertex with an edge of intersection between them. Every face of H can be seen as a patch in 5_v, which we will construct, and so, the local Euler formula (Theorem 3.1) can be

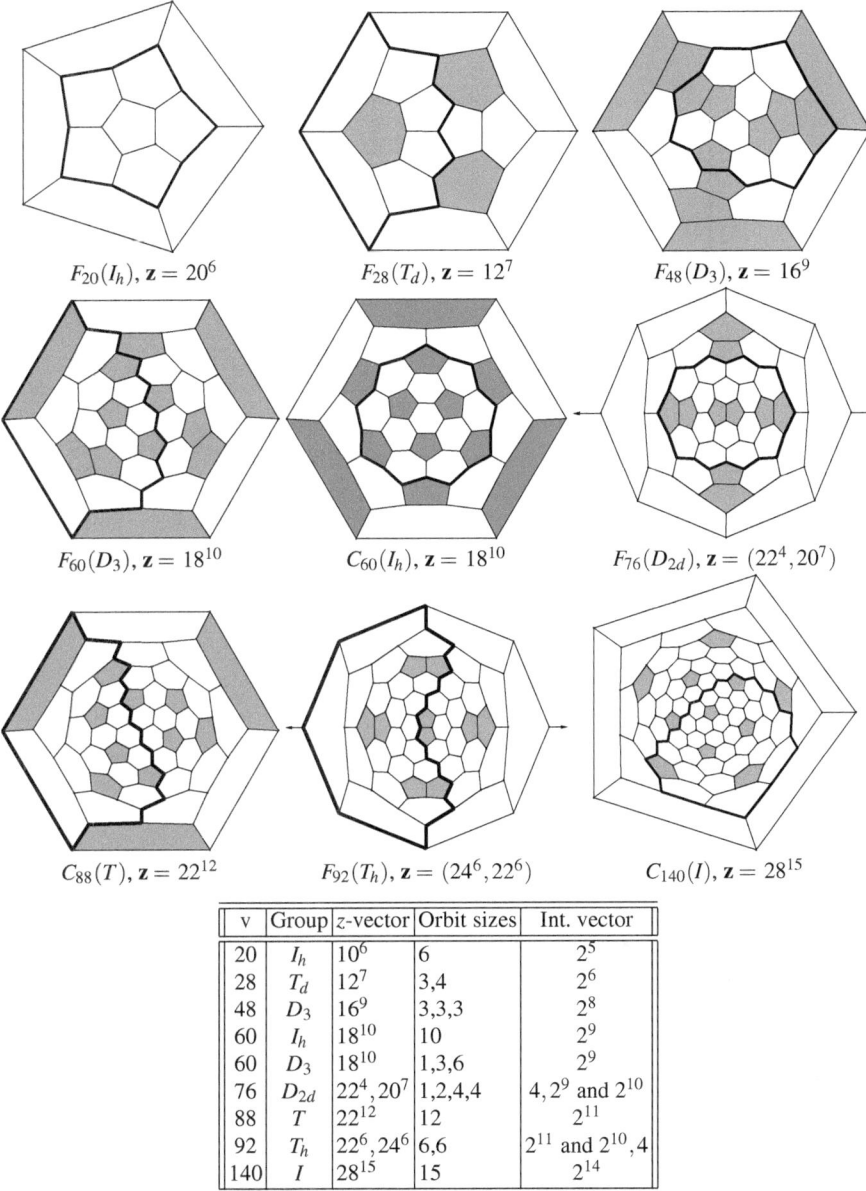

$F_{20}(I_h)$, $\mathbf{z} = 20^6$ $F_{28}(T_d)$, $\mathbf{z} = 12^7$ $F_{48}(D_3)$, $\mathbf{z} = 16^9$

$F_{60}(D_3)$, $\mathbf{z} = 18^{10}$ $C_{60}(I_h)$, $\mathbf{z} = 18^{10}$ $F_{76}(D_{2d})$, $\mathbf{z} = (22^4, 20^7)$

$C_{88}(T)$, $\mathbf{z} = 22^{12}$ $F_{92}(T_h)$, $\mathbf{z} = (24^6, 22^6)$ $C_{140}(I)$, $\mathbf{z} = 28^{15}$

v	Group	z-vector	Orbit sizes	Int. vector
20	I_h	10^6	6	2^5
28	T_d	12^7	3,4	2^6
48	D_3	16^9	3,3,3	2^8
60	I_h	18^{10}	10	2^9
60	D_3	18^{10}	1,3,6	2^9
76	D_{2d}	$22^4, 20^7$	1,2,4,4	$4, 2^9$ and 2^{10}
88	T	22^{12}	12	2^{11}
92	T_h	$22^6, 24^6$	6,6	2^{11} and $2^{10}, 4$
140	I	28^{15}	15	2^{14}

Fig. 2.8 All pure tight fullerenes 5_v with $v \leq 200$; no others are expected for larger v

(a)   **(b)**

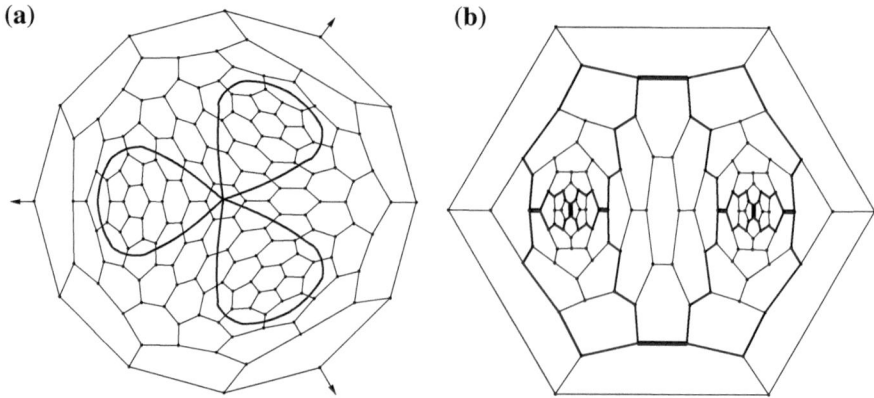

Fig. 2.9 Two particular fullerenes. **a** $5_{172}(C_{3v})$ with triple self-intersecting railroad (Trifolium). **b** Two simple zigzags Z, Z' in $5_{136}(D_{2h})$ with $|Z \cap Z'| = 8$ (8 underlined edges)

Table 2.9 Fullerenes F_n ($n \leq 60$) with self-intersecting railroads are listed below by position in the spiral lexicographic order, with the sizes of the orbits of zigzags and their r- (i.e., railroad) knots

Fullerene	Group	Orbits	z-vector	r-knots
50:13	C_{2v}	2,1,1	$20;\ \mathbf{24^2_{0,1}},\ 82_{14,5}$	0_1
52:94	D_{2d}	2,2	$\mathbf{38^2_{0,3}},\ \mathbf{40^2_{0,4}}$	4_1
54:13	C_2	2,2,1,1	$20^3;\ \mathbf{24^2_{0,1}},\ 54_{2,5}$	0_1
54:119	D_{3h}	3,2,1	$16^3;\ \mathbf{36^2_{3,0}},\ 42_{0,3}$	3_1
60:27	C_s	1,1,1,1	$14;\ \mathbf{26^2_{0,1}},$ $114_{18,14}$	0_1
60:34	C_s	1,1,1	$\mathbf{26^2_{0,1}},\ 128_{19,23}z$	0_1
60:207	C_s	1,1,1	$\mathbf{28^2_{0,1}},\ 124_{17,21}$	0_1
60:208	C_{2v}	2,1,1	$\mathbf{28^2_{0,1}},\ 60_{0,8},\ 64_{8,2}$	0_1
60:1379	C_{2v}	2,2	$\mathbf{28^2_{0,1}},\ 62^2_{2,6}$	0_1

A bold entry in a z-vector shows the origin of a significant knotted railroad. Knots are named in the Rolfsen notation [Rol76, Kaw96] and illustrated in Fig. 1.4

applied. Fix a face F of H; one can assign to every angle of F an angle (obtuse or acute), so that every 2-gon has one acute and one obtuse angle, while every 4-gon has two obtuse and two acute angles. See below the graph for the 1-st values of h.

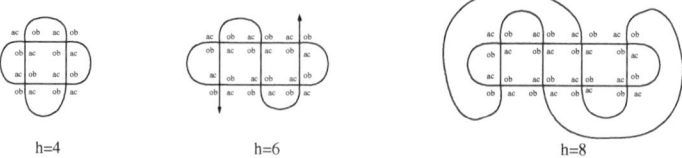

h=4   h=6   h=8

Table 2.10 IP fullerenes C_v ($v \leq 114$) having self-intersecting railroads

Fullerene	Group	Orbits	z-vector	r-knots
96:187	D_{6d}	2,2	$24^2; 120^2_{12,12}$	(0_1); $(4 \times 6)^*$
104:823	D_{3h}	3,3,2	$24^6; 84^2_{0,9}$	$(0_1)^3$; $9_{40} \equiv (3 \times 3)^*$
106:624	C_{3v}	3,1,1	$50^3_{0,1}, 84^2_{0,9}$	$9_{40} \equiv (3 \times 3)^*$
108:897	D_{3h}	6,2	$26^6; 84^2_{0,9}$	$9_{40} \equiv (3 \times 3)^*$
112:3341	D_2	2,2,2	$24^2; 64^2_{0,4}, 80^2_{0,8}$	$4_1, 8_{18} \equiv (2 \times 4)^*$
114:9	C_{2v}	2,2,1	$40^2_{0,1}, 86^2_{2,6}, 90_{3,6}$	0_1
114:1738	C_1	1,1,1,1	$50_{0,1}, 64^2_{0,4}, 82_{0,8}, 82_{1,7}$	4_1
114:2338	C_{2v}	1,1,1,1	$42_{0,1}, 80^2_{0,8}, 140_{14,12}$	$8_{18} \equiv (2 \times 4)^*$
114:3419	C_2	1,1,1	$64^2_{0,4}, 214_{13,46}$	4_1

Where the Rolfsen notation is not available, the Conway graph notation [Kaw96] is used

So, 2-gonal patches will contain three pentagons, while 4-gonal patches will contain only 6-gons. We replace 2- and 4-gonal faces of H by patches depicted below.

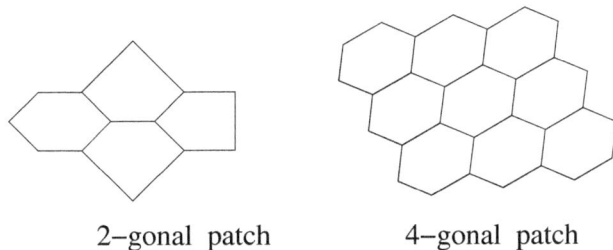

2–gonal patch   4–gonal patch

The obtained graph has $p_6 = 9(h - 2) + 4$, and so, $v = 18h - 8$ vertices. The symmetry group is the same as the one of H, except for the 1-st values $h = 2, 4$. See Fig. 2.9b for the corresponding graph with $h = 8$. □

2.6 Tight Fullerenes

Theorem 2.2 *The number of zigzags of a tight fullerene is at most* 15.

Proof For a zigzag z, we have two sides and the condition that it is tight implies that on each side there is at least one pentagon.

For $1 \leq i \leq 4$, denote by n_i the number of sides incident to exactly i pentagons. Denote by n_5 the number of sides incident to at least 5 pentagons. We denote by n_{sec}, n_{sec2}, and n_{sec3} the number of sides in the configurations depicted on Fig. 2.10 with n_{sec2} and n_{sec3} excluding n_{sec}.

If we have a lonely side, i.e., a side that contains only a single 5-gon, then we have the configuration of Fig. 2.11. The dashed side will eventually close itself, but

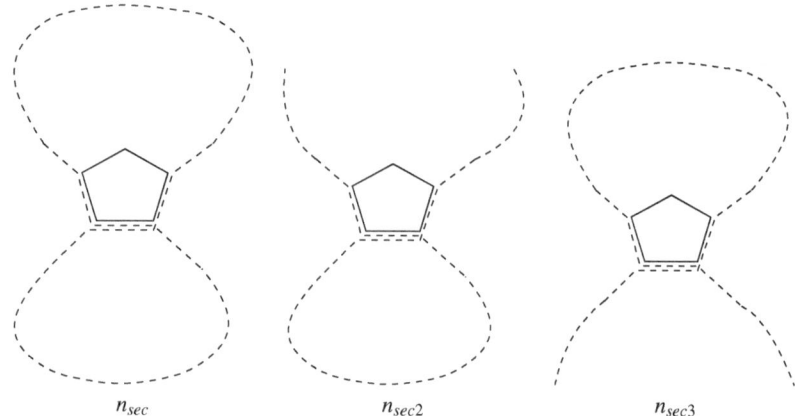

$$n_{sec} \qquad\qquad n_{sec2} \qquad\qquad n_{sec3}$$

Fig. 2.10 Possible local configurations corresponding to a lonely side

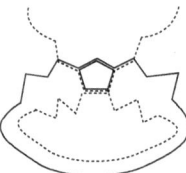

Fig. 2.11 Possible local configurations corresponding to a fourth configuration

it may or may not have another pentagon. Therefore, we are either in configuration n_{sec} or n_{sec2} and this gives the following equation:

$$n_1 = n_{sec} + n_{sec2}.$$

We denote by n_{4sec}, n_{4sec2} the number of sides that contain the configurations of Fig. 2.12. Let us take a configuration n_{sec} or n_{sec3} and add the opposing side passing by the 5-gon, according to Fig. 2.13. When doing this we get a sequence of 6-gon coming from the 5-gon. This sequence has to end on a 5-gon. Therefore, we get a configuration n_{4sec} or n_{4sec2}. This gives the equality:

$$n_{sec} + n_{sec3} = \frac{1}{2}n_{4sec} + n_{4sec2}.$$

The following other inequalities are easy to prove:

$$\begin{cases} n_1 \leq 12 \\ n_{4sec} + n_{4sec2} + n_{sec2} + n_{sec3} \leq n_3 + n_4 + n_5 \\ n_1 + 2n_2 + 3n_3 + 4n_4 + 5n_5 \leq 60 \\ n_{sec} \leq n_2 \\ n_{4sec2} \leq n_5 \\ n_{4sec} \leq n_4 \end{cases}$$

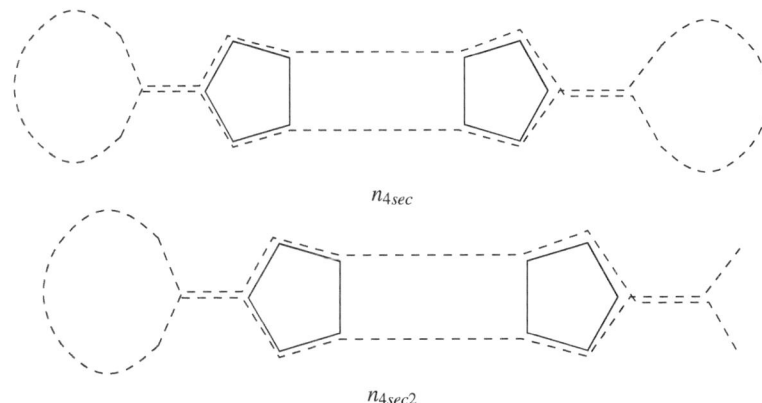

n_{4sec}

n_{4sec2}

Fig. 2.12 A lonely side in overlined drawing and the corresponding side

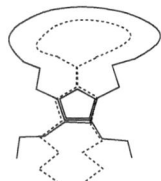

Fig. 2.13 A second configuration and the corresponding side

The objective function giving the number of zigzags is $\frac{1}{2}(n_1 + n_2 + n_3 + n_4 + n_5)$ and the maximum found value of the linear programming problem is 15 as obtained from [Fu]. $\qquad\square$

2.7 Disk-Fullerenes

This section is based partly on [DDS13a, DDS13b].

Given a ($\{5, 6, c\}$,3)-sphere with $c \neq 5, 6$ and $p_c = 1$, call it a *c-disk-fullerene* (or $c - DF$) if the c-gon has no self-intersections (i.e., it can be seen as the boundary of a disk) and call it a *c-multidisk-fullerene* (or $c - MDF$), otherwise. Denote such spheres $c - DF_v(G)$ or $c - MDF_v(G)$ if they have v vertices and the group G of symmetry. So, the fullerenes can be seen as $c - DF$'s with $c = 5, 6$. The unique c-gon is a negatively curved face if and only if $c \geq 7$.

In general, a $c - DF$ and a c-disk, obtained from it by deleting the c-gon, are different geometrical objects; but they have the same skeleton. Since we consider only their combinatorial properties, we will, by abuse of language, treat them as the same object. We will use the term *disk* but present a picture for sphere. Also, a "group" means here the group of a sphere, not only of its skeleton. Call a $c - DF$ with $c \neq 5, 6$ *tight* if it has no *belts*, i.e., simple railroads.

Call a *c-disk-nearfullerene* (or $c - DNF$) a c-disk, in which vertices have degree 3, interior faces are 5- or 6-gons, boundary faces have gonality within $\{2, 3, 4, 5, 6\}$, and at least one boundary face is neither 5, nor 6-gon. Any $c - MDF$ has (only) 1-connected graph—a tree of following possible components, connected by *bridges* (edges between components, i.e., edges of self-intersection of the c-gon):

1. vertex having (i.e., from which emanate) 3 bridges;
2. empty disk having 5 or 6 bridges;
3. disk-fullerene, having bridges on some (at least one) of its boundary 5-gons;
4. disk-nearfullerene, having $5 - t$ or $6 - t$ bridges on boundary t-gons, $2 \leq t \leq 5$.

One can slightly generalize $c - DNF$, in order to incorporate cases 2. and 3. in it. Also, a theory of c-multidisk-fullerenes with infinite c will be of interest.

Call a $c - MDF$ *extensible* if among faces, adjacent to the boundary c-gon, there is a 5-gon; call it *nonextensible*, otherwise, i.e., if all boundary faces are 6-gons.

Let D be a component $c' - DF$ or $c' - DNF$ of a $c - MDF$. One can check that $c' > 6 - t$ if there is a t-gon among the boundary faces of D; so, $c' \geq 2$. Hence, a $c - MDF$ exists if and only if $c \geq 8$, since $c \geq (2 + 1) + 2 + (2 + 1) = 8$. See a minimal $8 - MDF_v$ on Fig. 2.14. Four pictures there are "dumbbells" of two $2 - DF_{38}$, of $2 - DF_{38}$ and $3 - DF_{34}$, of two F_{20} and of two $6 - DNF_{22}$. This $6 - DNF_{22}$ comes from F_{24} by deleting edge connecting a 5-gon with the 6-gon nonadjacent to it.

It is easy to check that $(\{a, b, c\}, k)$-sphere with $p_c = 1$ and $p_b = 0$ has $c = a$, i.e., it is the k-regular map $\{a, k\}$ on the sphere. We conjecture that a $(\{a, b, c\}, k)$-sphere with $p_c = p_b = 1 < a, c$ has $c = b$, i.e., it is a $(\{a, b\}, k)$-sphere with $p_b = 2$.

Theorem 2.3 *The possible symmetry groups of a c-disk-fullerene with $c \neq 5, 6$ or a c-multidisk-fullerene are C_k, C_{kv} with $k \in \{1, 2, 3, 5, 6\}$ and k dividing c.*

Proof Any symmetry of such sphere should stabilize unique c-gon. So, the possible groups are only C_k and C_{kv} with k ($1 \leq k \leq c$) dividing c. Moreover, $k \in \{1, 2, 3, 5, 6\}$, since the axis has to pass by a vertex, edge, or face. Remind that $C_s = C_{1v}$.   □

Let $m(c, G)$ denote the smallest value of p_6 in a $c - DF(G)$ if $c = 1, 2$ and, in a 3-connected $c - DF(G)$, if $c \geq 3$; Table 2.11 gives $m(c, G)$ for small c and, in parenthesis, such values for $c - MDF(G)$, when they exist, i.e., for $c \geq 8$.

Proposition 2.2 [DDS13b]

(i) *The skeleton of any $c - DF$ is 2-connected if and only if $c > 1$.*
(ii) *The skeleton of any $c - DF$ is 3-connected if and only if $3 \leq c \leq 7$.*

Clearly, for $c = 1, 2$, the skeleton S of a $c - DF$ is only c-connected. For $c \geq 3$, it can be only 2- but not 3-connected, when the intersection of the c-gonal boundary with a face is not connected. Then the group of $c - DF$ can be a proper subgroup of $Aut(S)$. Also, the plane realization can be not unique; see, for example, Fig. 2.15.

Such 2-connected skeleton admits a convex realization in $\mathbb{R}^2$, but in $\mathbb{R}^3$ it can be realized only as a convex polyhedral surface with nonplanar boundary.

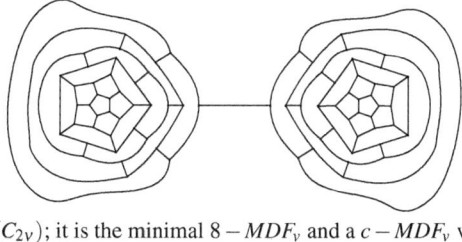

a $8 - MDF_{78}(C_{2v})$; it is the minimal $8 - MDF_v$ and a $c - MDF_v$ with smallest c

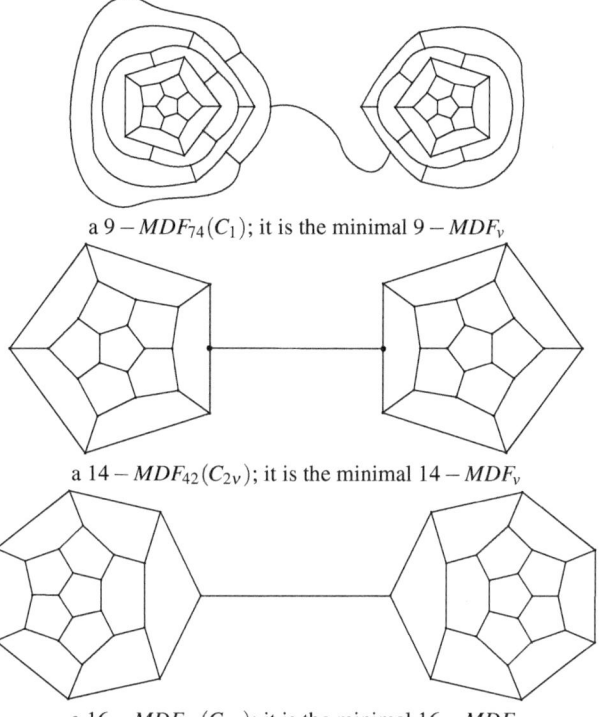

a $9 - MDF_{74}(C_1)$; it is the minimal $9 - MDF_v$

a $14 - MDF_{42}(C_{2v})$; it is the minimal $14 - MDF_v$

a $16 - MDF_{46}(C_{2v})$; it is the minimal $16 - MDF_v$

Fig. 2.14 Some *minimal*, i.e., with minimal v, c-multidisk-fullerenes $c - MDF_v$

Clearly, any $c - DF$ or $c - MDF$ has $p_5 = c + 6$, $v = 2(p_6 + c + 5)$ vertices and there is an infinity of $c - DF$'s for any $c \geq 1$ and of $c - MDF$'s for any $c \geq 8$.

A $(\{5, 6\}; 3)$-*polycycle* (cf. [DeDu08]) is the following relaxation of a $c - DF$: some—say, v_2'—of the vertices of the c-gon have degree 2. Clearly, $p_5 = c + 6 - 2v_2'$.

Let $m(c)$, $m_3(c)$, and $m_2(c)$ denote the smallest value of p_6 in a any $c - DF$, in a $c - DF$ whose skeleton is 3-connected and, respectively, 2- but not 3-connected.

Theorem 2.4 *(i) There are eight $c - DF$ with $p_6 \leq 3$ (all are 3-connected): unique ones $5 - DF_{20}$, $6 - DF_{24}$, $6 - DF_{26}$, $4 - DF_{22}$, $3 - DF_{22}$, $7 - DF_{30}$, and two $6 - DF_{28}$; their p_6 is 0, 1, 2, 2, 3, 3, 3, 3, respectively.*

Table 2.11 Minimal values of p_6 in a $c - DF(G)$ with small c not dividing 5 and 6; in parenthesis, minimum values of p_6 for $c - MDF(G)$, when they exist (i.e., if $c \geq 8$), are given

G	C_1	C_s	C_2	C_{2v}	C_3	C_{3v}
$m(1, G)$	18	14	–	–	–	–
$m(2, G)$	10	9	8	6	–	–
$m(3, G)$	7	5	–	–	9	3
$m(4, G)$	5	3	4	2	–	–
$m(7, G)$	5	3	–	–	–	–
$m(8, G)$	6(?)	5(?)	6(?)	4(26)	–	–
$m(9, G)$	7(23)	6(?)	–	–	12(?)	6(?)
$m(11, G)$	8(10)	8(9)	–	–	–	–
$m(13, G)$	5(5)	5 (4)	–	–	–	–
$m(14, G)$	5(4)	4(3)	5(6)	4(2)	–	–
$m(16, G)$	5(3)	4(3)	4(4)	6(2)	–	–
$m(17, G)$	5(4)	5(3)	–	–	–	–
$m(19, G)$	6(6)	6(5)	–	–	–	–

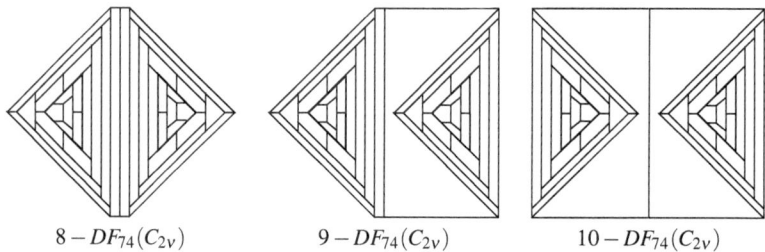

$$8 - DF_{74}(C_{2v}) \qquad 9 - DF_{74}(C_{2v}) \qquad 10 - DF_{74}(C_{2v})$$

Fig. 2.15 Three disk-fullerenes with isomorphic skeletons; also each of them is tight and contains two simple disjoint zigzags of length 10 nonintersecting the boundary

(ii) If $c \leq 11$, then all $c - DF$ realizing $m(c), m_2(c)$, and $m_3(c)$ are 23, 6, and 19 $c - DF$'s given for $c \leq 11$ in the Table 2.12, where the number of $c - DF$'s is given in parentheses if it is not 1; see also Figs. 2.16 and 2.17.

(iii) If $c \geq 12$, then:

(iii1) $m_2(c) = 4$ and $m(c) = m_2(c)$ for $c \equiv 4, 5, 6 \pmod{10}$;

(iii2) $m_2(c) = 5$ for $c \equiv 2, 3, 7, 8 \pmod{10}$;

(iii3) $m_2(c) = 6$ and $m(c) = m_3(c)$ for $c \equiv 0, 1, 9 \pmod{10}$;

(iii3) $4 \leq m_3(c) \leq 6$; we conjecture that $m_3(c) = 6$ and is realized only by the c-pentatube (it is checked it for $12 \leq c \leq 21$).

For $c \geq 12$, the *c-pentatube* is the 3-connected $c - DF_{2(c+11)}$—of symmetry C_2, C_s for even and odd c, respectively—depicted in the end of Fig. 2.16 for $c = 12, 13$. For any c, it consists of the same left and right subgraphs separated by $c - 12$ pentagons. A c-pentatube with c not divisible by 5 has z-vector 10^4; $6c + 26_{3c-3,0}$.

Table 2.12 Minimal values of p_6 in a c-disk-fullerene

c	1	2	3	4	5	6	7	8	9	10	11
$m(c)$	14	6	3	2	0	1	3	4	6 (2)	7 (3)	8 (10)
$m_2(c)$	–	6	–	–	–	–	–	23	17 (2)	10	8 (2)
$m_3(c)$	–	–	3	2	0	1	3	4	6 (2)	7 (3)	8 (8)
c	12	13	14	15	16	17	18	19	20	21	$c \geq 22$
$m(c)$	5	5	4	4	4	5	5	6	6	6	$\min(m_2, m_3)$
$m_2(c)$	5	5	4	4	4	5	5	6	6	6	Theorem 2.4
$m_3(c)$	6	6	6	6	6	6	6	6	6	6	$4 \leq m_3(c) \leq 6$

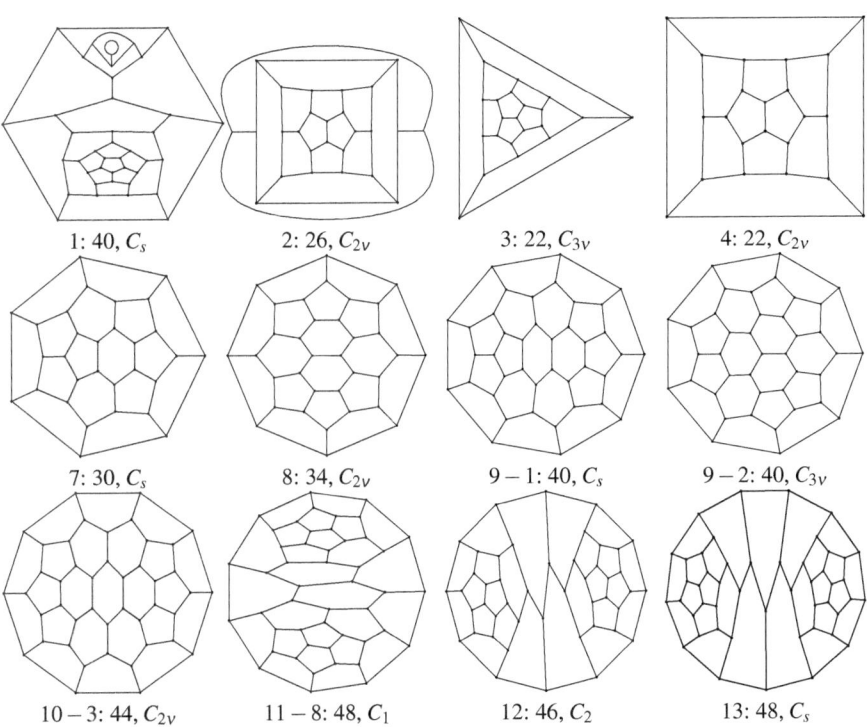

1: 40, C_s 2: 26, C_{2v} 3: 22, C_{3v} 4: 22, C_{2v}

7: 30, C_s 8: 34, C_{2v} 9 − 1: 40, C_s 9 − 2: 40, C_{3v}

10 − 3: 44, C_{2v} 11 − 8: 48, C_1 12: 46, C_2 13: 48, C_s

Fig. 2.16 Minimal $c - DF$ with $c = 1, 2$ and minimal 3-connected $c - DF$ with $3 \leq c \leq 13$, $c \neq 5, 6$; for $c = 10$ and 11 only one of 3 and, respectively, one of 8 is given

Conjecture 2.2 *A $c - DF_v$ exists—except the cases $(c, v) = (1, 42), (3, 24), (5, 22)$—if and only if v is even and $v \geq 2(m(c) + c + 5)$.*

Call c-*thimble* any 2-nd condition is necessary: see Fig. 2.21. It exists if and only if $c \geq 5$ and its skeleton is always 3-connected. One can consider also c-*multi-thimble*,

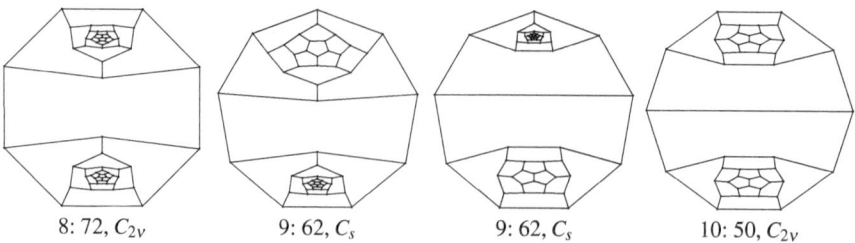

8: 72, C_{2v} 9: 62, C_s 9: 62, C_s 10: 50, C_{2v}

Fig. 2.17 Minimal only 2-connected $c - DF$ with $3 \leq c \leq 10$

i.e., $c - MDF$ such that the c-gon is adjacent only to 5-gons and only once to each of them; one example is $16 - MDF_{46}(C_{2v})$ from Fig. 2.14.

For $c \geq 5$, let $m_t(c)$ denote the smallest value of p_6 in a c-thimble.

Proposition 2.3 [DDS13a] *and* [DDS13b] *It holds* $c - 6 \leq m_t(c) \leq \left\lfloor \frac{3(c-5)}{2} \right\rfloor$.

We expect that $m_t(c) = \left\lfloor \frac{3(c-5)}{2} \right\rfloor$. It holds for $5 \leq c \leq 10$: all minimal $c - DF$ with $5 \leq c \leq 9$, as well as the 3-rd minimal $10 - DF_{44}$ (see Fig. 2.16), are c-thimbles.

A *fullerene c-patch* is a partial subgraph of fullerene bounded by a c-gon; so, it has $p_5 \leq 12$. Such $(\{5, 6\}; 3)$-polycycle is a $c - DF$ if and only if $c \in \{5, 6\}$. Let us cut a fullerene with a belt along the middle line of such railroad. Then each 6-gon splits into two 5-gons, and the fullerene splits into two thimbles. These thimbles are not necessarily identical, and not fullerene patches.

Remark 2.1 In any $c - DF$ it holds the following:

 (i) If two zigzags intersect, then each connected component of their intersection consists of a single edge. In each of them the intersection is transverse, that is, the segments of the zigzags near their common edge are on different sides.
 (ii) Each of c pairs of adjacent edges on the boundary generates a zigzag. These are the only zigzags which intersect the boundary; their number is at most c, while exactly 7 zigzags of length 20 in a $7 - DF_{72}$ on Fig. 2.18a intersect the boundary.
(iii) If an edge does not belong to the boundary, but both ends of the edge belong to it, then any zigzag containing this edge is self-intersecting. If a simple zigzag intersects the boundary, then any two boundary edges in it alternate with at least two non-boundary edges.
(iv) A shortest simple zigzag is the one bounding the half of Dodecahedron; it consists of 10 edges. So, if all zigzags are simple, then there are at most $\frac{e}{5} = \frac{3v}{10}$ of them, since each edge belongs to at most two zigzags.
 (v) Adjacent edges of a face in a $c - DF$ cannot belong to its boundary, since the graph is 3-regular. So, a 6-gon has at most three (and a 5-gon has at most two) edges on the boundary. Each 6-gon in a $c - DF$ belongs to at most three railroads.

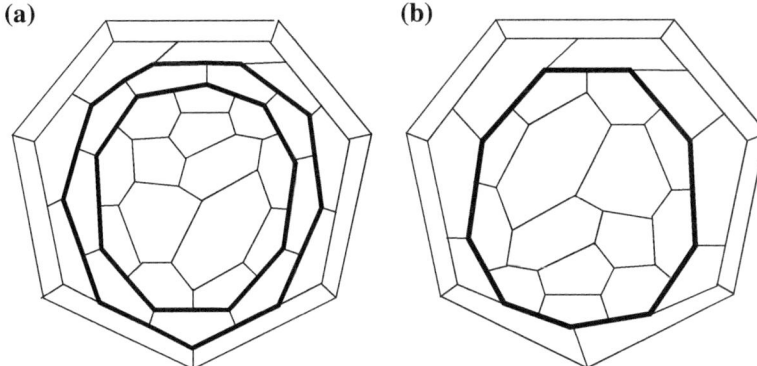

Fig. 2.18 Reduction of a $7 - DF_{72}$ (unique pure $c - DF$ with $p_6 \leq 25$ and $c \leq 15$, $c \neq 5, 6$) to a tight $7 - DF_{54}$; all its zigzags, except two of the length 18 intersect pairwise in two edges. **a** A reducible pure $7 - DF_{72}(C_1)\mathbf{z} = 20^9$, 18^2. **b** One of its reductions, $7 - DF_{54}(C_1)\mathbf{z} = 18$; $144_{24,39}$

(vi) The number of faces adjacent to a simple zigzag on any side is equal to half its length. There is one exception: when counting the number of faces adjacent to a zigzag on the exterior side, the boundary edges of the zigzag are not considered.

Proposition 2.4 *The intersection of a face in a c-disk-fullerene with a simple zigzag is connected.*

Proof A face in a disk-fullerene cannot have just one common edge with a zigzag. A nonempty intersection of a 5-gon with a simple zigzag contains exactly two adjacent edges. Let us prove by contradiction that two opposite pairs of adjacent edges of a 6-gon cannot belong to a simple zigzag. We first prove this for a fullerene. If two opposite pairs of adjacent edges of a 6-gon belong to a simple zigzag, then the face adjacent to the 6-gon along an edge, which does not belong to the zigzag, has a disconnected intersection with the zigzag. So, this face is another 6-gon. There is a third 6-gon which is adjacent to the second and has a disconnected intersection with the zigzag, and so on. All possible ways of termination of this process (in any direction) give an m-gon with $m \neq 5, 6$, i.e., are not possible for a fullerene.

In the case of a disk-fullerene the proof is similar, but the following modification is needed. Consider the 1-st 6-gon which intersects the zigzag in two opposite pairs of adjacent edges. If this 6-gon is on the interior side of the zigzag, then the proof is the same. If the 1-st 6-gon is on the exterior side, then we choose a 2-nd 6-gon on the same side as the part of the zigzag which is interior with respect to the cycle consisting of the other part of the zigzag and an edge of the 1-st 6-gon not belonging to the zigzag. This 2-nd 6-gon is adjacent to a 3-rd 6-gon which also has a disconnected intersection with the zigzag, and so on. As in the case of a fullerene, this process (in only one direction) cannot terminate at a 6-gon. ☐

Lemma 2.1 *All the boundary 5-gons of a thimble form a cylinder, one of whose edges is the boundary of the thimble and the other is a simple zigzag.*

Proof Consider the cyclic sequence of 5-gons adjacent to the boundary of a thimble. The 5-gons in it are all different. Indeed, each 5-gon has a single edge on the boundary. Neighboring 5-gons in the sequence have a common edge with an endpoint on the boundary of the thimble. We will prove that non-neighboring 5-gons do not intersect. Assume that they do intersect, and consider the following two cases:

Case 1. Two non-neighboring 5-gons have a common vertex A which is opposite to the boundary edge of each 5-gon. The two 5-gons have a common edge AB, since the graph is 3-regular. The star of the vertex B contains a third face, which has a common edge BC with one 5-gon and a common edge BD with the other 5-gon. The vertices C and D belong to the boundary of the thimble. Since there are only 5-gons adjacent to the boundary, the third face in the star of B is a 5-gon, which has a disconnected intersection with the boundary. This is a contradiction.

Case 2. Two neighboring 5-gons have a common edge AB, where the vertex A is opposite to the boundary edge of one 5-gon, and the vertex B is opposite to the boundary edge of the other 5-gon. The star of B contains a 3-rd face, which has a common edge with the 1-st 5-gon of the sequence. One of the ends of this edge belongs to the boundary of the thimble. Hence, the 3-rd face in the star of B is a 5-gon. The 3-rd and the 2-nd 5-gons have a common edge, say BC. The star of C consists of the 2-nd 5-gon, the 3-rd 5-gon, and some 4-th 5-gon. The 4-th and 3-rd 5-gons have a common edge CD, and so on. The sequence $ABCD\ldots$ can be continued indefinitely. (Indeed, two sequences of 5-gons, one indexed by even numbers and the other by odd numbers, cannot close up into a cylinder with two boundary components, but they also cannot close up into a Möbius band with only one boundary component.) Again, we end up with a contradiction.   □

In a thimble, pairs of edges of boundary 5-gons with a common vertex, which is opposite to a boundary edge of a 5-gon, form a simple zigzag, because all faces adjacent to the boundary of a thimble are different 5-gons. If all the boundary faces of an $c - DF$ are 5-gons, but the $c - DF$ is not a thimble (see one in Fig. 2.21a), then these pairs of edges form a self-intersecting zigzag. But if all the boundary faces of an $c - DF$ are 5-gons forming a cylinder whose second boundary component is a simple zigzag, then this $c - DF$ is a thimble.

Proposition 2.5 *A disk D cut out from a c-disk-fullerene by a simple zigzag Z contains exactly six pentagons.*

Proof The degrees of the vertices of Z in D alternate between 3 and 2. We attach a 5-gon to each pair of boundary edges of D with a common vertex of degree 3; the number of the attached 5-gons equals half the length of the zigzag. By identifying edges of the 5-gons with vertices of degree 2 in D we obtain a thimble. Besides the boundary 5-gons, this thimble contains exactly six 5-gons belonging to D.   □

Lemma 2.2 *If a disk-fullerene has a simple zigzag, such that all the adjacent faces on one of its sides are 6-gons, then these 6-gons form a belt.*

Proof A 6-gon adjacent to a zigzag contains a pair of zigzag edges with a common vertex (see Proposition 2.4). The number of such 6-gons is half the number of edges in the zigzag; see Remark 2.1 (vi). Take two neighboring 6-gons in the cyclic sequence. They have a common edge. One end of this edge belongs to the zigzag. Denote the other end by A and assume that A belongs to a 3-rd 6-gon which is adjacent to the zigzag along the two edges with common vertex opposite to A. This 3-rd 6-gon is adjacent to the first two 6-gons. Let AB be the common edge of the 3-rd and 2-nd 6-gons. The vertex B belongs to some 4-th face, because the graph is 3-regular. Since one of the ends of the common edge of the 4-th and 3-rd faces belongs to the zigzag, the 4-th face is a 6-gon adjacent to the zigzag. Let BC be the common edge of the 4-th and 2-nd 6-gons. The vertex C belongs to some 5-th face, which is a 6-gon adjacent to the zigzag (by the same reason as for the 4-th face). Let CD be the common edge of the 5-th and 4-th 6-gons. Then the vertex D belongs to a 6-th 6-gon, which is adjacent to the zigzag, and so on.

If we now swap the 2-nd and the 3-rd hexagons, then we obtain two sequences of 6-gons, one indexed by even numbers and the other by odd numbers, that are adjacent along the middle polygonal line $ABCD \ldots$, which is infinite. Indeed, it cannot close up, for suppose that it does close up. It has either an even or an odd number of edges. However, both cases are impossible. In the first case two sequences of 6-gons cannot close up into a cylinder, which has two boundary components. In the second case the two sequences cannot close up into a (forbidden, see above) Möbius band, even though it has one boundary component. We got a contradiction.

So, adjacent 6-gons on one side of the zigzag cannot intersect unless they are neighboring in the cyclic sequence. It follows that these 6-gons form a belt. □

Lemma 2.1 can be viewed as a particular case of Lemma 2.2: one can cut the belt along its middle line and get a thimble.

Proposition 2.6 *If a fullerene has two disjoint simple zigzags, then*

(i) the fullerene has at least one belt
(ii) and both zigzags have the same length.

Proof (i) Two simple zigzags Z_1 and Z_2 cut the fullerene into two disks and an *annulus* (a region bounded by two concentric zigzags). Each disk contains six 5-gons, so the annulus contains only 6-gons. By Lemma 2.2, there exists a belt.

(ii) Since there are only 6-gons between Z_1 and Z_2, we can apply Lemma 2.2. If the first boundary component of the belt is Z_1 and the second boundary component is not Z_2, then there is another belt. If the second boundary component of the second belt is not Z_2, then there is a third belt, and so on. After a finite number of steps we get a belt whose second boundary component is Z_2. Then all the boundary components of all these belts have the same length. □

By Proposition 2.5, a simple zigzag intersects the boundary of a thimble in $s \leq 3$ connected components (pairs of edges); the disk cut by this zigzag has $2s$ boundary 5-gons. See Fig. 2.1 for $s = 0, 1$ and Fig. 2.19b for $s = 0, 2, 3$. On Fig. 2.19a, three longest zigzags, i.e., those of length 72 and 66, have $s = 0$, while each of remaining 15 zigzags has $s = 2$.

(a)   **(b)**

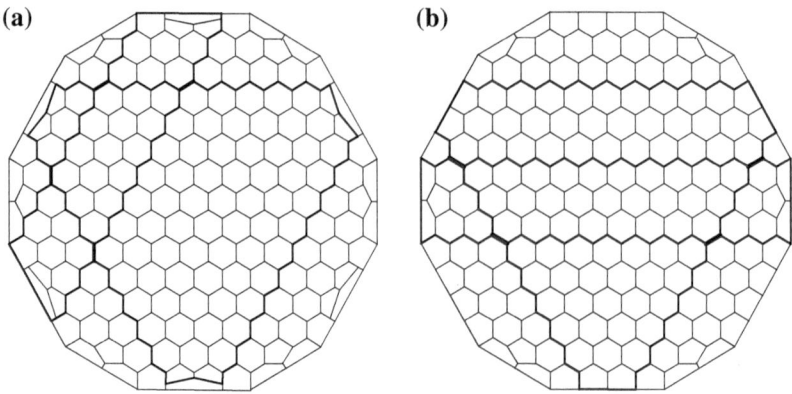

Fig. 2.19 Two special disk-fullerenes. **a** a tight pure $30 - DF_{36}(C_{6v})$ with $\mathbf{z} = 72, 66^2, 58^6, 56^3, 48^6$. **b** a tight 36-thimble (C_{6v}) with two simple zigzags having 2 or 3 boundary components

Proposition 2.7 *If two simple zigzags in a c-thimble do not intersect the boundary and do not intersect each other, then:*

(i)   the thimble has at least one belt
(ii) and the two zigzags have the same length.

Proof (i) By Proposition 2.5, each simple zigzag cuts a disk containing six 5-gons off the thimble. They are all not adjacent to the boundary, because the zigzag does not intersect the boundary. If the disks cut off from the thimble by the two zigzags do not intersect, then the thimble contains 12 rather than 6 non-boundary 5-gons, which is impossible since $p_5 = c + 6$. So, one disk is contained in the other, and all six 5-gons are contained in the smaller disk. Since there are only 6-gons between the zigzags, the existence of a belt follows from Lemma 2.2.

(ii) If the second boundary component of the belt does not coincide with the second zigzag, then there are only 6-gons adjacent to the belt from the other side. These 6-gons form another belt, and so on. After a finite number of steps, we end up at a belt whose boundary component is the second given zigzag. Both boundary components of each belt have the same length.   □

Any disk-fullerene can be turned into a tight disk-fullerene by a finite number of the following *reductions steps*. If a $c - DF$ has a belt, call *z-reduction* of this $c - DF$ the operation of deleting the interior of such railroad with subsequent identification of its boundary zigzags. This step is not uniquely defined.

If the belt does not intersect the boundary of the $c - DF$, then after deleting the interior of such a belt with inner boundary Z_1 and outer boundary Z_2 we obtain a disk with boundary Z_1 and an annulus with inner boundary Z_2. They can be united into a new $c - DF$ in z different ways, where z is the number of edges of the simple zigzag Z obtained after identifying the zigzags Z_1 and Z_2.

If Z_2 intersects the boundary of the $c - DF$, then we attach an c-gon to this boundary, obtaining a sphere. By removing from it the interior of the belt, we obtain two disks with boundaries Z_1 and Z_2. We identify them in one of z ways and remove the attached c-gon. As a result, we obtain a new $c - DF$. The number of simple zigzags can increase under z-reduction (the number of simple central circuits *decreases* under cc-reduction; see Sect. 4.1. But p_6 always decreases. So, any disk-fullerene can be reduced to a tight one in a finite number of steps.

See z-reduction of a pure $7 - DF$ on Fig. 2.18. See also first and seventh fullerenes on Fig. 2.3: a 5-thimble $F_{30}(D_{5h})$ (with $\mathbf{z} = 10^2, 70_{15,10}$ and a belt of five 6-gons) reduces to 5-thimble $F_{20}(I_h)$ with $\mathbf{z} = 10^6$ and pure 6-thimble $F_{36}(D_{6h})$ (with $\mathbf{z} = 12^2, 14^6$ and 6-belt) reduces to 6-thimble $F_{24}(D_{6d})$ with $\mathbf{z} = 12; 60_{12,12}$.

A disk-fullerene with a simple zigzag Z can be extended by cutting it into two parts along Z and inserting a belt joining them. The *extending* operation is the reverse of the reduction operation and is also not uniquely defined. A disk-fullerene without simple zigzags cannot be extended or reduced. Any tight thimble can be extended along its unique simple zigzag not intersecting the boundary.

Lemma 2.3 *A c-thimble is tight if and only if at least one of its non-boundary 5-gons is adjacent to a boundary 5-gon.*

Proof A c-thimble has c boundary 5-gons. Since we have a non-boundary 5-gon adjacent to a boundary 5-gon, there is a set of at least $c + 1$ pentagons whose union is connected. If the thimble had a belt, all these pentagons would be on the same side of the belt. Then on the other side of the belt there would be at most five pentagons, which contradicts Proposition 2.5. If none of the non-boundary 5-gons is adjacent to a boundary 5-gon, then the thimble has a belt by Lemma 2.2.   □

The next Proposition follows from Proposition 2.6.

Proposition 2.8 *(i) Any two simple zigzags of a tight fullerene intersect.*
(ii) A non-tight disk-fullerene has at least two simple zigzags of the same length.

Lemma 2.4 *If a fullerene or thimble is not tight, then all its belts with pairwise nonintersecting interiors form a cylinder. Both boundary components of this cylinder are adjacent to 5-gons.*

Proof In a fullerene, there are two disks on the different sides of the belt, each containing six 5-gons (see Proposition 2.5). Any other belt, which has no common interior point with the 1-st belt, is inside one of these disks. If two belts have no common boundary points, then there is a cylinder (an annulus) between them consisting only of 6-gons. By Lemma 2.2, this cylinder contains a sequence of belts in which two neighbors are adjacent along a zigzag. The belts in this sequence fill another cylinder. If there are only 6-gons adjacent to this cylinder from outside, then they form a belt by Lemma 2.2. We join this belt to the cylinder. After a finite number of steps, we obtain a cylinder with only 5-gons adjacent to its boundary from outside.

In the case of a c-thimble, all c boundary 5-gons are on the same side of any belt in the thimble, because they form an connected annulus homeomorphic to the lateral surface of a round cylinder. We prove by contradiction that the remaining six 5-gons are all on the other side of the belt. Assume that there are fewer than six 5-gons there. Consider the middle edge cycle splitting the belt into two cylinders. It cuts out a disk from the thimble and this disk is a new thimble. The number of 5-gons in this new thimble must be equal to the number of boundary 5-gons plus six. However, by assumption there are fewer than six old 5-gons inside the new thimble. This is a contradiction. Therefore, there are exactly six 5-gons on the other side of the belt. All of them are non-boundary faces of the thimble.

The same six 5-gons are on the disk side of another belt, because only these 5-gons are not adjacent to the boundary. Indeed, if the two disks did not intersect, then the thimble would contain 12 non-boundary 5-gons, which is excluded by $p_5 = c+6$. So, one disk is inside the other, and all six 5-gons are in the smaller disk. The complement of the smaller disk in the larger one is a cylinder.

There are only 6-gons between the belts in the cylinder. The cylinder has a sequence of belts in which any two neighbors are adjacent along a zigzag. These belts fill the cylinder. (The cylinder can consist of only two original belts if they are adjacent.) There are 6 non-boundary 5-gons on one side of the cylinder, and c boundary 5-gons on the other side. If there are only 6-gons adjacent to the boundary of the cylinder from the outside, then they form a belt by Lemma 2.2. We attach this belt to the cylinder. After a finite number of steps, we obtain a cylinder with only 5-gons adjacent to its boundary components from the outside. There is at least one non-boundary 5-gon adjacent to a boundary component, and all c boundary 5-gons are adjacent to the other boundary component. So, all belts constructed above form a cylinder with only 5-gons adjacent to its boundary from the outside.   □

We note that all the belts in a thimble are pairwise nonintersecting and belong to a single cylinder. Each 6-gon in a thimble belongs to at most one belt. If there is a belt containing a given 6-gon, then when trying in one of the two other possible ways to construct a belt containing this 6-gon and corresponding to each of the two other pairs of edges opposite to it, we come to one of the c boundary 5-gons of the thimble. Any other $c - DF$ can contain several cylinders.

Lemma 2.1 and Remark 2.1 (ii) imply that the number of simple zigzags in a tight c-thimble is within $[1, c + 1]$:

Proposition 2.9 *(i) Any c-thimble has at least one simple zigzag nonintersecting its boundary.*
(ii) A tight c-thimble has exactly one such zigzag, and so, the number of its simple zigzags is at most $c + 1$.

Above boundary is best possible: a tight pure 5-thimble $F_{20}(I_h)$ and tight pure 6-thimble $F_{28}(T_d)$ have $\mathbf{z} = 10^6$ and $\mathbf{z} = 12^7$, respectively. A pure 6-thimble $F_{36}(D_{6h})$, given as 7-th on Fig. 2.3, have $\mathbf{z} = 12^2, 14^6$ but it is not tight.

But a tight disk-fullerene, which is not a tumble, can have several simple disjoint zigzags nonintersecting the boundary; see example on Fig. 2.15.

Lemma 2.5 *In a tight fullerene or tight c-thimble, each simple zigzag has at least one adjacent 5-gon on each side.*

Proof If a simple zigzag in a thimble intersects the boundary, then it has at least two adjacent 5-gons on each side. Therefore, we may assume that the zigzag does not intersect the boundary.

We prove the statement by contradiction. If there are no adjacent 5-gons on some side of a simple zigzag, then all faces adjacent to it are 6-gons. By Lemma 2.2 they would form a belt, implying that the fullerene or thimble is non-tight.   □

Each of five pairs of adjacent edges in a 5-gon belongs to a zigzag, possibly self-intersecting. The number of zigzags adjacent to 5-gons in a fullerene is at most $5p_5 = 60$. By Lemma 2.5, there are at least two 5-gons adjacent to each simple zigzag of a tight fullerene. So, such a fullerene has at most 30 simple zigzags. But this bound is not the best possible.

Each of c pairs of adjacent edges on the boundary of a thimble generates a zigzag, possibly self-intersecting. This zigzag intersects the boundary and so has at least two adjacent 5-gons on each side. A tight thimble has no belts; hence, it has only one simple zigzag nonintersecting the boundary.

Lemma 2.6 *Assume that a fullerene has a simple zigzag with only one adjacent 5-gon on one of its sides. Then this fullerene also has a self-intersecting zigzag.*

Proof Assume that there is only one adjacent 5-gon on one of the sides of a simple zigzag Z. All the other adjacent faces are 6-gons, which form an open chain. By Proposition 2.4, each 6-gon in this chain has only two edges in Z (equivalently, all the 6-gons in the chain are different).

Let us prove that non-neighboring 6-gons in the chain do not intersect (in boundary points). Two neighboring 6-gons in the chain have a common edge. Only one end of this edge belongs to Z (see above). If the other end of the edge (denote it by A) belongs to a third 6-gon, which is adjacent to Z along two edges with a common vertex opposite to A. Then these three 6-gons split Z into two parts.

We can assume without loss of generality that the 2-nd 6-gon is adjacent to that part of the zigzag which is adjacent only to the 6-gons from the open chain. The 3-rd 6-gon is adjacent to each of the 1-st two. Let the edge AB be the intersection of the 3-rd and 2-nd 6-gons. Since all vertices are of degree 3, the vertex B belongs to a 4-th face. But one end of the common edge of the 3-rd and 4-th faces belongs to Z. Hence, the 4-th face is a 6-gon adjacent to Z. Let the edge BC be the intersection of the 4-th and 2-nd 6-gons. Then the vertex C belongs to a 5-th face which is a 6-gon adjacent to Z (as in the case of the 4-th face). Let the edge CD be the intersection of the 5-th and 4-th 6-gons. Then, as above, the vertex D belongs to a sixth 6-gon which is adjacent to the zigzag, and so on. If we swap the 2-nd and 3-rd 6-gons, then we obtain two chains of 6-gons, one indexed by even numbers and the other by odd numbers. These two chains are adjacent along the polygonal line $ABCD\ldots$, which must be infinite. This is a contradiction.

Therefore, any two 6-gons in the open chain do not have common boundary points unless they are neighboring in the chain.

In the 6-gons of the open chain, pairs of adjacent edges that are opposite to pairs of adjacent edges belonging to Z form a polygonal line Λ which is a proper part of a zigzag containing an edge passed through twice that is in the original 5-gon and is opposite to the 5-gon's vertex at which the two of its edges belonging to Z are adjacent. (The open polygonal line Λ together with this edge forms a closed polygonal contour which is not a zigzag but is part of the zigzag.)   □

The following alternative is a trivial corollary of Lemma 2.6.

Corollary 2.1 *If a fullerene has only simple zigzags, then among the faces adjacent to any zigzag from one side there are either at least two 5-gons, or only 6-gons.*

So, Lemma 2.6 justifies Proposition 8 (iii) from [DDF04]. Note that Theorem 2.2 prove, by another way, a more general result than (ii) in Theorem 2.5.

Theorem 2.5 *If all zigzags in a fullerene are simple and pairwise intersecting, then:*

(i) there are at least two 5-gons adjacent to any zigzag on each side;
(ii) the number of all zigzags is at most 15 (this bound is the best possible).

Proof (i) Clearly, there are no belts. By Corollary 2.1, there are two 5-gons adjacent to any zigzag on each side.

(ii) Each 5-gon is adjacent to at most five zigzags. So, the number of zigzags is at most $5p_5 = 60$. By (i), there are at least four 5-gons adjacent to any simple zigzag. Therefore, the fullerene has at most $\frac{5p_5}{4} = 15$ zigzags.   □

All the pairwise nonintersecting belts in a fullerene form a cylinder which has only 5-gons adjacent to its boundary components from the outside. It gives:

Corollary 2.2 *In a fullerene, the number of cylinders formed by pairwise noninter-secting belts is at most 15. (Recall that in a thimble, the number of cylinders formed by all its belts is at most 1.)*

Theorem 2.6 *If a $c - DF$ is tight and all the zigzags not intersecting its boundary are simple, then:*

(i) there are at least two 5-gons adjacent to each side of any such zigzag;
(ii) the number of such zigzags is at most $\frac{5(c+6)}{4}$.

The proof is similar to the proof of Theorem 2.5. As regards a non-tight $c - DF$, if only two 5-gons are adjacent to such a zigzag and all the zigzags are simple, then both 5-gons are on the same side of the zigzag.

Remark 2.2 (i) If all the zigzags in an $c - DF$ are simple, then their number is at most $c + \frac{5p_5}{4} = c + \frac{5(c+6)}{4}$. This follows from Theorem 2.6 (ii) and Remark 2.1 (ii).

(ii) For any $c = 4k + 12$, $v = 20k + 54$, where $k \in \mathbb{N}$, there is a tight $c - DF_v$ with $2k + 1$ simple and 4 self-intersecting zigzags. For $k = 5$, it is shown in Fig. 2.20.

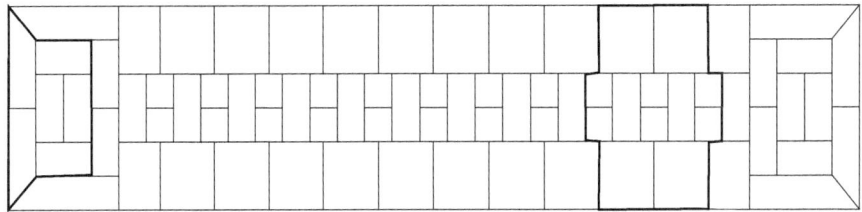

Fig. 2.20 The case $k = 5$ of a tight $c - DF_v(C_{2v})$ with $c = 4k+12$, $v = 20k+54$ and many simple zigzags (both types of them are shown): $\mathbf{z} = 10^2$, 16^{2k-1}; $36^2_{0,3}$, $(14k+30)_{k+1,0}$, $(14k+56)_{k+4,4}$

The number of simple zigzags grows with k. So, this number in a tight disk-fullerene can be arbitrarily large. It would be interesting to know for which c there exist tight $c - DF$'s with an arbitrarily large number of simple zigzags.

Examples of other, besides disk-fullerenes, interesting $(\{5, 6, c\}, 3)$-maps follow.

- Haeckel, 1887: $(\{5, 6, c\}, 3)$-spheres with $c = 7, 8$ representing skeletons of radiolarian zooplankton *Aulonia hexagona*.
- *G-fulleroids*, i.e., $(\{5, c\}, 3)$-spheres with $c > 6$ and symmetry G.
- *Azulenoids*, i.e., $(\{5, 6, 7\}, 3)$-$\mathbb{T}^2$; such tori have $p_5 = p_7$.
- *Schwartzits*, i.e., $(\{5, 6, c\}, 3)$-maps on minimal surfaces of constant negative curvature (of genus $g \geq 2$) with $c = 7, 8$.

Also, *plane fullerenes* (or *nanocones*) are $(\{5, 6\}, 3)$-$\mathbb{E}^2$. Such planes have $0 \leq p_5 \leq 6$. Their number is 1 if $p_5 = 0, 1$ and infinity if $2 \leq p_5 \leq 6$; they are just *nanotubes* (see Sect. 2.4) if $p_5 = 6$. The number of their equivalence (isomorphism up to a finite induced subgraph) classes is [KlBa06] 2, 2, 2, 1 for $p_5 = 2, 3, 4, 5$.

[DDD10, DD12] considered *space fullerenes*, i.e., tilings of $\mathbb{E}^3$ by fullerenes, while [DeSt99] treated *fullerene manifolds*, i.e., $(d - 1)$-dimensional d-valent compact connected manifolds whose 2-faces are 5- and 6-gons; they exist for $2 \leq d \leq 5$ (Fig. 2.21).

(a)   **(b)**

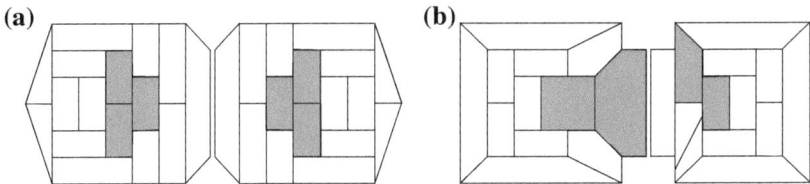

Fig. 2.21 Two only 2-connected $c - DF$ split in two parts with the same p_6 and the number of simple zigzags of length 10 in each; all those zigzags are adjacent to the boundary in two edges. **a** $20 - DF_{62}(C_{2v})$: not a 20-thimble, but all its 18 boundary faces are 5-gons; $\mathbf{z} = 10^2$; $30^2_{0,3}$, $40_{2,0}$, $66_{4,5}$. **b** $15 - DF_{48}(C_1)$ split in two parts with $p_6 = 2$ and 2 simple zigzags of length 10 each; $\mathbf{z} = 10^4$; $40_{0,4}$, $64_{4,8}$

References

[DeSt99]   Deza, M., Shtogrin, M.: Three, four and five-dimensional fullerenes. Southeast Asian Bull. Math. **23**, 9–18 (1999)

[DeSt03]   Deza, M., Shtogrin, M.: Octahedrites, special issue "Polyhedra in Science and Art", symmetry: culture and science. The Quarterly of the International Society for the Interdisciplinary Study of Symmetry **11/1-4**, 27–64 (2003)

[DuDe04]   Dutour, M., Deza, M.: Goldberg-Coxeter construction for 3- or 4-valent plane graphs. Electron. J. Comb. **11–1**, R20 (2004)

[DeDu08]   Deza, M., Dutour Sikirić, M.: Geometry of Chemical Graphs: Polycycles and Two-Faced Maps. Encyclopedia of Mathematics and Its Applications, vol. 119. Cambridge University Press, Cambridge (2008)

[DD12]   Dutour Sikirić, M., Deza, M.: Space fullerenes: computer search for new Frank-Kasper structures II. Struct. Chem. **23–4**, 1103–1115 (2012)

[DDE96]   Dresselhaus, M.S., Dresselhaus, G., Eklund, P.C.: Science of Fullerenes and Carbon Nanotubes. Academic Press, New York (1996)

[DDF04]   Deza, M., Dutour, M., Fowler, P.W.: Zigzags, railroads and knots in fullerenes. J. Chem. Inf. Comput. Sci. **44**, 1282–1293 (2004)

[DDD10]   Dutour Sikirić, M., Delgado-Friedrichs, O., Deza, M.: Space fullerenes: computer search for new Frank-Kasper structures. Acta Crystallogr. **A66–5**, 602–615 (2010)

[DDS13a]   Deza, M., Dutour Sikirić, M., Shtogrin, M.: Fullerene-like spheres with faces of negative curvature. In: Diudea, M.V., Nagy, C.L. (eds.) Diamond D5 and Related Nanostructures. Carbon Materials: Chemistry and Physics, vol. 6, pp. 251–274. Springer, Dordrecht (2013)

[DDS13b]   Deza, M., Dutour Sikirić, M., Shtogrin, M.: Fullerene and disk-fullerenes. Uspechi Matemat. Nauk **68–4**, 69–128 (2013)

[Fu]   Fukuda, K.: The CDD program, http://www.ifor.math.ethz.ch/~fukuda/cdd_home/cdd.html

[Fö77]   Föppl, L.: Stabile Anordnungen von Elektronen in atom. J. Reine Angew. Math. **141**, 251–301 (1912)

[FoMa95]   Fowler, P.W., Manolopoulos, D.E.: An Atlas of Fullerenes. Clarendon Press, Oxford (1995)

[Gold34]   Goldberg, M.: The isoperimetric problem for polyhedra. Tohoku Math. J. **40**, 226–236 (1934)

[Gold37]   Goldberg, M.: A class of multisymmetric polyhedra. Tohoku Math. J. **43**, 104–108 (1937)

[Kaw96]   Kawauchi, A.: A Survey of Knot Theory. Birkhäuser, Basel (1996)

[KlLi92]   Klein, D.J., Liu, X.: Theorems for carbon cages. J. Math. Chem. **11**, 199–205 (1992)

[KlBa06]   Klein, D.J., Balaban, A.T.: The eight classes of positive-curvature graphitic nanocones. J. Chem. Inf. Model. **46**, 307–320 (2006)

[KHCS85]   Kroto, H.W., Heath, J.R., Curl, R.F., Smalley, R.E.: C_{60}: Buckminsterfullerene. Nature **318**, 162–163 (1985)

[Rol76]   Rolfsen, D.: Knots and Links, Mathematics Lecture Series 7, Publish or Perish, Berkeley, 1976; second corrected printing: Publish or Perish, Houston (1990)

[ScZJ04]   Schaeffer, G., Zinn-Justin, P.: On the asymptotic number of plane curves and alternating knots. Exp. Math. **13–4**, 483–493 (2004)

Chapter 3
Zigzags and Railroads of Spheres 3_v and 4_v

We consider the zigzag and railroad structures of 3-regular plane graphs and, especially, graphs a_v, i.e., $v - vertex$ ($\{a, 6\}$, 3)-spheres, where $a = 2, 3$, or 4. The case $a = 5$ has been treated in the previous chapter.

Every a_v, where $a = 3, 4$ or 5, except one infinite series of only 2-connected ones ($G_n, n \geq 1$; see Fig. 3.2) among 3_v, are polyhedral since they are 3-connected.

Every graph 2_v is described by the Goldberg–Coxeter construction, denoted $GC_{k,l}$; see Theorem 6.1. All graphs 3_v, 4_v, and 5_v with maximal pair of symmetry (T or T_d, O or O_h, I or I_h, respectively) are of the form $GC_{k,l}(G_0)$ with $0 \leq l \leq k$ and G_0 being tetrahedron, cube, or dodecahedron, respectively; see Theorem 6.2.

The graphs 3_v admit a combinatorial description by Grünbaum–Motzkin which allows describing them simply and proving a number of specific results of algebraic nature. For example, we prove that all 3_v are tight if and only if $\frac{v}{4}$ is a prime number. The 4_v do not have such description, but their growth rate is very moderate. We completely determine the tight 4_v with simple zigzags and prove that tight 4_v have at most 9 zigzags. We conjecture that the sharp upper bound is 8 which is true for $v \leq 400$. Finally, we give the list of small railroads occurring in graphs 4_v.

3.1 General Results for Plane Graphs

It was conjectured in [DDF04] that any 3-regular plane graph, which is z-knotted, has an odd number of edges of type I. The condition of 3-regularity is necessary: for example, the 3-bipyramid (i.e., $Prism_3^*$) has z-vector $18_{6,3}$.

Clearly, any graph with 1, 2, or 3 zigzags is z-balanced. The smallest z-unbalanced one among 3-regular graphs and among graphs 4_v and 5_v is given in Fig. 3.1.

Denote by $(G_n)_{n \geq 1}$, the $4(n + 1)$-vertex sphere ($\{3, 6\}$, 3)-$\mathbb{S}^2$, whose faces are (organized in pairs of adjacent ones) triangles and hexagons. Any graph G_n is 2- but not 3-connected, its z-vector is $(4^{n+1}, (4n + 4)^2)$; its group is D_{2d} or D_{2h}, if n is even or odd, respectively. The first occurrences are depicted in Fig. 3.2.

© Springer India 2015
M. Deza et al., *Geometric Structure of Chemistry-Relevant Graphs*,
Forum for Interdisciplinary Mathematics 1, DOI 10.1007/978-81-322-2449-5_3

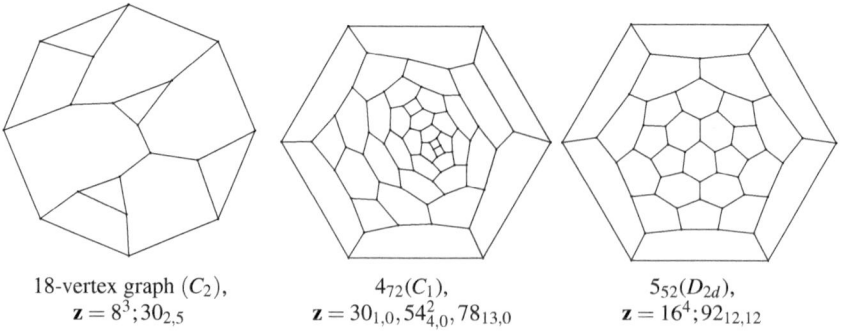

| 18-vertex graph (C_2),
$\mathbf{z} = 8^3; 30_{2,5}$ | $4_{72}(C_1)$,
$\mathbf{z} = 30_{1,0}, 54_{4,0}^2, 78_{13,0}$ | $5_{52}(D_{2d})$,
$\mathbf{z} = 16^4; 92_{12,12}$ |

Fig. 3.1 Smallest z-unbalanced 3-regular, 4_v, and 5_v graphs

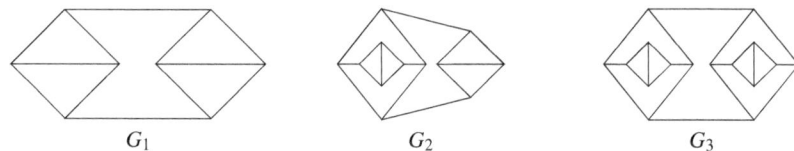

G_1 G_2 G_3

Fig. 3.2 The first 3 graphs of series $(G_n)_{n \geq 1}$ of $(\{3, 4, 5, 6\}, 3)$-spheres, which are not 3-connected

Proposition 3.1 [DeDu05] *Any* $(\{3, 4, 5, 6\}, 3)$-$\mathbb{S}^2$, *except* $G_n, n \geq 1$, *is 3-connected.*

Now we give a local Euler formula for zigzags. Let G be a plane 3-regular graph. Consider a patch A in G, which is bounded by t arcs (i.e., paths of edges) belonging to zigzags (different or coinciding).

We admit also 0-gonal patch A, i.e., just the interior of a simple zigzag. Let the patch A is *regular*, i.e., the continuation of any of its bounding arcs (on the zigzag, to which it belongs) lies outside of the patch (see Fig. 3.3). Let $p'(A) = (\ldots, p_i', \ldots)$ be the p-vector enumerating the faces of G, which are contained in the patch A.

There are two types of intersections of arcs on the boundary of a regular patch: either intersection in an edge of the boundary, or intersection in a vertex of the boundary. Let us call these types of intersections *obtuse* and *acute*, respectively (see Fig. 3.3); denote by t_{ob} and t_{ac}, the respective number of obtuse and acute intersections. Clearly, $t_{ob} + t_{ac} = t$, where t is the number of arcs forming the patch.

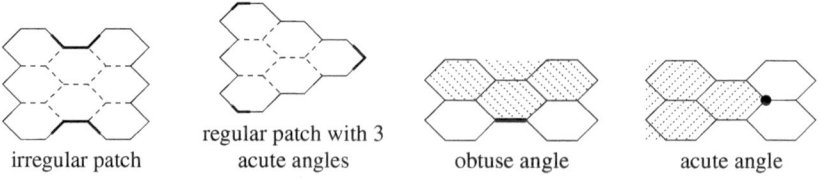

irregular patch   regular patch with 3   obtuse angle   acute angle
                         acute angles

Fig. 3.3 Examples of patches and their angles

Theorem 3.1 *Let G be a 3-regular plane graph (1- and 2-gonal faces are permitted) and A be a regular patch with p-vector $(\ldots, p'_i, \ldots)$ in it. Then it holds*

$$6 - t_{ob} - 2t_{ac} = \sum_{i \geq 1}(6 - i)p'_i .$$

Proof Let $P(A)$ be the induced plane subgraph of G formed by the patch A. The vertices of $P(A)$ have degree 2 or 3; we denote their respective number by v_2 and v_3. The exterior face is l-gonal, where l is the length of the boundary of A. Direct enumeration gives the following expressions for the number $|E|$ of edges of $P(A)$

$$|E| = \frac{1}{2}(3v_3 + 2v_2) = \frac{1}{2}(l + \sum_{i \geq 1} ip'_i) .$$

Euler's formula, applied to the plane graph $P(A)$, yields

$$2 = (v_2 + v_3) - |E| + (1 + \sum_{i \geq 1} p'_i) .$$

The above two expressions of $|E|$ give

$$1 = -\frac{v_3}{2} + \sum_{i \geq 1} p'_i = (v_2 + v_3) - \frac{l}{2} + \sum_{i \geq 1}(1 - \frac{i}{2})p'_i .$$

Eliminating v_3 yields $3 = v_2 - \frac{l}{2} + \sum_{i \geq 1}(3 - \frac{i}{2})p'_i$.

For example, if the patch A is a 0-gon, then $l = 2v_2$ and we get $3 = \sum_{i \geq 1}(3 - \frac{i}{2})p'_i$.

Denote by v'_2 and v'_3 the numbers of vertices on the boundary of A, having degree 2 and 3, respectively. Clearly, $l = v'_2 + v'_3$ and $v'_2 = v_2$. Now, using that A is a regular patch and specifying acute and obtuse types of arc intersections on its boundary, we get $v'_2 = v'_3 + t_{ob} + 2t_{ac}$, which gives $6 - t_{ob} - 2t_{ac} = \sum_{i \geq 1}(6 - i)p'_i$. □

Corollary 3.1 *In the conditions of Theorem 3.1 with p-vector $(\ldots, p'_i, \ldots)$, we have:*

(i) If G has no q-gonal faces with $q > 6$, then A is formed by $t \leq 6$ arcs.
(ii) If G is a bifaced sphere a_v, $a \in \{3, 4, 5\}$, then

$$6 - t_{ob} - 2t_{ac} = (6 - a)p'_a .$$

Proof (i) follows directly from the above theorem;
(ii) also follows by direct application with p-vector $(\ldots, p'_i, \ldots)$. □

3.2 Graphs 3_v

We present in Table 3.1 the numbers $N_3(v)$ of 3_v and $N_3^t(v)$ of tight 3_v for $v \leq 512$.

In [GrünMo63] it was shown that each zigzag in a graph 3_v is simple and a general construction of all 3_v was given; we sketch this construction with slightly different notation. If a zigzag contains two edges, say, e_1 and e_2, of a triangle, then its third edge, say, e_3 defines a pseudo-road (see Sect. 1.3) PR that finish on another triangle. Denote by s one fourth of the length of the zigzag; then $s - 1$ is the number of hexagons of PR. Consider the sequence of concentric belts, which, possibly, goes around the patch depicted in Fig. 3.4; let m denote the number of zigzags and $m - 1$ be the number of corresponding railroads. The same patch (with the same number $s - 1$ of hexagons between the other two triangles and with the same number $m - 1$ of concentric belts, going around it) occurs on the other side of the sphere. As in [GrünMo63], we get, by direct computation, the equality $p_6 = 2(sm - 1)$ and $v = 4sm$. See Fig. 3.5c, d for two examples with $s = 2$ and $m = 4$.

Take a triangle T and an edge e of this triangle; e belongs to a sequence of adjacent 6-gons, which is concluded by another triangle T'. If one considers the three edges e_1, e_2, e_3, then we obtain three triangles T_1, T_2, and T_3, respectively, that may be distinct or not.

Theorem 3.2 *Any graph 3_v has exactly one of the following forms:*

  (i) *Tetrahedron or Truncated Tetrahedron (the only case, when a hexagon is adjacent to more than two triangles).*
 (ii) *There are two hexagons, each of which is adjacent to two triangles on opposite edges; there are one or two such graphs 3_v depending on $v \equiv 8$ (mod 16) or $v \equiv 0$ (mod 16), respectively. Their symmetry groups are, respectively, D_2 if $v \equiv 8$ (mod 16) and D_{2h}, D_{2d} if $v \equiv 0$ (mod 16).*
(iii) *There are four hexagons (in two adjacent pairs), each of which is adjacent to two triangles on nonopposite and nonadjacent edges; there is exactly one such graph 3_v for every $v \geq 16$, $v \equiv 0$ (mod 4).[1] The symmetry group is D_{2h} or D_{2d} if $\frac{v}{4}$ is even or odd, respectively.*
 (iv) *Each hexagon is adjacent to at most one triangle; there is an one-to-one correspondence between such 3_v with isolated triangles, and IPT fullerenes 5_v (i.e., those having their 12 pentagons organized in 4 triples of mutually adjacent ones).*

Proof For given graph 3_v G, denote by $t(G)$ the maximal number of triangles, which are adjacent to a hexagon. The only graph 3_v with $t(G) = 0$ is Tetrahedron. The case $t(G) = 4$ corresponds to the 2- but not 3-connected 6-vertex plane graph G_1 (see Fig. 3.2). The only graph 3_v with $t(G) = 3$ is the Truncated Tetrahedron. If $t(G) = 2$, then either two triangles are adjacent to an hexagon on opposite edges, or they are adjacent to an hexagon on nonopposite and nonadjacent edges.

---

[1] The two graphs 3_{16} are cube truncated at 4 nonadjacent vertices or at 4 vertices of two opposite edges. The second one is unique, which is of type (ii) and (iii).

Table 3.1 Numbers $N_3(v)$ of graphs 3_v and $N_3^t(v)$ of tight 3_v for $v \leq 512$

v	N_3	N_3^t	v	N_3	N_3^t	v	N_3	N_3^t	v	N_3	N_3^t	v	N_3	N_3^t	v	N_3	N_3^t	v	N_3	N_3^t
12	1	1	84	7	2	156	11	3	228	15	4	300	23	3	372	23	6	444	27	7
16	2	0	88	6	0	160	19	0	232	15	0	304	26	0	376	24	0	448	48	0
20	1	1	92	4	4	164	7	7	236	10	10	308	17	8	380	21	9	452	19	19
24	2	0	96	14	0	168	17	0	240	33	0	312	29	0	384	50	0	456	41	0
28	2	2	100	6	3	172	8	8	244	11	11	316	14	14	388	17	17	460	25	11
32	4	0	104	7	0	176	16	0	248	16	0	320	37	0	392	29	0	464	37	0
36	3	1	108	8	2	180	15	2	252	20	3	324	22	5	396	28	5	468	33	6
40	3	0	112	12	0	184	12	0	256	26	0	328	21	0	400	40	0	472	30	0
44	2	2	116	5	5	188	8	8	260	15	6	332	14	14	404	17	17	476	25	13
48	7	0	120	13	0	192	27	0	264	25	0	336	43	0	408	37	0	480	69	0
52	3	3	124	6	6	196	11	7	268	12	12	340	19	8	412	18	18	484	23	17
56	4	0	128	14	0	200	16	0	272	23	0	344	22	0	416	39	0	488	31	0
60	5	1	132	9	2	204	13	3	276	17	4	348	21	5	420	35	3	492	29	7
64	8	0	136	9	0	208	19	0	280	25	0	352	34	0	424	27	0	496	40	0
68	3	3	140	9	3	212	9	9	284	12	12	356	15	15	428	18	18	500	27	13
72	7	0	144	19	0	216	21	0	288	39	0	360	41	0	432	52	0	504	54	0
76	4	4	148	7	7	220	13	5	292	13	13	364	21	11	436	19	19	508	22	22
80	9	0	152	10	0	224	24	0	296	19	0	368	30	0	440	37	0	512	48	0

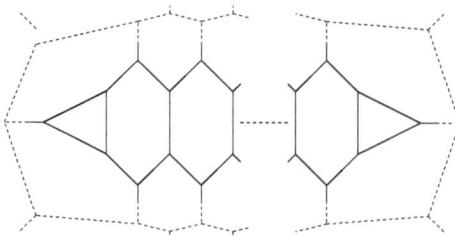

Fig. 3.4 A pseudo-road in a graph 3_v and a circular railroad

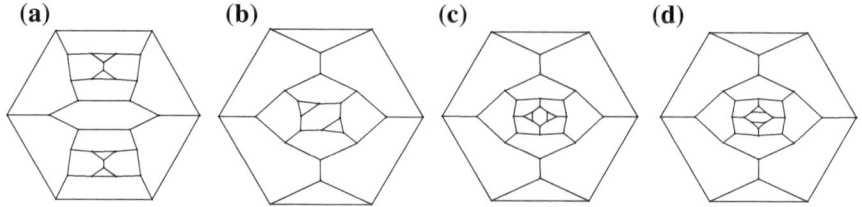

Fig. 3.5 Graphs 3_{32} with hexagons adjacent to two triangles **a** 3_{32}, case (iii), D_{2h}, **b** 3_{24}, case (ii), D_2, **c** 3_{32}, case (ii), D_{2d}, **d** 3_{32}, case (ii), D_{2h}

The first case corresponds to the case $s = 2$ of the Grünbaum–Motzkin construction. The divisibility of v by 8 follows from the formula $v = 4sm$. At the last step of the Grünbaum–Motzkin construction, namely, after adding railroads, we add a pair of triangles in one of two possible ways. If $v \equiv 8 \pmod{16}$, then both ways give isomorphic graphs, both of symmetry D_2, for $v \geq 8$ (see Fig. 3.5b), for this possibility). If $v \equiv 0 \pmod{16}$, then we get two nonisomorphic graphs, one of symmetry D_{2h}, the other of symmetry D_{2d} (see Fig. 3.5c, d, for those two possibilities).

Let our graph be of type (iii), i.e., two hexagons H_1 and H_2 are adjacent to two triangles T_1 and T_2 on nonopposite and nonadjacent edges, as in the picture below.

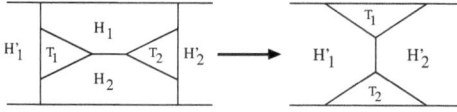

These two hexagons H_1 and H_2 are adjacent, respectively, to two hexagons H_1' and H_2'. By replacing the patch H_1, H_2, T_1, T_2 by two triangles, one gets a graph 3_{v-4}, which is also of type (iii). By induction, one gets, for any $v \equiv 0 \pmod{4}$ with $v \geq 16$, an unique graph. A graph of type (iii) has two pairs of hexagons, each pair being adjacent to two triangles; if one does the operation, depicted above, to the two pairs, then the symmetry group remains the same. So, one has a periodicity of order 8 for the symmetry groups and the result on groups follows.

(iv). If the graph has $t(G) = 1$, then the operation, depicted in picture below, on all triangles realize a one-to-one correspondence.   □

See Fig. 3.5a, c, d, for all 3_{32} with hexagons adjacent to two triangles.

The *graph of curvatures* of a plane graph is the graph (possibly, with loops and multiple edges) having as vertex-set all its faces of positive curvature, two of them being *adjacent* if there exists a pseudo-road connecting them. Clearly, the graph of curvatures of a graph a_n, $a \in \{3, 4, 5\}$, is a-regular.

Proposition 3.2 *The graph of curvatures of a 3_v is one of three following graphs:*

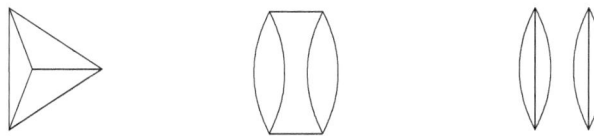

Proof Take a triangle, say, T_1 and an edge e_1 of T_1. The pseudo-road, which is defined by T_1 and e_1, establishes an edge between T_1 and, say, T_2; this pseudo-road is bounded by one zigzag Z_1. This zigzag belongs to a sequence of m concentric zigzags $Z_1, Z_2, \ldots, Z_m$. The zigzag Z_m defines a pseudo-road between the remaining triangles T_3 and T_4, i.e., an edge connecting T_3 to T_4 in the graph of curvatures. Since any vertex of the graph of curvatures has degree 3, we are done. □

Conjecture 3.1 *The graph of curvatures of any tight graph 3_v is K_4.*

Fix a graph G 3_v. Denote by T_i, $1 \le i \le 4$, the triangles in G, by $(e_i)_{1 \le i \le 3}$ the edges of T_1, by PR_i the pseudo-road defined by T_1, e_i and by $(4s_i)_{1 \le i \le 3}$ the lengths of corresponding zigzags. Without loss of generality, we can suppose that $s_1 \le s_2 \le s_3$. Denote by $m_1 - 1$, $m_2 - 1$, $m_3 - 1$, respectively, the corresponding number of concentric railroads around each of three patches. Denote by $Z_{i,k}$, where $1 \le i \le 3$ and $1 \le k \le m_i$, the zigzags of G.

Theorem 3.3 *Let G be a graph 3_v, then the following properties hold:*

(i) $\mathbf{z}(G) = (4s_1)^{m_1}, (4s_2)^{m_2}, (4s_3)^{m_3}$; the number of railroads is $m_1 + m_2 + m_3 - 3$.

(ii) G has at least three zigzags with equality if and only if it is tight.

(iii) If G is tight, then $\mathbf{z}(G) = v^3$ (so, each zigzag is a Hamiltonian circuit).

(iv) G is z-balanced and $|Z_{i,k} \cap Z_{j,l}| = \begin{cases} 0 & \text{if } i = j, \\ \frac{v}{2m_i m_j} & \text{if } i \ne j. \end{cases}$

Proof If Z is a zigzag of G, then it belongs to a sequence of m_i concentric zigzags of a patch defined by two triangles and a pseudo-road. (i) follow immediately; (ii) is a corollary of (i), since G is tight if and only if all m_i are equal to 1.

Let G be a tight 3_v; the formula $v = 4sm$ becomes $v = 4s_1 = 4s_2 = 4s_3$. So, $\mathbf{z} = v^3$.

Clearly, any two zigzags $Z_{i,k}$ and $Z_{j,l}$ are disjoint if $i = j$. Moreover, the size of the intersection $Z_{i,k} \cap Z_{j,l}$ depends only on i and j. Denote by β_{ij} the size of the pairwise intersections $|Z_{i,1} \cap Z_{j,1}|$. We obtain the linear system

$$\begin{cases} 4s_1 = \beta_{13}m_3 + \beta_{12}m_2 \\ 4s_2 = \beta_{23}m_3 + \beta_{12}m_1 \\ 4s_3 = \beta_{23}m_2 + \beta_{13}m_1 \end{cases},$$

which has the unique solution $\beta_{ij} = \frac{v}{2m_i m_j}$. If one writes this intersection size as $\frac{8}{v}s_i s_j$, then it is easy to see that G is z-balanced.   □

Theorem 3.4 *All graphs 3_v are tight if and only if $\frac{v}{4}$ is odd prime.*

Proof If $\frac{v}{4}$ is a odd prime, let G be a graph 3_v with parameters s_i, m_i defined before Theorem 3.3. If $\frac{v}{4} = s_i m_i$, then either $s_i = 1$, or $m_i = 1$. The first case corresponds to a 2-connected but not 3-connected plane graph; so, $m_1 = m_2 = m_3 = 1$ and G has no railroad. The second case $m_i = 1$ means also the absence of a railroad.

Assume that $\frac{v}{4}$ is not prime; if $v = 4pq$ for $q > 1$, then, using the Grünbaum–Motzkin construction, we can construct a graph of type 3_v with a system of $q > 1$ concentric zigzags of length $4p$ and so, with at least $q - 1 \geq 1$ railroads.   □

Remark 3.1 In Table 3.1, the number of 3_v for prime $\frac{v}{4}$ is a nondecreasing function.

Theorem 3.5 *None of the graphs 3_v is tight if and only if $\frac{v}{4}$ is even.*

Proof The unique (for every integer $\frac{v}{4} \geq 4$) graph G, defined in Theorem 3.2 (iii), is tight if $\frac{v}{4}$ is odd; we represent this graph in Fig. 3.6a for $v = 28$.

Suppose now that $\frac{v}{4}$ is even and that G is tight. We will use the necessary conditions of Theorem 3.3 to get a contradiction.

Consider the patch, formed by a pseudo-road PR between two triangles, say, T_1 and T_2. Since G is assumed to be tight, there are $\frac{v}{4} - 1$ hexagons in PR. Moreover, since there are no railroads, any of the remaining two triangles T_3 and T_4 will be adjacent to the hexagons of PR. Let i be the position of T_3 in PR. The choice of position for T_3 and T_4 determines our 3_v; so, we need to show that every choice of i leads to a nontight graph 3_v. If we find a shorter pseudo-road, then we are done.

Define a pseudo-road PR' by starting with T_1 and taking the upper edge if i is even and the lower edge if i is odd; (see Fig. 3.6b). PR' intersects with PR. From the choice of PR', the position p of the first hexagon of intersection, of PR' with PR, satisfies $p \equiv 0 \pmod{4}$. Moreover, all positions of hexagons of intersection satisfy this condition. So, PR' is shorter than $\frac{v}{4} - 1$ and the graph is *not* tight.   □

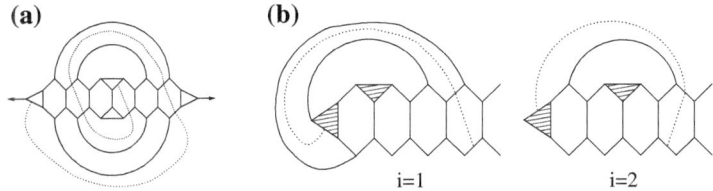

Fig. 3.6 The pseudo-road construction

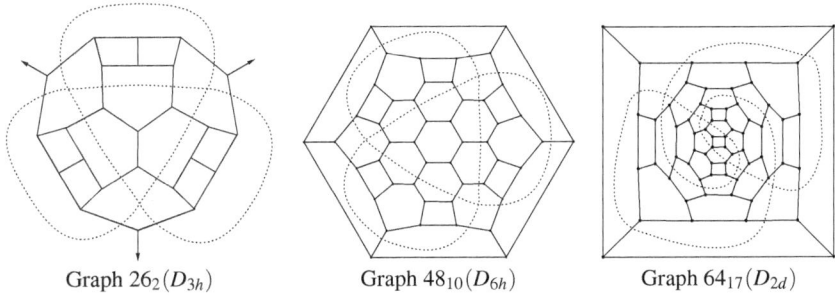

Graph $26_2(D_{3h})$ Graph $48_{10}(D_{6h})$ Graph $64_{17}(D_{2d})$

Fig. 3.7 Graphs 4_{26}, 4_{48}, and 4_{64} with self-intersecting railroads

3.3 Graphs 4_v

Since any graph 4_v is bipartite, we can apply Proposition 1.1: there exists an orientation of zigzags, with respect to which each edge has type I. We will always work with such orientation. Any self-intersection of a railroad is of type I (see Fig. 3.7). See on Fig. 3.7 several self-intersecting railroads.

We present in Table 3.2, the numbers $N_4(v)$ of 4_v and $N_4^t(v)$ of tight 4_v for $v \leq 260$.

Conjecture 3.2 *(i) z-knotted 4_v exists if and only if $v \geq 30$, $v \equiv 2$ (mod 4).*
(ii) Tight 4_v exists if and only if $v \geq 8$, $v \neq 10, 14$.
(iii) Every tight 4_v has at most eight zigzags.

The smallest tight 4_v with eight zigzags is $4_{88}(D_{2h})$ given on Fig. 1.14.

Theorem 3.6 *Every tight 4_v has at most 9 zigzags.*

Proof Let G be a tight 4_v with zigzags $Z_1, \ldots, Z_l$. We obtain $2l$ sides, since every zigzag has two sides. A side S is called *lonely* if it is incident to exactly one square sq. We can define a pseudo-road PR, parallel to S, which will begin and finish on sq. On the other side of the pseudo-road, there will be a side S' of a zigzag Z', which will be incident two times to sq. Moreover, if S' is incident exactly two times to sq, then it defines another lonely side S''; (see Fig. 3.8b).

Denote by n_{1a} the number of lonely sides in first case and n_{1b} the number of lonely sides in second case. Also, let n_2 be the number of sides incident to exactly two squares (identical or not) and n_3 be the number of sides incident to at least three squares (identical or not). Obviously, $l = \frac{1}{2}(n_{1a} + n_{1b} + n_2 + n_3)$.

Table 3.2 Numbers $N_4(v)$ of graphs 4_v and $N_4^t(v)$ of tight 4_v for $v \leq 260$

v	N_4	N_4^t	v	N_4	N_4^t	v	N_4	N_4^t	v	N_4	N_4^t	v	N_4	N_4^t	v	N_4	N_4^t
8	1	1	52	13	10	94	46	27	136	165	104	178	291	176	220	646	403
12	1	1	54	10	6	96	59	45	138	110	89	180	298	215	222	426	276
14	1	0	56	23	16	98	65	38	140	220	134	182	356	190	224	776	469
16	1	1	58	12	8	100	70	50	142	150	96	184	388	246	226	584	361
18	1	1	60	19	15	102	48	34	144	164	116	186	259	174	228	567	413
20	3	2	62	21	11	104	99	62	146	189	109	188	479	285	230	684	357
22	1	1	64	22	17	106	65	43	148	207	145	190	352	203	232	754	481
24	3	3	66	16	12	108	79	60	150	142	106	192	352	264	234	489	331
26	3	1	68	36	24	110	89	53	152	265	167	194	418	230	236	894	517
28	3	3	70	21	12	112	97	63	154	190	109	196	463	310	238	681	421
30	2	2	72	29	23	114	68	47	156	202	146	198	303	207	240	663	471
32	8	5	74	31	16	116	133	80	158	237	121	200	559	338	242	781	463
34	3	2	76	34	24	118	90	58	160	262	162	202	419	274	244	867	533
36	7	5	78	24	14	120	99	75	162	175	123	204	424	313	246	566	381
38	7	4	80	53	32	122	115	67	164	330	205	206	498	260	248	1016	594
40	7	5	82	32	21	124	127	81	166	239	140	208	553	355	250	790	466
42	5	4	84	42	33	126	86	58	168	249	193	210	365	251	252	751	541
44	14	7	86	47	25	128	171	103	170	288	154	212	663	400	254	902	487
46	6	5	88	50	36	130	118	79	172	319	221	214	500	309	256	1006	629
48	12	9	90	35	27	132	133	100	174	209	135	216	483	358	258	661	459
50	12	8	92	75	48	134	152	82	176	397	250	218	580	332	260	1173	675

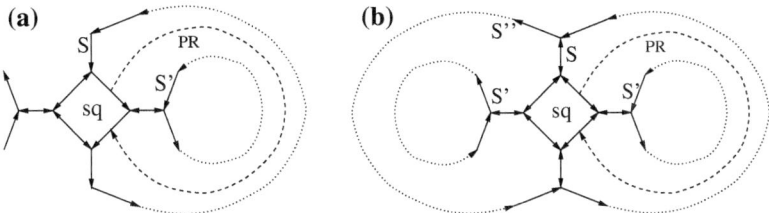

Fig. 3.8 Two cases of lonely sides **a** One lonely side S. **b** Two lonely sides S and S''

In case (a) of lonely side, S' is incident to at least three squares; so, $n_{1a} \leq n_3$, while in case (b) S' is incident exactly two times to sq; so, $n_{1b} \leq n_2$. Every square sq can be incident to 0, 1, or 2 lonely sides; so, one has the inequality $n_{1a} + \frac{n_{1b}}{2} \leq 6$. By an enumeration of incidences, one gets $n_{1a} + n_{1b} + 2n_2 + 3n_3 \leq 4 \times 6 = 24$. The 4-dimensional polyhedron defined by above inequalities and nonnegativity inequalities $n_{1a}, \ldots, n_3 \geq 0$ is denoted by $\mathscr{P}$. We can maximize the linear function $\frac{1}{2}(n_{1a} + n_{1b} + n_2 + n_3)$ over P using the cdd program (see [Fu]) and obtain the upper bound 9, which is attained for the unique 4-uple $(n_{1a}, n_{1b}, n_2, n_3) = (0, 12, 6, 0)$. □

The following theorem is in sharp contrast with Theorem 3.3 (iv) for 3_v and Theorem 2.1 for 5_v. But the same holds for $(\{2, 3\}, 6)$-spheres, implying Theorem 5.18: classification, similar to Corollary 3.2, of such *weakly tight* pure graphs.

Theorem 3.7 *The intersection of every two simple zigzags of a graph 4_v, if nonempty, has one of the following forms (and so, its size is 2, 4, or 6).*

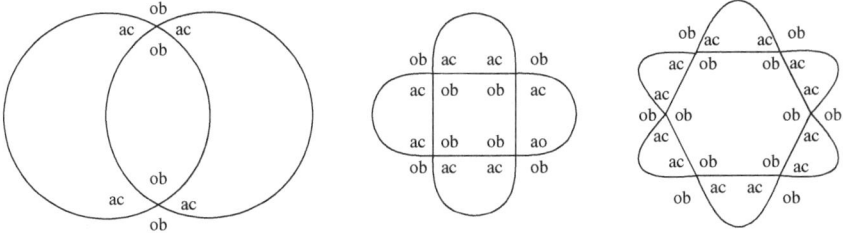

Proof Let H be the graph, whose vertices are edges of intersection between simple zigzags Z and Z', with two vertices being adjacent if they are linked by a path belonging to Z or Z'. Then H is a plane 4-regular graph and Z, Z' define two central circuits in H. Since Z and Z' are simple, the faces of H are t-gons with even t.

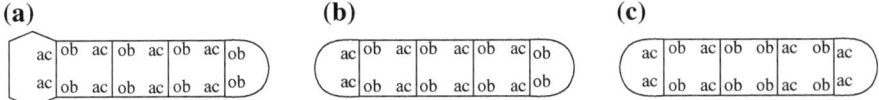

Fig. 3.9 Three cases for sequence of 4-gons

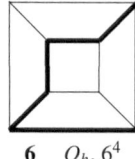   

6   $O_h, 6^4$ 24 $O_h, 12^6$

Fig. 3.10 The tight 4_v with simple zigzags

Applying Theorem 3.1 to a t-gonal face F of H, we obtain that the number p_4' of 4-gons in F satisfies $6 - t_{ob} - 2t_{ac} = 2p_4'$. So, the numbers t_{ob} and t_{ac} are even, since $t = t_{ob} + t_{ac}$. Also, $6 - t_{ob} - 2t_{ac} \geq 0$. So, $t \leq 6$.

We obtain the following five possibilities for the faces of H: 2-gons with two acute angles, 2-gons with two obtuse angles, 4-gons with four obtuse angles, 4-gons with two acute and two obtuse angles, and 6-gons with six obtuse angles.

Take an edge e of a 6-gon in H and consider the sequence (possibly, empty) of adjacent 4-gons of H emanating from this edge. This sequence will stop at a 2-gon or a 6-gon; the case-by-case analysis of angles yields that it
stops at a 2-gon; (see Fig. 3.9a).

Take an edge of a 2-gon in H and consider the same construction. If the angles are both obtuse, then the sequence will terminate at a 2-gon or a 6-gon. If the angles are both acute, then cases (b) and (c) of Fig. 3.9 are possible.

In the first case, all 4-gons contain two obtuse angles and two acute angles; so, the pseudo-road finishes with an edge of two obtuse angles. In the second case, there is a 4-gon, whose angles are all obtuse; this 4-gon is unique in the sequence and its position is arbitrary. Every pair of opposite edges of a 4-gon belongs to a sequence of 4-gons considered above. So, all angles of a 4-gon are the same, i.e., obtuse. This fact restricts the possibilities of intersections to the three cases of the theorem.   □

Corollary 3.2 *The tight pure 4_v are cube and truncated octahedron $= GC_{1,1}(Cube)$.*

Proof By Theorem 3.7, every two simple zigzags intersect in at most six edges. Since, by Theorem 3.6, there are at most 9 zigzags, we obtain the upper bound $(9-1)6 = 48$ on the length of every zigzag. This yields the upper bounds $\frac{9}{2}48 = 216$ on the number of edges of G and $\frac{2}{3}216 = 144$ on the number of its vertices. But, our exhaustive computation in this range of values gave cube and truncated octahedron as the only solutions (Figs. 3.10 and 3.11).   □

3.4 Railroads and Pseudo-Roads

The set of railroads of a graph G can be seen as a plane graph H with degree 4 or 6 of its vertices, accordingly to double or triple points of intersection or self-intersection of the representing curves. Every t-gonal face of H can be seen as a regular patch bounded by t arcs. If G has $p_t = 0$ for $t > 6$, Corollary 3.1 (i) implies $t \leq 6$.

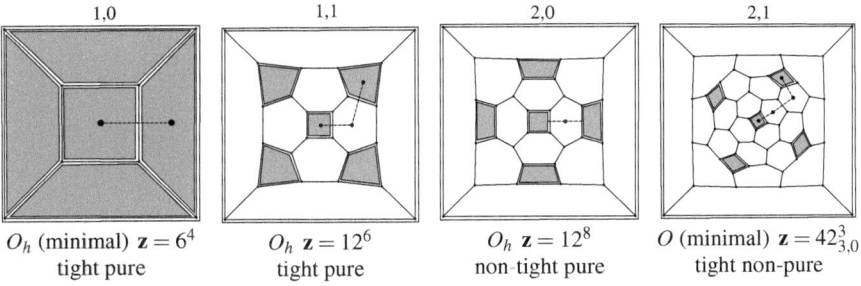

1,0   1,1   2,0   2,1

O_h (minimal) $\mathbf{z} = 6^4$ O_h $\mathbf{z} = 12^6$ O_h $\mathbf{z} = 12^8$ O (minimal) $\mathbf{z} = 42^3_{3,0}$
tight pure   tight pure   non-tight pure   tight non-pure

Fig. 3.11 $GC_{k,l}(Cube)$ for $(k, l) = (1, 0), (1, 1), (2, 0), (2, 1)$

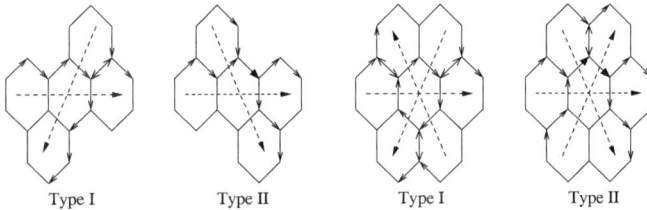

Type I   Type II   Type I   Type II

Fig. 3.12 Types of self-intersection of a railroad

A railroad R is bounded by two zigzags Z_1 and Z_2, which have the same length and signature. Each edge of self-intersection of Z_1, Z_2 corresponds to an hexagon of self-intersection of R. Since Z_1 can self-intersect in an edge of type I or type II, there are two types of double self-intersections of railroad. Simple analysis yields two types for triple self-intersections of railroad: either Z_1 self-intersects in three edges of type I or it self-intersects in one edge of type I and two edges of type II. See Fig. 3.12 for all possibilities.

In order to illustrate the notion of railroads, we give in the remainder of this section two lists: those having, or, respectively, not having, triple self-intersections of the curves representing railroads in bifaced v-vertex polyhedra 4_v for small v. In Tables 3.3 and 3.4, we present the first cases of such curves; the first column being the name of the curve. The smallest graph 4_v is listed in the column "First appearance". If the curve appears twice in this graph; we mark that by putting "twice" in parenthesis. The third and fourth column give the number m of intersection points and the number p_i of i-gons, $1 \leq i \leq 6$. Columns "Group" and "z-vector" give the point group and the z-vector of the corresponding graph 4_v.

The computation of all railroads in the class of polyhedra 4_v, $v \leq 88$, with at least two self-intersections, all being double self-intersections, produced a list of projections of alternating knots presented in Table 3.3. We use Rolfsen's notation, see [Rol76], for knots with at most 10 crossings and those of Thistlethwaite [Thi], for other knots. We denote by 15_y and 18_y the 15- and 18-crossing alternating knots, appearing in a graph 4_{80}, which are 80_{29} and 80_{28}, respectively. (The knot 15_y comes from $APrism_3$ or from knot 9_{40} by inscribing consecutively 3 or 2 triangles,

Table 3.3 Alternating knots in railroads of graphs 4_v, $v \leq 88$

Knot	*First appearance*	m	$p_1, \ldots, p_6$	Group	*z*-vector
0_1	20_3	0		D_{3d}	6^3; $42_{12,0}$
3_1	26_2	3	0,3,2,0,0,0	D_{3h}	$24^2_{3,0}$, $30_{3,0}$
4_1	$64_{17}(twice)$	4	0,2,4,0,0,0	D_{2d}	$48^4_{4,0}$
5_2	74_{24}	5	0,3,2,2,0,0	C_{2v}	18^4; $48^2_{5,0}$, $54_{3,0}$
6_3	74_{21}	6	0,2,4,2,0,0	C_2	$54^2_{6,0}$, $114_{27,0}$
7_4	$80_{30}(twice)$	7	0,4,2,2,0,1	D_{2h}	$60^4_{7,0}$
9_{23}	62_{16}	9	0,2,6,2,0,1	C_{2v}	$60^2_{9,0}$, $66_{9,0}$
9_{40}	56_{19}	9	0,8,3,0,0,0	D_{3h}	12^4; $60^2_{9,0}$
11_{332}	88_{44}	11	0,2,4,7,0,0	C_{2v}	24^4; $84^2_{11,0}$
15_y	80_{29}	15	0,0,8,6,0,0	D_{3d}	12^6; $84^2_{12,0}$
18_y	80_{28}	18	0,0,14,0,6,0	D_3	24^2; $96^2_{18,0}$

respectively.) All curves, representing railroads with triple self-intersections in the graphs 4_v, with $v \leq 142$ are given in Table 3.4. The notation $i - j$, given in first column, means that i is the number of triple points of the curve and j is order of appearance (among curves with i triple points) in this table.

Remark 3.2 Let Z be a zigzag of signature (α_1, α_2) bounding a railroad R with m_2 double and m_3 triple self-intersections. Then one has $m_2 + 3m_3 = \alpha_1 + \alpha_2$.

Proposition 3.3 *Let G be a graph 4_v, having railroads $R_1, \ldots, R_p$; let H be a plane graph formed by the curves representing those railroads. Then it hold:*

(i) *Every t-gonal face of H with $t = 0, 1$ contains exactly $3 - t$ 4-gons of G; every 2-gonal face of H contains one or two 4-gons of G.*
(ii) *If q_t is the number of t-gonal faces of H, then one has the inequality*

$$3q_0 + 2q_1 + q_2 \leq 6 .$$

Proof Every t-gonal face F of H can be viewed as a regular patch with $t = t_{ob} + t_{ac}$ (obtuse and acute) intersections of arcs. Let p'_4 be the number of 4-gons inside F. We will apply Theorem 3.1.

(i) If $t = 0$, then $2p'_4 = 6$ and we are done.
    If $t = 1$, then $2p'_4 = 6 - t_{ob} - 2t_{ac} \geq 4$ and $2p'_4 = 6 - t_{ob} - 2t_{ac} < 6$. Since $2p'_4$ is even, we get $2p'_4 = 4$, i.e., every 1-gon has exactly two 4-gons.
    If $t = 2$, then $2p'_4 = 6 - t_{ob} - 2t_{ac} \geq 2$ and $2p'_4 = 6 - t_{ob} - 2t_{ac} < 6$. This yields $p'_4 = 1$ or $p'_4 = 2$.
(ii) Any graph 4_v has $p_4 = 6$; so, the result follows. $\square$

Table 3.4 Curves with triple self-intersections in railroads of graphs 4_v, $v \leq 142$

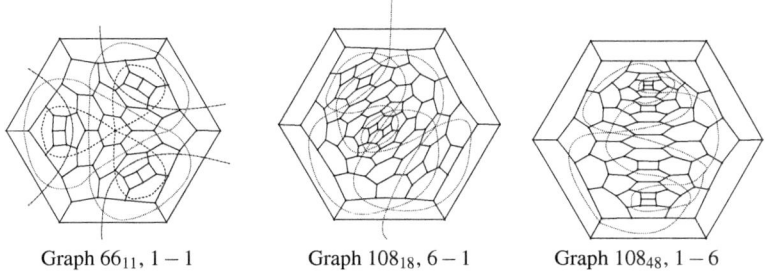

Graph 66_{11}, $1-1$ Graph 108_{18}, $6-1$ Graph 108_{48}, $1-6$

Graph 4_{66} and two graphs 4_{108} with triple self-intersecting railroads

Curve	First appearance	m	$p_1,\ldots,p_6$	Group	z − vector
$1-1$	66_{11} (twice)	1	$3,0,1,0,0,0$	D_{3h}	$36^4_{3,0}, 54_{3,0}$
$1-2$	126_{76}	7	$0,3,4,3,0,0$	D_{3h}	$18^6; 90^3_{9,0}$
$1-3$	106_{53}	3	$1,2,3,0,0,0$	C_s	$18^4; 30_{1,0}, 42_{1,0}, 54_{1,0}, 60^2_{5,0}$
$1-4$	86_{39}	4	$1,2,3,1,0,0$	C_1	$60^2_{6,0}, 138_{33,0}$
$1-5$	114_{36} (twice)	5	$0,4,3,0,1,0$	C_{2v}	$54^3_{3,0}, 72^4_{7,0}$
$1-6$	108_{48} (twice)	7	$0,2,7,0,1,0$	C_{2v}	$78^3_{9,0}, 90_{9,0}$
$1-7$	142_{82}	5	$0,3,4,2,0,0$	C_1	$24^3; 60_{1,0}, 90^2_{8,0}, 114_{14,0}$
$1-8$	114_{61}	11	$0,2,8,5,0,0$	C_1	$102^2_{13,0}, 138_{19,0}$
$2-1$	140_{68}	2	$2,2,2,0,0,0$	C_2	$24^2; 72^2_{6,0}, 114^2_{13,0}$
$2-2$	90_{30}	5	$0,6,0,3,0,0$	D_{3h}	$18^4; 54_{3,0}, 72^2_{9,0}$
$2-3$	122_{21}	5	$2,2,2,3,0,0$	C_2	$42_{1,0}, 78^2_{5,0}, 84^4_{9,0}$
$2-4$	126_{39}	6	$0,4,4,2,0,0$	C_2	$36^2_{1,0}, 90^2_{10,0}, 126_{21,0}$
$2-5$	134_{130}	6	$0,3,6,1,0,0$	C_{2v}	$36^2; 42^2_{2,0}, 54_{3,0}, 96^2_{11,0}$
$2-6$	122_{104}	7	$0,4,4,3,0,0$	C_{2v}	$30^4; 54_{3,0}, 96^2_{11,0}$
$2-7$	124_{100}	7	$0,3,6,2,0,0$	C_2	$90^2_{9,0}, 96^2_{11,0}$
$2-8$	128_{64}	7	$0,2,6,3,0,0$	C_s	$54_{3,0}, 96^2_{11,0}, 138_{17,0}$
$2-9$	110_{24}	8	$0,4,4,4,0,0$	C_2	$90^2_{12,0}, 150_{33,0}$
$2-10$	110_{74}	8	$0,2,8,2,0,0$	C_2	$90^2_{12,0}, 150_{33,0}$
$2-11$	134_{50}	10	$0,4,4,6,0,0$	C_2	$114^2_{14,0}, 174_{29,0}$
$2-12$	134_{149}	14	$0,2,8,8,0,0$	C_2	$126^2_{18,0}, 150_{21,0}$
$2-13$	126_{57}	14	$0,0,12,6,0,0$	D_3	$126_{9,0}, 126^2_{18,0}$
$4-1$	72_{17}	4	$0,6,4,0,0,0$	D_2	$24; 48_{4,0}, 72^2_{12,0}$
$6-1$	108_{18}	6	$0,6,8,0,0,0$	D_3	$24^3; 36; 108^2_{18,0}$
$6-2$	138_{102}	21	$0,1,20,6,2,0$	C_2	$78_{3,0}, 168^2_{33,0}$
$6-3$	158_{150}	24	$0,0,24,6,0,2$	D_3	$102_{9,0}, 186^2_{36,0}$
$8-1$	144_{151}	8	$0,6,12,0,0,0$	D_2	$36; 108_{12,0}, 144^2_{24,0}$

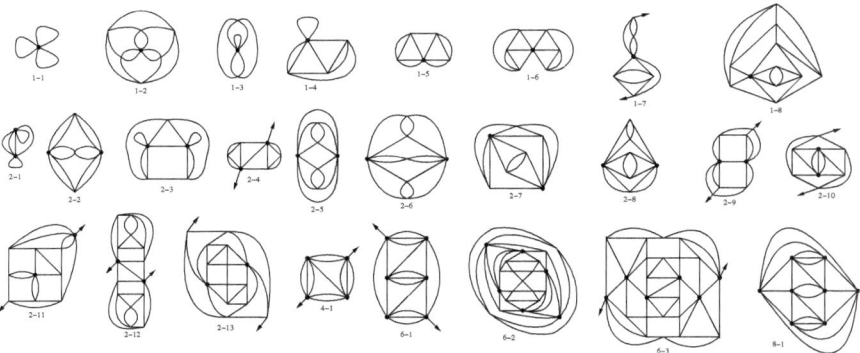

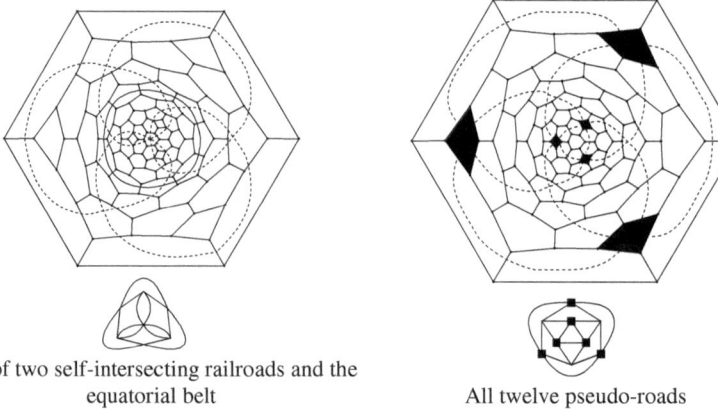

One of two self-intersecting railroads and the
equatorial belt

All twelve pseudo-roads

Fig. 3.13 Road-decomposition of the graph $4_{126}(D_{3h})$

Given an hexagon in a graph 3_v, 4_v, or 5_v, any pair of opposite edges belongs to a railroad or a pseudo-road; so, we exactly have a triple covering of the set of all hexagons by the set of all railroads and pseudo-roads. Every non 6-gonal face, i.e., t-gonal face with $t < 6$, has t adjacencies with the system of pseudo-roads.

For the special case, when our graph is 4_v, any pseudo-road arriving on a 4-gonal face can be extended on the opposite edge. So, the set of such *extended pseudo-roads*, together with the set of railroads, can be identified with the set of central circuits of the dual of our graph. Hence, extended pseudo-roads can be seen (in the same way as it was done for railroads) as projections of a Jordan curve on the plane with double and, eventually, triple points of self-intersection. Those notions are illustrated in Fig. 3.13. This graph $4_{126}(D_{3h})$ has the following *road decomposition* : five concentric belts (we indicate only the central, *equatorial* one), two self-intersecting railroads (they differ only by their opposite position on the sphere; we present only one of them), and all 12 pseudo-roads.

References

[DDF04]   Deza, M., Dutour, M., Fowler, P.W.: Zigzags railroads and knots in fullerenes. J. Chem. Inf. Comput. Sci. **44**, 1282–1293 (2004)

[DeDu05] Deza, M., Dutour, M.: Zigzag structure of simple two-faced polyhedra. Comb. Probab. Comput. **14**, 31–57 (2005)

[Fu]   Fukuda, K.: The cdd program. http://www.ifor.math.ethz.ch/fukuda/cdd_home/cdd.html

[GrünMo63] Grünbaum, B., Motzkin, T.S.: The number of hexagons and the simplicity of geodesics on certain polyhedra. Can. J. Math. **15**, 744–751 (1963)

[Rol76]   Rolfsen, D.: Knots and Links, Mathematics Lecture Series 7, Publish or Perish, Berkeley, (1976); second corrected printing: Publish or Perish, Houston (1990)

[Thi]   Thistlewaite, M.: Homepage. http://www.math.utk.edu/morwen

Chapter 4
ZC-Circuits of 4-Regular and Self-dual $\{2, 3, 4\}$-Spheres

For a 4-regular plane graph without loops, Euler formula $v - e + f = 2$ takes the form

$$\sum_{i \geq 2} (4 - i) p_i = 8. \tag{4.1}$$

Call an i-*hedrite* any ($\{2, 3, 4\}$, 4)-sphere with $p_2 + p_3 = i$. So, an v-vertex i-hedrite has $v = i - 2 + p_4$ vertices and p-vector $(p_2, p_3, p_4) = (8 - i, 2i - 8, p_4)$. Clearly, $(i; p_2, p_3) = (8; 0, 8)$, $(7; 1, 6)$, $(6; 2, 4)$, $(5; 3, 2)$, $(4; 4, 0)$ are all possibilities. The 8-hedrites were introduced (in [DeSt03]) as *octahedrites* and further studied in [DeDuSh03]. A i-*self-hedrite* is a self-dual plane graph with faces of size 2, 3 and 4 and $p_2 = 4 - i$. All possibilities are $(i; p_2, p_3) = (4; 0, 4)$, $(3; 1, 2)$ and $(2; 2, 0)$. Such graphs were first studied in [DuDe11].

In this chapter we give an account of those works that is relevant to the study of zigzags and central circuits. We first consider the k-connectivity property and give the full list of possible symmetry groups. Systematic enumeration methods of i-hedrites and i-self-hedrites are developed. For i-hedrites this gives the full list up to 70 vertices and for i-self-hedrites, up to 40 vertices. Sharp upper bound on the number of central circuits of tight i-hedrites are given and those of them with only simple central circuits are completely enumerated. For i-self-hedrites the corresponding circuit notion is zigzags and we do similar study of them. In addition, we determine the self-dual spheres with faces of size 1, 3, and 4 such that $p_1 = 2$.

4.1 Central Circuits of i-hedrites

The following Proposition is a local version (for "parts" of the sphere) of the Euler formula 4.1 for p-vector of any 4-regular plane 3-connected graph P.

For any 4-regular plane graph P, a *patch* A is a region of P bounded by q arcs (paths of edges) belonging to central circuits (different or coinciding), such that all q arcs form together a circle. A patch can be seen as a q-gon; we admit also 0-gonal A, i.e.,

© Springer India 2015
M. Deza et al., *Geometric Structure of Chemistry-Relevant Graphs*,
Forum for Interdisciplinary Mathematics 1, DOI 10.1007/978-81-322-2449-5_4

just the interior of a simple central circuit. Suppose that the patch A is *regular*, i.e., the continuation of any of bounding arc (on the central circuit to which it belongs) lies outside of the patch. See below two examples of patch.

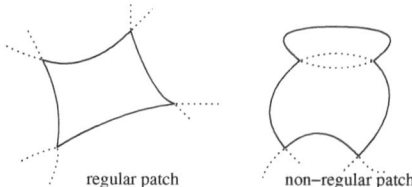

regular patch   non–regular patch

Let $p'(A) := p'_1, \ldots$ be the p-vector enumerating the faces of the patch A. The *curvature of the patch* A is defined as $c(A) = \sum_{k \geq 1}(4 - k)p'_k$. So, the k-gon is, respectively, positively curved, flat, or negatively curved, if $k < 4$, $k = 4$, or $k > 4$; such curvature is proportional to the one defined in Sect. 1.4.

Proposition 4.1 *(i) If A be a regular patch, then $c(A) = 4 - q$, moreover:*
(i1) $c(A) = 4$ if and only if A is bounded by a simple central circuit.
(i2) $c(A) = 0$ if and only if A is a rectangle formed by 4-gons put together.
(ii) Any patch A is the union of regular patches $A_1, \ldots, A_p$; one has $c(A) = (4 - q_1) + \cdots + (4 - q_p)$, where each patch A_i is bounded by q_i arcs.
(iii) If a graph G is the union of patches $A_1, \ldots, A_p$, then $c(A_1) + \cdots + c(A_p) = 8$.

Proof (i) is a restatement in our terms of Theorem 1 of [DeSt03], (i1) and (i2) are easy consequences. The properties (ii) and (iii) follow from the definition of the curvature of a patch and from Euler formula.   □

Proposition 4.2 *(i) Any 4-regular plane graph, whose faces are k-gons with k even, has only simple central circuits.*
(ii) At least one central circuit of a 7-hedrite self-intersects.

Proof In fact, if a central circuit in the case (i) self-intersects, then we have an 1-gonal regular patch. The equality of above Proposition becomes $\sum_i(4 - i)p'_i = 3$, an impossibility since the left-hand side is even.

For a 7-hedrite, take a central circuit containing an edge of the unique 2-gon, then the sequence (possibly empty) of adjacent 4-gons will necessarily finish by a 3-gon, or this 2-gon; both cases yield a self-intersection.   □

The graph of curvatures (see Sect. 3.2) of an i-hedrite is the i-vertex multigraph having as vertex-set all 2- and 3-gons of with two vertices being adjacent if there is a pseudo-road connecting them. Clearly, in this graph, the vertices corresponding to 2- and 3-gons, have degree 2 and 3, respectively.

Proposition 4.3 *Two central circuits C_1, C_2 of an i-hedrite are disjoint if and only if they are simple and there exist a ring of 4-gons separating them.*

Proof If both C_1 and C_2 are simple circuits, is evident: the curvature of the interior of a patch is 4 and so, two circuits are separated by 4-gons only. Suppose that C_1 is self-intersecting. Then it has at least three regular patches and each of them has curvature at most 3. The circuit C_2, being disjoint with C_1, lies entirely inside one of those patches, say, A. So, all its 3- and 2-gons, except, possibly, those from its exterior patch, lie in A. So, $c(A) \geq 5$, since the exterior patch of C_2 has curvature at most 3. It contradicts to the fact that A has curvature at most 3.   □

Remark 4.1 Consider a 4-regular plane graph G having only one central circuit; then, the set of faces of G can be partitioned into two classes $\mathscr{C}_1$, $\mathscr{C}_2$ in chess manner. Every vertex V is contained in two faces F and F'. Also, the unique central circuit can be given an orientation, which induces an orientation on the set of edges.

The vertex V is incident to two edges of F, e_1 and e_2, and to two edges of F', e'_1 and e'_2. If e_1, e_2 have both arrow pointing to the vertex or both arrow pointing out of the vertex, then the same is true for e'_1 and e'_2, and then, we say that V *belongs to Class I*. Class I and its complement, Class II, form a bipartition of the set of vertices of the knot; reversing orientation of the central circuit or interchanging C_1 and C_2 does not change the bipartition.

If the graph consists of p, $p \geq 2$, central circuits $C_1, \ldots, C_p$, then, one can put orientations on every central circuit and get a bipartition of the set of vertices. But in that case the bipartition will depend on the chosen orientations.

The *deleting of a central circuit C* in an i-hedrite G consists of removal of all edges and vertices contained in C. It produces a 4-regular plane graph P' having only k-gonal faces with $k \leq 4$. But since cases $k = 0, 1$ are possible, we do not always obtain an i-hedrite.

The *cutting of an i-hedrite G* consists of adding another central circuit to it. The faces of the new i'-hedrite G' with $8 \geq i' \geq i$ comes from the cutting of faces of G. This operation is only partially defined, since arbitrary cutting can produce k-gons with $k > 4$. The cutting of a 4-gon in several 4-gons (two, if the face is traversed only once) is possible only if the 4-gon is traversed on opposite edges. This corresponds to the notion of *shore zone* in [DeSt03]. A cutting changes partition of edge-set of an i-hedrite into central circuits only in the following way: new central circuit C is added and all others central circuits remain unchanged, except that the length of each of them increases by one for any intersection with C.

A railroad of 4-gons is bounded by two "parallel" central circuits. The deleting of one of them (i.e., collapsing railroad into one central circuit) is called *cc-reduction*. The cutting produces a railroad if and only if it replaces C by (thin enough) railroad. A *t-inflation along a central circuit C* is replacing C by $t-1$ parallel (thin enough) railroads. A *t-inflation of an i-hedrite* is new i-hedrite obtained from the original one by simultaneous t-inflation along all of its central circuits. A t-inflation of G is G if $t = 1$, and it is the chamfering of G if $t = 2$.

Remark 4.2 Let C be a central circuit of G with $\mathbf{cc}(G) = (\ldots, a_i^{\alpha_i}, \ldots; \ldots, b_j^{\beta_j}, \ldots)$, and let $\mathbf{Int}(C) = (c_0; c_1^{\gamma_1}, \ldots, c_r^{\gamma_r})$. Let G^t denote the t-inflation of G and let C' be one of t parallel copies of C. Then $\mathbf{cc}(G^t) = (\ldots, ta_i^{t\alpha_i}, \ldots; \ldots, tb_j^{t\beta_j}, \ldots)$ and $\mathbf{Int}(C') = (c_0; c_1^{t\gamma_1}, \ldots, c_r^{t\gamma_r}, (2c_0)^{t-1})$.

4.2 Connectivity and Symmetries of *i*-hedrites

For any integer $m \geq 2$, let us denote:

by $I_{6,2m}$ the $2m$-vertex 6-hedrite, such that each 2-gon is adjacent to two 3-gons;

by $I_{5,2m+1}$ the $(2m + 1)$-vertex 5-hedrite, such that two 2-gons share a vertex and the remaining 2-gon is adjacent to two 3-gons;

by $I_{4,2m+2}$ the $(2m + 2)$-vertex 4-hedrite, such that four 2-gons are organized into two pairs sharing a vertex;

by $J_{4,2m}$ the m-inflation of only one central circuit of 4-hedrite **2-1**; it is a projections of *composite* alternating links $2_1^2 \# 2_1^2 \# 2_1^2 \ldots \# 2_1^2$ (m times).

See in Fig. 4.1 the first occurrences (for $2 \leq m \leq 5$) of those graphs, followed by their symmetry groups and *cc*-vectors.

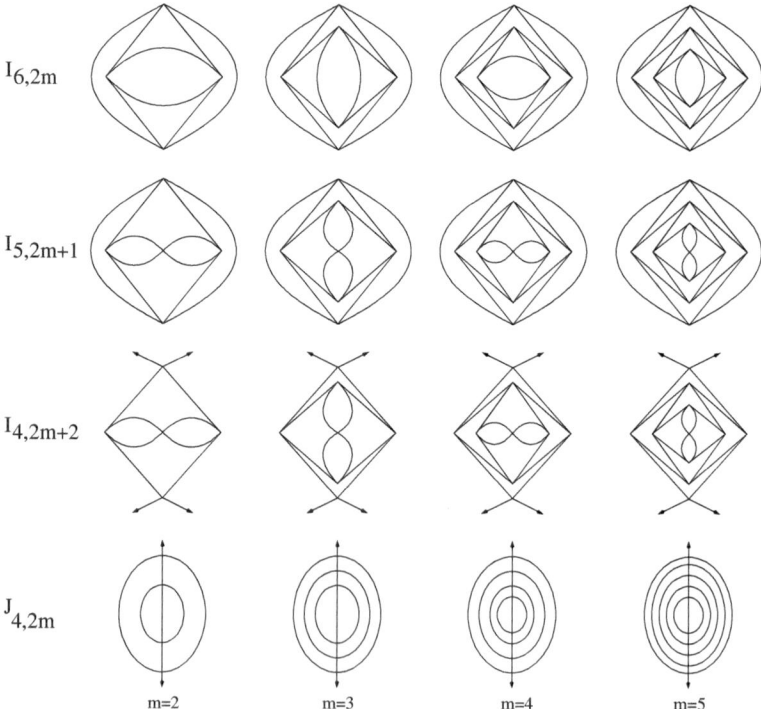

$I_{6,2m}$

$I_{5,2m+1}$

$I_{4,2m+2}$

$J_{4,2m}$

m=2   m=3   m=4   m=5

Fig. 4.1 The first four graphs of all series of *i*-hedrites, which are not 3-connected

Table 4.1 The infinite series of *i*-hedrites, which are not 3-connected	*i*-hedrite	Group	*cc*-vector
	$I_{6,2m}$, *m* even	D_{2d}	$4m$
	$I_{6,2m}$, *m* odd	D_{2h}	$(2m)^2$
	$I_{5,2m+1}$	C_{2v}	$4m+2$
	$I_{4,2m+2}$, *m* even	D_{2d}	$(2m+2)^2$
	$I_{4,2m+2}$, *m* odd	D_{2h}	$(2m+2)^2$
	$J_{4,2m}$	D_{2h}	2^m, $2m$

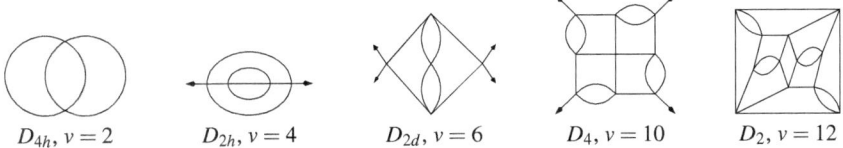

$D_{4h}, v = 2$ $D_{2h}, v = 4$ $D_{2d}, v = 6$ $D_4, v = 10$ $D_2, v = 12$

Fig. 4.2 Minimal representatives for the symmetry groups of 4-hedrites

In [DeDuSh03], we proved:

Theorem 4.1 *(i) Any i-hedrite is 2-connected.*
(ii) All not 3-connected i-hedrites are $I_{6,2m}$, $I_{5,2m+1}$, $I_{4,2m+2}$, $J_{4,2m}$ for some $m \geq 2$.

Theorem 4.2 *(i) If G is an i-hedrite with two adjacent 2-gons, then this is a 4-hedrite **2-1** or a $J_{4,2m}$ with $m \geq 2$.*
*(ii) If G is an i-hedrite with two 2-gons sharing a vertex, then it is either 4-hedrite **4-1**, or an $I_{4,2m+2}$, or 5-hedrite **3-1**, or an $I_{5,2m+1}$ with $m \geq 2$ (Table 4.1).*

Theorem 4.3 *(i) The symmetry groups of 4-hedrites are the point subgroups of D_{4h}, containing D_2, i.e., D_{4h}, D_4, D_{2d}, D_{2h}, D_2.*
(ii) The symmetry groups of 5-hedrites are: D_{3h}, D_3, C_{2v}, C_2, C_s, C_1.
(iii) The symmetry groups of 6-hedrites are the point subgroups of D_{2d} or D_{2h}, i.e., D_{2d}, D_{2h}, D_2, C_{2h}, C_{2v}, C_2, C_i, C_s, C_1.
(iv) The symmetry groups of 7-hedrites are the point subgroups of C_{2v}, i.e., C_{2v}, C_2, C_s, C_1.
(v) The symmetry groups of 8-hedrites are: C_1, C_s, C_i, C_2, C_{2v}, C_{2h}, S_4, D_2, D_{2h}, D_{2d}, D_3, D_{3h}, D_{3d}, D_4, D_{4h}, D_{4d}, O, O_h.
(vi) Minimal representatives for each symmetry group in (i)-(v) above are given in Figs. 4.2, 4.3, 4.4, 4.5 and 4.6, respectively.

In general, the D_4, D_{4h} 4-hedrites, D_3,D_{3h} 5-hedrites, and O, O_h 8-hedrites come as $GC_{k,l}(G_{0,i})$ with $G_{0,i}$ the smallest *i*-hedrite. See Theorem 6.2 for a full list of such graphs obtained by Goldberg-Coxeter construction.

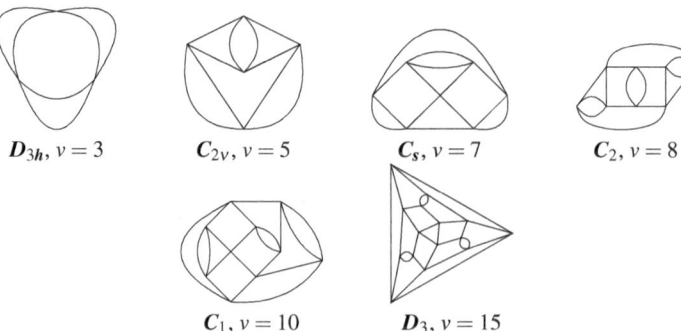

Fig. 4.3 Minimal representatives for the symmetry groups of 5-hedrites

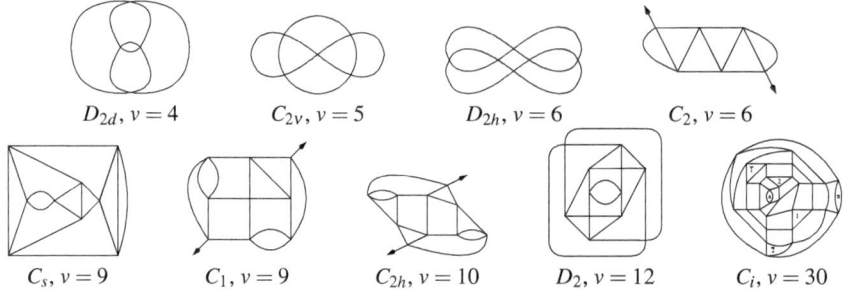

Fig. 4.4 Minimal representatives for the symmetry groups of 6-hedrites

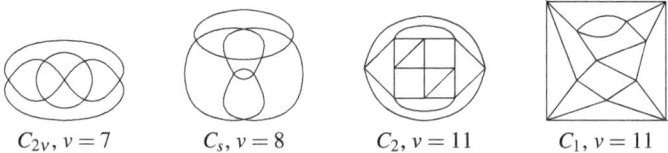

Fig. 4.5 Minimal representatives for the symmetry groups of 7-hedrites

4.3 Tight *i*-hedrites and All Pure Tight *i*-hedrites

Remind (see Sect. 1.3) that an *i*-hedrite is called *tight* if it contains no railroad.

Theorem 4.4 [DeDuSh03] *Any tight i-hedrite has at most $i - 2$ central circuits and equality is attained for each i, $4 \le i \le 8$.*

Proof For $i = 8$, the Theorem is proved in [DeSt03]. We will show, using suitable cutting, that this result implies the Theorem for $i < 8$.

For $i = 7$, consider a simple circuit S in its curvature graph, which contains the vertex corresponding to unique 2-gon. Remind that the vertices in the curvature graph

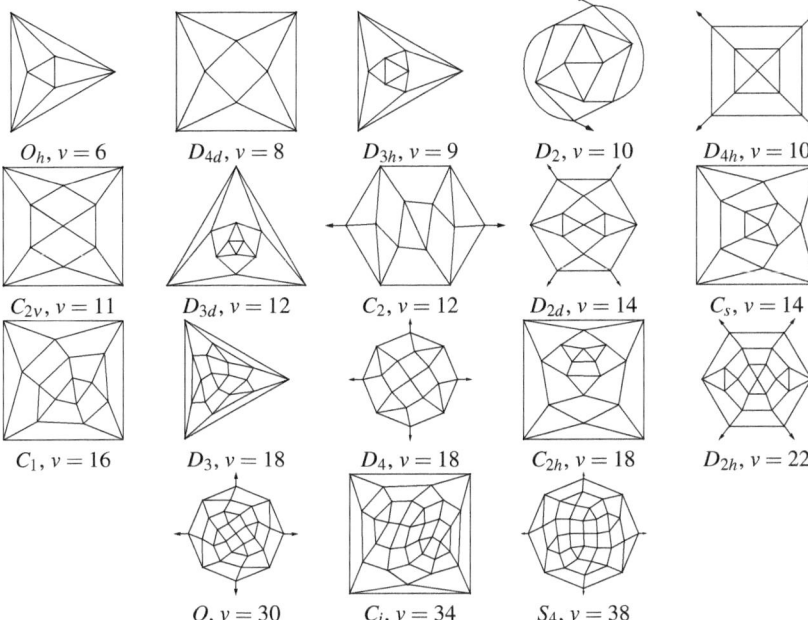

$O_h, v = 6$ $D_{4d}, v = 8$ $D_{3h}, v = 9$ $D_2, v = 10$ $D_{4h}, v = 10$

$C_{2v}, v = 11$ $D_{3d}, v = 12$ $C_2, v = 12$ $D_{2d}, v = 14$ $C_s, v = 14$

$C_1, v = 16$ $D_3, v = 18$ $D_4, v = 18$ $C_{2h}, v = 18$ $D_{2h}, v = 22$

$O, v = 30$ $C_i, v = 34$ $S_4, v = 38$

Fig. 4.6 Minimal representatives for the symmetry groups of 8-hedrites

Fig. 4.7 Schematic of
central circuit in a 7-hedrite;
see Theorem 4.4

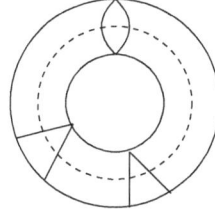

correspond to 2- or 3-gons, while edges correspond to pseudo-roads. So, the simple
circuit S corresponds to the circuit of faces of G, containing our 2-gon, some 4-gons
and, possibly, some of six 3-gons; see Fig. 4.7.

Suppose that the 7-hedrite G is tight and has k central circuits. By adding the
central circuit C, which is shown by dotted line on Fig. 4.7, we produce an 8-hedrite
(since, the 2-gon is cut by C in two 3-gons), which is still tight and has $k + 1$ central
circuits. So, $k + 1 \leq 6$, by Theorem 3 of [DeSt03].

For remaining cases of *i*-hedrites with $i = 4, 5, 6$, the proof is similar. We consider
all possible distributions of 2-gons by simple circuits in the graph of curvatures and,
for each such circuit, we add suitable number of new central circuits.   □

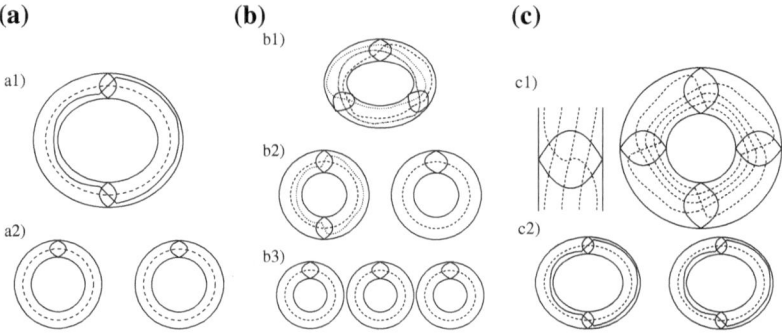

Fig. 4.8 The construction of Theorem 4.4 for 6-, 5- and 4-hedrites. **a** 6-hedrite: 2 cases, **b** 5-hedrite: 3 cases, **c** 4-hedrite: 2 cases

Fig. 4.9 A tight 7-hedrite with the maximum number of central circuits

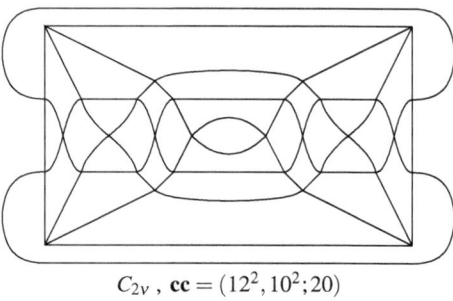

$$C_{2v}, \mathbf{cc} = (12^2, 10^2; 20)$$

Fig. 4.10 No *pure* tight *i*-hedrite can be obtained by cutting of the above 4-hedrite

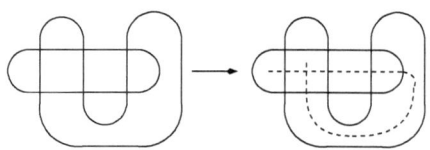

All possibilities are presented in Fig. 4.8: two for $i = 6$, three for $i = 5$ and two for $i = 4$. In the last case, there are no 3-gons and so, simple circuits in the curvature graph contain only even number of 2-gons by local Euler formula of Theorem 4.1. Note that the case of 4-hedrites is obvious by Theorem 4.5 of [DeSt03].

See in Fig. 4.9 an example of a tight 7-hedrite with 5 central circuits (Fig. 4.10).

For all other i, there is a tight i-hedrite with $i - 2$ central circuits (see Theorem 4.6), which is, moreover, pure: 4-hedrite **2-1** ($\mathbf{cc} = 2^2$), 5-hedrite **6-2** ($\mathbf{cc} = 4^3$), 6-hedrite **14-20** ($\mathbf{cc} = (6^2, 8^2)$) and, 7-th in Fig. 4.11, 8-hedrite with $\mathbf{cc} = 10^6$.

Conjecture 4.1 *A tight i-hedrite cannot be obtained from another i'-hedrite by a cutting if and only if it has $i - 2$ central circuits.*

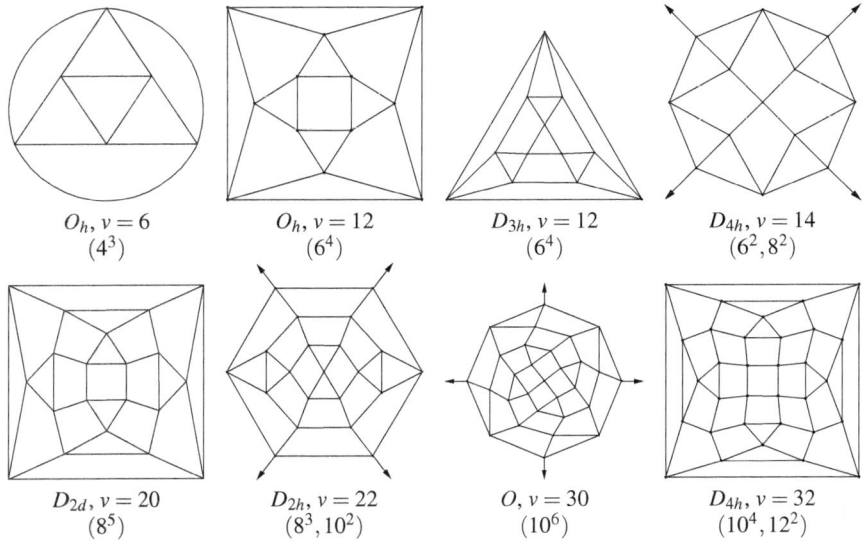

O_h, $v = 6$	O_h, $v = 12$	D_{3h}, $v = 12$	D_{4h}, $v = 14$
(4^3)	(6^4)	(6^4)	$(6^2, 8^2)$

D_{2d}, $v = 20$	D_{2h}, $v = 22$	O, $v = 30$	D_{4h}, $v = 32$
(8^5)	$(8^3, 10^2)$	(10^6)	$(10^4, 12^2)$

Fig. 4.11 The pure tight 8-hedrites with their symmetries and cc-vectors. First four are **6-1**, **12-4**, **12-5**, **14-7**; last four are all pure tight with at least 5 central circuits

The easiest case, $i = 4$, of i-hedrites admits following complete characterization:

Theorem 4.5 *(i) Any 4-hedrite can be obtained from some 4-hedrite with two central circuits by simultaneous t_1- and t_2-inflation (repetition of) along those circuits; it is tight if and only if $t_1 = t_2 = 1$.*

(ii) Any 4-hedrite with two central circuits is defined by its number of vertices v and by shift j, $0 \leq j \leq \frac{v}{4}$ with $gcd(\frac{v}{2}, j) = 1$, vertices between the pair of boundary 2-gons on the horizontal circuit and the remaining pair of 2-gons.

(iii) Any 4-hedrite is balanced.

Proof (i) and (ii) are proved in [DeSt03], while (iii) is obvious for 4-hedrite with two central circuits and remain true under t_1- and t_2-inflation. See, as an example of (ii), 4-hedrite **8-1**. Different values of shift can yield the same graph.

The shift $j = 0$ corresponds to **2-1** and its only on one circuit m-inflations $J_{4, v=2m}$. The shift $j = 1$ corresponds to **4-1** and $I_{4, v=2m+2}$. Denote by $K_{4, 4m}$, for any $m \geq 2$, any $4m$-vertex 4-hedrite obtained from **4-1** by m-inflation of only one its central circuit; so, its cc-vector is $(4^m, 4m)$, its symmetry is D_{2d} and it is not tight. Clearly, any $K_{4, v=4m}$ has the maximal shift $j = \frac{v}{4}$.   □

No unbalanced 5-hedrite or 7-hedrite exists for $v \leq 15$. The first unbalanced 6-hedrite is **12-12**. The first unbalanced 8-hedrite is **14-7** (4th on Fig. 4.11) inflated along a central circuit of length 8. Remind (see Sect. 1.3) that an i-hedrite is called *balanced*, if all its central circuits of same length have the same intersection vector and *pure* if it has no self-intersecting central circuits.

Clearly, any pure i-hedrite has an even number of vertices; in fact, any vertex in this case belong to the intersection of exactly two central circuits.

Theorem 4.6 *Any pure tight i-hedrite is, either any 4-hedrite with two central circuits, or a 5-hedrite* **6-2**, *or one of* 6-hedrites **8-5**, **14-20** *(see all three on two pictures below), or one of the eight pure* 8-hedrites, *given of Fig.* 4.11.

Proof Let G be a pure tight i-hedrite having r central circuits. If one deletes a central circuit, then, in general, 1-gon can appear. It does not happen for G, since it would imply a self-intersection of a central circuit. So, the result of deletion of a central circuit from G produces a pure tight i-hedrite with $r - 1$ central circuits.

If $r = 2$, then, by Theorem 5 from [DeSt03], such G are exactly 4-hedrites with two central circuits; all of them are classified in Theorem 4.5. We prove the Theorem by systematic analysis of all possible ways to add to G (for $r = 2, 3, 4, 5$) a central circuit, in order to get a pure tight i-hedrite with $r + 1$ central circuits.

Let $r = 2$. Then G can be only one of two smallest 4-hedrites. In fact, if G is another 4-hedrite, then, because of classification Theorem 4.5, it has a form as in Fig. 4.10. New central circuit should cut both 2-gons on opposite edges, since, otherwise, there is a railroad. But Fig. 4.10 shows, on example for $v = 6$, that a self-intersection appears if two central circuits intersect in more than four vertices.

So, the only possible 4-hedrites with *two* central circuits, which can be cut in order to produce tight pure i-hedrite are **2-1** and **4-1**. All cases are indicated below.

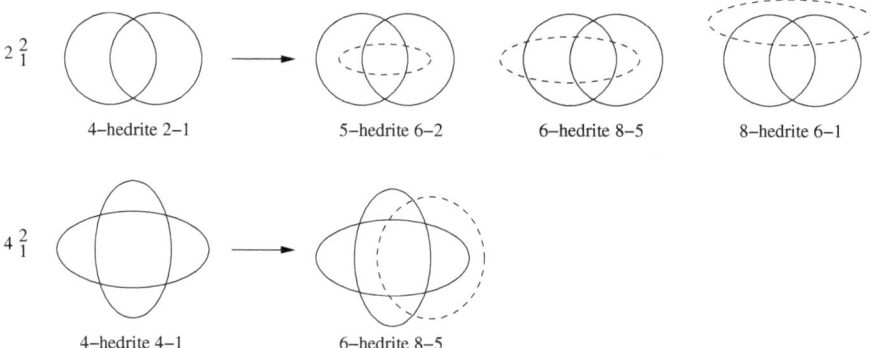

Now, all tight pure i-hedrites with 3 central circuits are 5-hedrite **6-2**, 8-hedrite **6-1**, and 6-hedrite **8-5** (projections of links 6_1^3, 6_2^3, and 8_6^3); first two have cc-vector 4^3. We apply the same procedure to those three i-hedrites; see picture below:

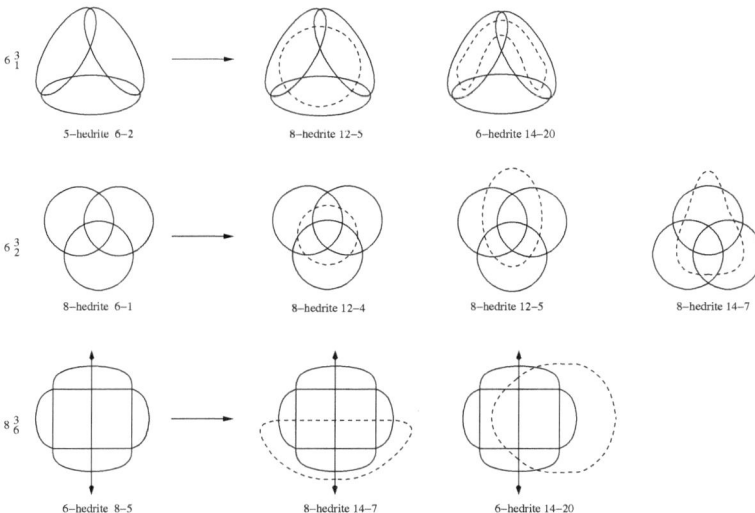

Next, all tight pure *i*-hedrites with 4 central circuits are 8-hedrites **12-4**, **12-5**, **14-7**, and 6-hedrite **14-20**(D_{2h}); last two have the same *cc*-vector (6^2, 8^2).

By the same method, one can see that there are exactly two pure tight *i*-hedrites with 5 central circuits and two with 6 central circuits: last four in Fig. 4.11.   □

Remark 4.3   Any pure *i*-hedrite comes from a pure *tight* *i*-hedrite with, say, *j* central circuits by simultaneous $t_1-, \ldots, t_j$-inflation along those circuits; it is tight if and only if $t_1 = \cdots = t_j = 1$.

4.4 Enumeration and Generation of *i*-hedrites

The enumeration of 8-hedrites was done by using the program ENU (see [Heid98, BrHaHe03]) by Heidemeier, that enumerates classes of 4-regular graphs with constraint on the size of their faces, fairly efficiently.

Theorem 4.7   *(i) an v-vertex 4-hedrite exists if and only if $v \geq 2$, even.*
(ii) an v-vertex 5-hedrite exists if and only if $v \geq 3$, $v \neq 4$.
(iii) an v-vertex 6-hedrite exists if and only if $v \geq 4$.
(iv) an v-vertex 7-hedrite exists if and only if $v \geq 7$.
(v) an v-vertex 8-hedrite exists if and only if $v \geq 6$, $v \neq 7$ (see [Grün67]).

Here and below we use Rolfsen's notation [Rol76] for links with at most 9 crossings and knots with 10 crossings. We write $\sim$ if the projection in the pictures and Table below is different from the one given in corresponding cases above.

We give on the pictures below all *i*-hedrites with at most 10 vertices, indicating under picture of each its symmetry, *cc*-vector and corresponding alternating link. If an *i*-hedrite is 2-connected but not 3-connected, then we add a symbol ∗ just after

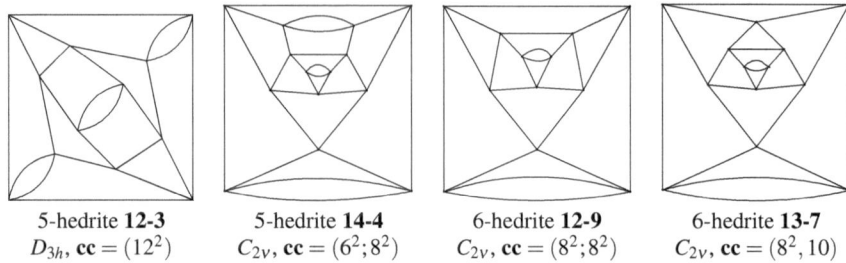

5-hedrite **12-3**	5-hedrite **14-4**	6-hedrite **12-9**	6-hedrite **13-7**
D_{3h}, **cc** $= (12^2)$	C_{2v}, **cc** $= (6^2; 8^2)$	C_{2v}, **cc** $= (8^2; 8^2)$	C_{2v}, **cc** $= (8^2, 10)$

Fig. 4.12 The i-hedrites with at most 15 vertices and a self-intersecting railroad

the number. If an i-hedrite is not tight, i.e., has a railroad, then we add mention "r" (reducible). All i-hedrites with 12 vertices are given in Sect. 9 of [DeDuSh03]. For the ones with 13–15 vertices, Table 2 of [DeDuSh03] gives essential information.

All four non-tight i-hedrites with $v \leq 15$ and a self-intersecting railroad are given in Fig. 4.12.

Below, we give the list of small i-hedrites. For a v-vertex i-hedrite G, $Med(G) = GC_{1,1}(G)$ is a $2v$-vertex i-hedrite, in which all 2- and 3-gonal faces are *isolated*, i.e., adjacent to only 4-gons. The medial of the smallest ones, 8-hedrite **6-1**, 7-hedrite **7-1**, 6-hedrite **4-1**, 5-hedrite **3-1**, and 4-hedrite **2-1**—are, respectively, 8-hedrite **12-4** (Cuboctahedron), 7-hedrite **14-9** (C_{1v}, **cc** $= (6^2; 16)$), 6-hedrite **8-3**, 5-hedrite **6-2** and 4-hedrite **4-1**.

4.4.1 8-hedrites

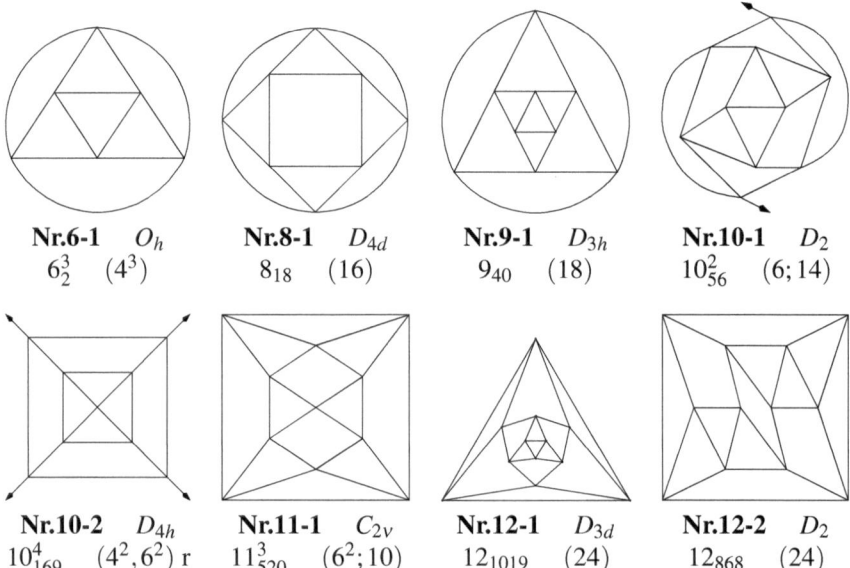

Nr.6-1 O_h	**Nr.8-1** D_{4d}	**Nr.9-1** D_{3h}	**Nr.10-1** D_2
6_2^3 (4^3)	8_{18} (16)	9_{40} (18)	10_{56}^2 $(6; 14)$

Nr.10-2 D_{4h}	**Nr.11-1** C_{2v}	**Nr.12-1** D_{3d}	**Nr.12-2** D_2
10_{169}^4 $(4^2, 6^2)$ r	11_{520}^3 $(6^2; 10)$	12_{1019} (24)	12_{868} (24)

4.4.2 4-hedrites

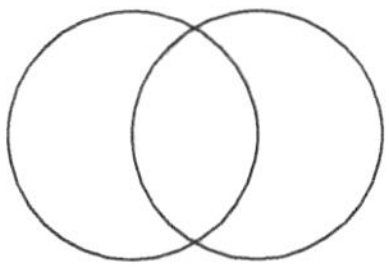

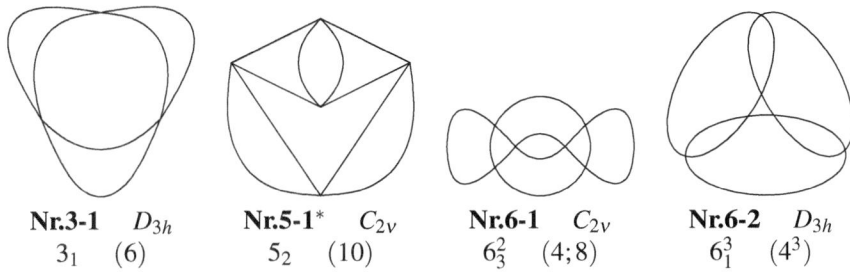

Nr.2-1 D_{4h}
2_1^2 (2^2)

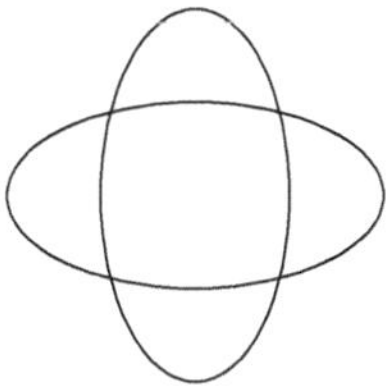

Nr.4-1 D_{4h}
4_1^2 (4^2)

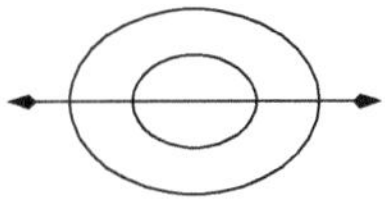

Nr.4-2* D_{2h}
$2 \times 2_1^2$ $(2^2, 4)$ r

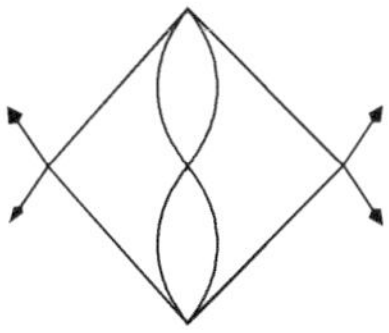

Nr.6-1* D_{2d}
6_2^2 (6^2)

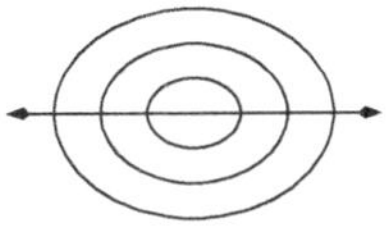

Nr.6-2* D_{2h}
$3 \times 2_1^2$ $(2^3, 6)$ r

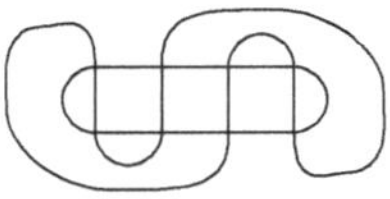

Nr.8-1* D_{2h}
$\sim 8_4^2$ (8^2)

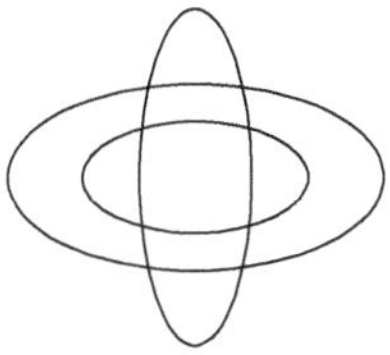

Nr.8-2 D_{2d}
8_4^3 $(4^2, 8)$ r

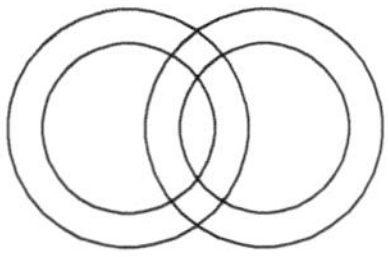

Nr.8-3 D_{4h}
8_1^4 (4^4) r

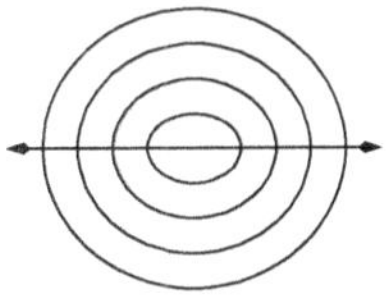

Nr.8-4* D_{2h}
$4 \times 2_1^2$ $(2^4, 8)$ r

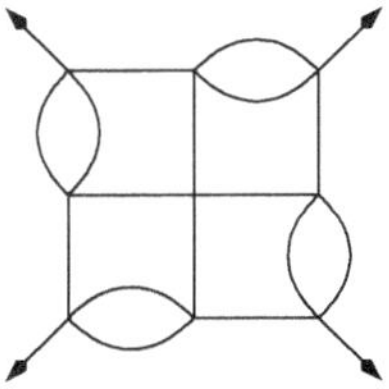

Nr.10-1 D_4
10_{121}^2 (10^2)

Nr.10-2* D_{2d}
$\sim 10_{120}^2$ (10^2)

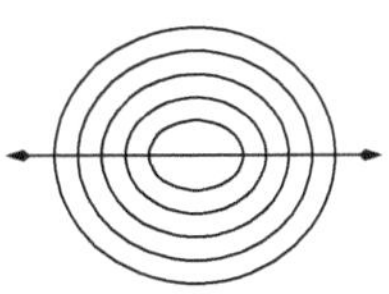

Nr.10-3* D_{2h}
$5 \times 2_1^2$ $(2^5, 10)$ r

4.4.3 5-hedrites

Nr.3-1 D_{3h}
3_1 (6)

Nr.5-1* C_{2v}
5_2 (10)

Nr.6-1 C_{2v}
6_3^2 $(4; 8)$

Nr.6-2 D_{3h}
6_1^3 (4^3)

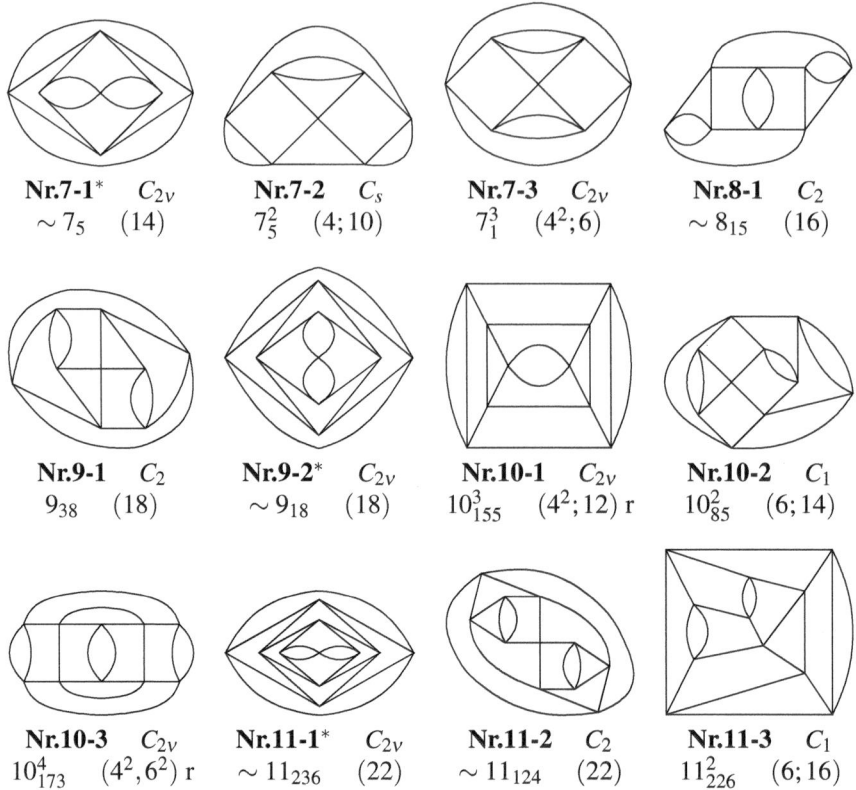

| **Nr.7-1*** C_{2v} | **Nr.7-2** C_s | **Nr.7-3** C_{2v} | **Nr.8-1** C_2 |
| $\sim 7_5$ (14) | 7_5^2 (4; 10) | 7_1^3 (4^2; 6) | $\sim 8_{15}$ (16) |

| **Nr.9-1** C_2 | **Nr.9-2*** C_{2v} | **Nr.10-1** C_{2v} | **Nr.10-2** C_1 |
| 9_{38} (18) | $\sim 9_{18}$ (18) | 10_{155}^3 (4^2; 12) r | 10_{85}^2 (6; 14) |

| **Nr.10-3** C_{2v} | **Nr.11-1*** C_{2v} | **Nr.11-2** C_2 | **Nr.11-3** C_1 |
| 10_{173}^4 ($4^2, 6^2$) r | $\sim 11_{236}$ (22) | $\sim 11_{124}$ (22) | 11_{226}^2 (6; 16) |

4.4.4 6-hedrites

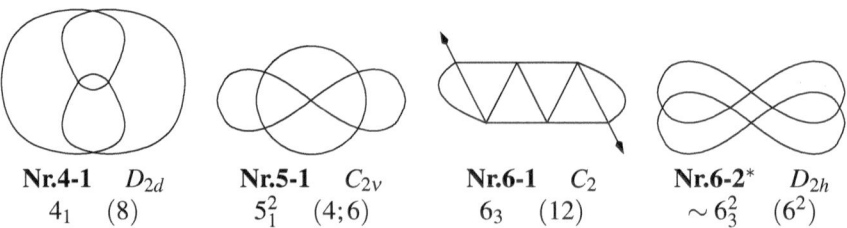

| **Nr.4-1** D_{2d} | **Nr.5-1** C_{2v} | **Nr.6-1** C_2 | **Nr.6-2*** D_{2h} |
| 4_1 (8) | 5_1^2 (4; 6) | 6_3 (12) | $\sim 6_3^2$ (6^2) |

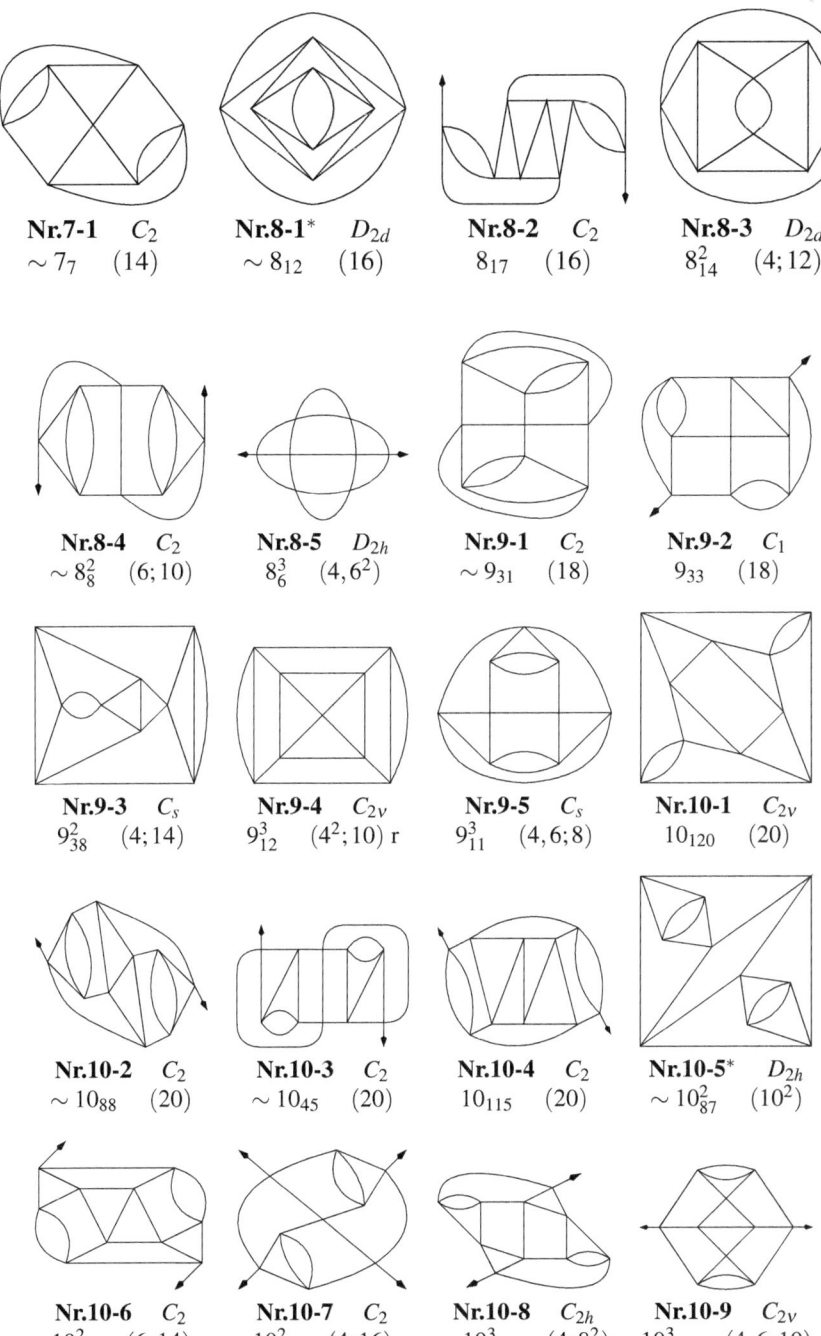

Nr.7-1 C_2	**Nr.8-1*** D_{2d}	**Nr.8-2** C_2	**Nr.8-3** D_{2d}
$\sim 7_7$ (14)	$\sim 8_{12}$ (16)	8_{17} (16)	8_{14}^2 (4; 12)
Nr.8-4 C_2	**Nr.8-5** D_{2h}	**Nr.9-1** C_2	**Nr.9-2** C_1
$\sim 8_8^2$ (6; 10)	8_6^3 $(4, 6^2)$	$\sim 9_{31}$ (18)	9_{33} (18)
Nr.9-3 C_s	**Nr.9-4** C_{2v}	**Nr.9-5** C_s	**Nr.10-1** C_{2v}
9_{38}^2 (4; 14)	9_{12}^3 $(4^2; 10)$ r	9_{11}^3 (4, 6; 8)	10_{120} (20)
Nr.10-2 C_2	**Nr.10-3** C_2	**Nr.10-4** C_2	**Nr.10-5*** D_{2h}
$\sim 10_{88}$ (20)	$\sim 10_{45}$ (20)	10_{115} (20)	$\sim 10_{87}^2$ (10^2)
Nr.10-6 C_2	**Nr.10-7** C_2	**Nr.10-8** C_{2h}	**Nr.10-9** C_{2v}
10_{86}^2 (6; 14)	10_{43}^2 (4; 16)	$\sim 10_{136}^3$ $(4; 8^2)$	10_{136}^3 (4, 6, 10)

4.4.5 7-hedrites

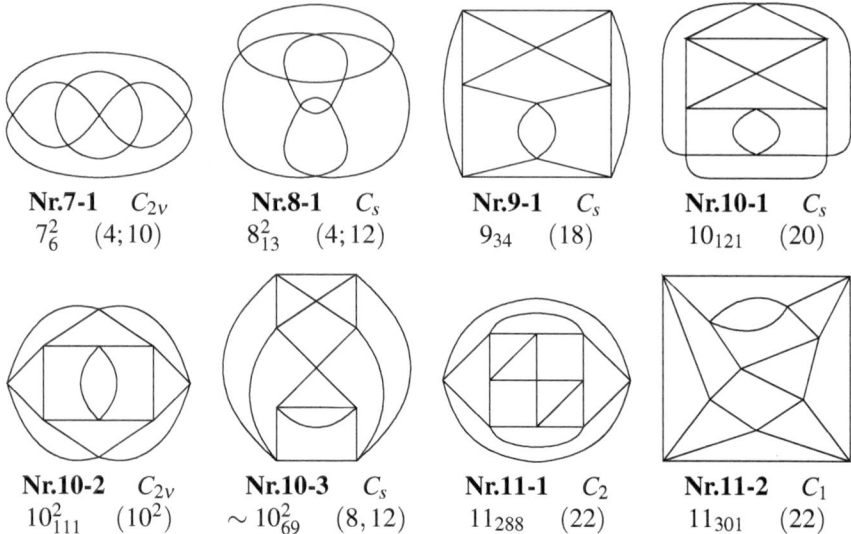

Nr.7-1 C_{2v}	Nr.8-1 C_s	Nr.9-1 C_s	Nr.10-1 C_s
7_6^2 (4; 10)	8_{13}^2 (4; 12)	9_{34} (18)	10_{121} (20)

Nr.10-2 C_{2v}	Nr.10-3 C_s	Nr.11-1 C_2	Nr.11-2 C_1
10_{111}^2 (10^2)	$\sim 10_{69}^2$ (8, 12)	11_{288} (22)	11_{301} (22)

The alternating knots, projections of which are i-hedrites with $11 \leq v \leq 15$ and such links with $v = 11$, are (in Dowker–Thistlewhaite's numbering, [Thi]):

- 5-hedrites: $\sim 11_{124}, 11_{226}^2, \sim 11_{236}, 11_{500}^3, 11_{547}^4, \sim 12_{431}, 13_{3097}, 13_{4054}, 14_{16368},$ $15_{20161}, 15_{54593}, 15_{83814}, 15_{83824}.$
- 6-hedrites: $\sim 11_{125}, 11_{297}, 11_{317}^2, 11_{332}, 11_{351}^2, \sim 12_{458}, \sim 12_{477}, \sim 12_{499}, \sim$ $12_{626}, 12_{1102}, 12_{1152}, 12_{1167}, \sim 13_{1345}, 13_{1485}, \sim 13_{1739}, \sim 13_{2957}, 13_{3586}, 13_{3811},$ $13_{3957}, 14_{5570}, 14_{8767}, 14_{17079}, 14_{17148}, 14_{17173}, 14_{17309}, 14_{17734}, \sim 15_{20975}, \sim$ $15_{39533}, 15_{45248}, \sim 15_{45357}, 15_{64488}, 15_{66949}, 15_{83008}.$
- 7-hedrites: $11_{150}^2, 11_{288}, 11_{301}, 11_{487}^3, \sim 12_{361}, 13_{3769}, 13_{3861}, 14_{5714}, 14_{10841},$ $14_{13725}, 14_{14207}, 15_{60207}, 15_{80242}, 15_{82225}.$
- 8-hedrites: $11_{520}^3, 12_{868}, 12_{1019}, 13_{3478}, 14_{17895}, 15_{82477}.$

4.4.6 Generation of i-hedrites

4-hedrites admit a simple explicit description, see Theorem 4.5. So, it remains to find efficient methods for the enumeration of 5-, 6-, and 7-hedrites. The program ENU cannot deal with 2-gons; so, we sought a method for reasonable enumeration of such graphs. See Table 4.2 for the number of i-hedrites with at most 70 vertices.

Take an i-hedrite G with $i \in \{5, 6, 7\}$. Then, if F is a face of size 2, we shrink it to a vertex by using the following *shrinking operation*:

Table 4.2 Number of *i*-hedrites, $4 \leq i \leq 8$, with $2 \leq \nu \leq 70$

ν	4	5	6	7	8
2	1	0	0	0	0
3	0	1	0	0	0
4	2	0	1	0	0
5	0	1	1	0	0
6	2	2	2	0	1
7	0	3	1	1	0
8	4	1	5	1	1
9	0	2	5	1	1
10	3	3	9	3	2
11	0	5	7	4	1
12	5	3	14	5	5
13	0	4	14	7	2
14	3	7	23	9	8
15	0	10	17	12	5
16	7	6	28	18	12
17	0	6	27	22	8
18	5	7	44	25	25
19	0	12	35	36	13
20	7	9	54	46	30
21	0	8	57	48	23
22	4	15	77	62	51
23	0	20	59	76	33
24	11	11	87	88	76
25	0	12	85	107	51
26	5	16	119	126	109
27	0	21	105	142	78
28	8	18	134	179	144
29	0	16	135	198	106
30	8	24	187	216	218
31	0	32	149	257	150
32	12	24	189	304	274
33	0	18	197	329	212
34	6	26	251	382	382
35	0	37	218	431	279
36	13	23	278	483	499
37	0	24	275	547	366
38	6	38	354	601	650
39	0	45	313	643	493
40	15	37	361	764	815
41	0	30	359	838	623
42	10	33	472	889	1083
43	0	52	405	998	800
44	11	44	480	1134	1305
45	0	34	511	1197	1020
46	7	56	609	1324	1653
47	0	69	519	1435	1261
48	21	45	613	1574	2045
49	0	40	614	1751	1554
50	10	54	771	1874	2505
51	0	66	704	1963	1946
52	13	58	771	2247	3008
53	0	48	788	2419	2322
54	12	66	989	2511	3713
55	0	92	849	2735	2829
56	18	68	938	3041	4354
57	0	49	1005	3187	3418
58	9	71	1175	3453	5233
59	0	98	1038	3659	4063
60	22	70	1215	3954	6234
61	0	63	1193	4315	4784
62	9	96	1440	4526	7301
63	0	104	1328	4674	5740
64	21	92	1378	5248	8514
65	0	74	1440	5600	6631
66	14	80	1751	5741	10103
67	0	122	1531	6159	7794
68	16	98	1675	6730	11572
69	0	72	1792	7005	9097
70	14	120	2066	7465	13428

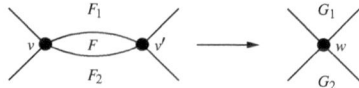

and get a graph denoted by $Red_F(G)$. During this operation the vertices v and v' are merged into one vertex w and the faces F_1 and F_2 are changed into G_1 and G_2 with one edge less. Thus, it is possible that G_1 and/or G_2 are themselves of size 2. We apply the shrinking operation whenever the shrunk graph is still an i-hedrite. Eventually, since every such operation diminish the vertex-set, one obtains a graph, denoted by $Red_\infty(G)$ for which we cannot apply the shrinking operation anymore.

We call a graph *unshrinkable* if we cannot apply to it any shrinking operation. If an unshrinkable graph G' has no faces of size 2, then it is an 8-hedrite. If G' has a face F of size 2, then denote by e_1, e_2 the two edges of F. Since G' is unshrinkable, F is adjacent on e_1 or e_2, say e_1, to another face of size 2.

If e_2 is incident to another face of size 2, then G' is actually 4-hedrite $2 - 1$, i.e., the unique graph with two vertices, and four faces of size 2. It is easy to see that e_2 cannot be incident to another face of size 3, but it can be incident to another face of size 4. In that case G' is not 3-connected and thus (see [DeDuSh03]) it belongs to one of the infinite families of 2-connected i-hedrites of Fig. 4.1. The only one that satisfies our requirement of unshrinkability is the infinite family $J_{4,2m}$.

Call *expansion operation* the reverse of the shrinking operation. The generation method of i-hedrites is to consider all unshrinkable i-hedrites and all possible ways of expanding them. For an unshrinkable graph G denote by $\mathscr{E}xp(G)$ the set of all possible i-hedrites that can be obtained by repeated application of the expansion operation. For the graphs of the infinite family $J_{4,2m}$ from Table 4.1, no expansion is possible and thus no i-hedrite is obtained from them. A priori, the set $\mathscr{E}xp(G)$ can be infinite but, as far as we know, for any 8-hedrite G the set $\mathscr{E}xp(G)$ is finite, although we have no proof of it. The set $\mathscr{E}xp(2_1)$ is infinite but it has a simple description.

Further generalization of i-hedrites are ({1, 2, 3, 4}, 4)-spheres. Then, besides i-hedrites (for which $p_1 = 0$), we get graphs with $(p_1, p_2, p_3) = (2, 1, 0)$, $(2, 0, 2)$, $(1, 2, 1)$, $(1, 1, 3)$, $(1, 0, 5)$.

The enumeration method is then to use i-hedrites and to add a 1-gon when we have a pair of 2-gon and 3-gon that are adjacent in all possible ways. This is similar to the strategy of squeezing of 2-gons used for the enumeration of i-hedrites.

Since an 8-hedrite with v vertices is 3-connected, the corresponding alternating link cannot be represented with less than v crossings. But it can happen that two 8-hedrites, that are not equivalent as graphs, give rise to equivalent alternating links.

4.5 Self-dual {1, 2, 3, 4}-Spheres

A plane graph G is called *self-dual* if it is isomorphic to its dual G^*. The Euler formula $v - e + f = 2$ for a self-dual plane graph is, clearly:

$$\sum_{j=1}^{\infty} p_j(4 - j) = 4. \tag{4.2}$$

Also, $e = 2(v - 1)$ in such sphere and so, the sum of lengths of all zigzags is $4(v - 1)$.

Call a self-dual {1, 2, 3, 4}-sphere (i.e., negatively curved faces are excluded) a *i-self-hedrite* if $p_2 + p_3 = i$. Clearly, $(p_1, p_2, p_3;\ i = p_2 + p_3) = (1, 0, 1;\ 1), (0, 2, 0;\ 2), (0, 1, 2;\ 3), (0, 0, 4;\ 4)$ are all possibilities. The smallest *i-self-hedrites* have $p_4 = 0$; they are: loop with pending edge, $\text{Bundle}_2 = 2 \times K_2$, triangle with one doubled edge and Tetrahedron, respectively. Also, if G is a *i-self-hedrite* with $i \geq 2$, then $Med(G)$ is a $2i$-hedrite. Easy to check that such a v-vertex *i-self-hedrite* exists exactly if $v \geq i$.

Proposition 4.4 *Any 4-self-hedrite is 2-connected.*

Proof Given a self-dual {3, 4}-sphere G, its medial $Med(G)$ is a 8-hedrite and so, is 3-connected. If G is not 2-connected, then there exists a vertex u such that $G - u$ has connected components $G_1, \ldots, G_m$ with $m \geq 2$. Since u is of degree at most 4, there is one components G_i, such that u is connected to vertices in G_i by at most two edges. Those edges correspond to vertices in $Med(G)$ and by removing them, we disconnect $Med(G)$, which contradicts 3-connectedness of 8-hedrites.   □

Figures 4.19, 4.20 and 4.21 provide examples, respectively, of only 2-connected 4-self-hedrite, only 2-connected 3-self-hedrite and only 1-connected 2-self-hedrite.

Theorem 4.8 *All 1-self-hedrites are the ones presented in Figs. 4.13, 4.14, and 4.15.*

Proof Let G be such a 1-self-hedrite. It necessarily has $v_1 = p_1 = 1$ and $v_3 = p_3 = 1$. Let u be the vertex of degree 1. It is connected to a vertex u' of degree 3 or 4 and it is contained in a face F with 3 or 4 edges. Denote by F' the face adjacent to F. If u' and F are of degree 3, then F' is a loop and G is the map A_2 of Fig. 4.13.

If u' is of degree 3 and F has 4 edges, we must add a vertex u'' on the face F. Then u'' is necessarily of degree 4 and F' is adjacent to F by two edges. Either u'' is connected to itself by a loop and we get the map A_3 of Fig. 4.13, or u'' is connected to another vertex u''' by two edges. In the second case we have to iterate the same choices and we get the infinite series A_i of Fig. 4.13.

If u' is of degree 4, then an edge exits from it and joins a vertex u'' of degree 3 or 4. The face F' has already 3 edges so, there is an edge connecting u'' to itself. If u'' is of degree 3, we get the map B_3 of Fig. 4.13. If u'' is of degree 4, we can iterate this procedure indefinitely and so obtain the infinite series B_i of Fig. 4.13.

If u' is of degree 4 and F has 4 edges, then all possible ways of add edges are:

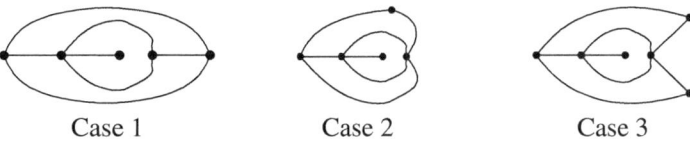

Case 1   Case 2   Case 3

In Case 1, the only way to extend the graph is to add two edges on each side and complete by adding two circumscribing edges, which is essentially Case 1 augmented. Therefore, this case does not lead to any map.

In Case 2, we can add a loop around the vertex of degree 2; it gives the map C_5 of Fig. 4.13. Or we can add an edge, with a vertex on it, between the vertices of degree 3 and 2. We then face the same choices leading to series C_i of Fig. 4.13.

In Case 3, if we add an edge between one of the two vertices of degree 2 and the vertex of degree 3, then we obtain the sporadic sphere D_5 of Fig. 4.14.

But if we connect two vertices of degree 2, then we need to add a vertex on at least one of the edges. This gives the following three choices:

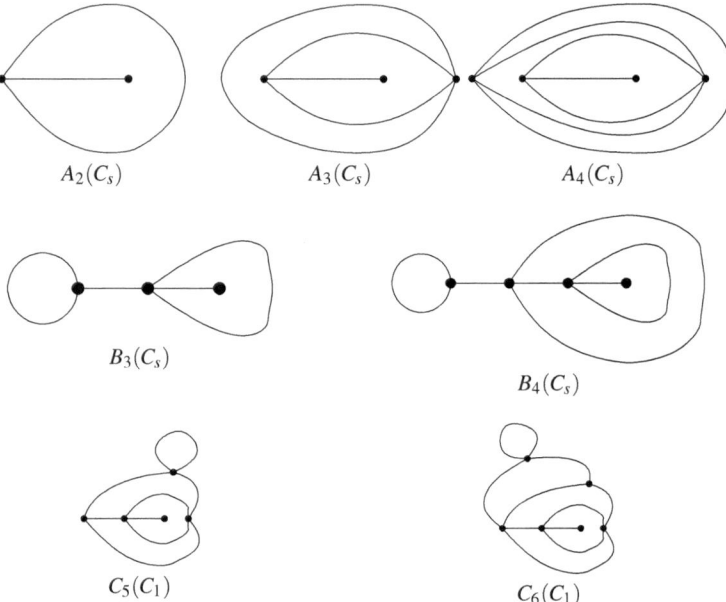

Fig. 4.13 First terms of a series $A_{i \geq 2}(C_s)$, $B_{i \geq 3}(C_s)$, and $C_{i \geq 5}(C_1)$ of 1-self-hedrites

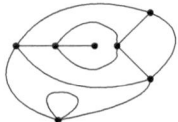

Fig. 4.14 Sporadic 1-self-hedrite of symmetry C_1

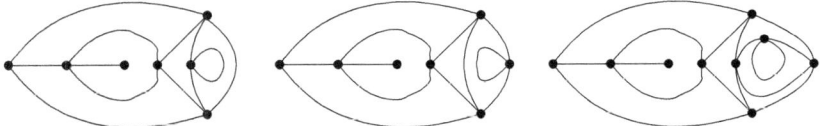

Fig. 4.15 First terms of a series $E_7'(C_1), E_{2i+1 \geq 7}(C_1)$ of 1-self-hedrites of symmetry C_1

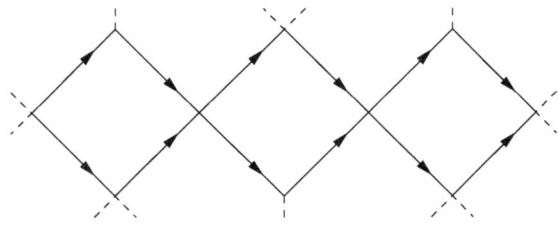

Fig. 4.16 Railroad in self-dual {2, 3, 4}-sphere. A sequence of 4-gons is bounded by two zigzags. The central vertices are of degree 4, while other vertices are of degree 2, 3, or 4

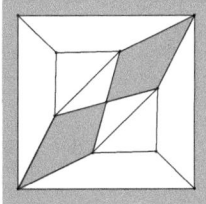

Fig. 4.17 Belt of three 4-gons in a minimal C_{2v} 4-self-hedrite, different from the one in Fig. 4.21; it has $\mathbf{z} = (6^2, 8, 10^2)$

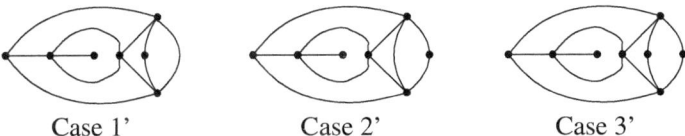

Case 1'   Case 2'   Case 3'

The first two cases give the maps of Fig. 4.15. In the third case, we are forced to add edges between the two vertices that have been inserted and we face the same set of choices. Therefore, we get the infinite series of Fig. 4.15.   □

Note that the medial of a 1-self-hedrite is a ({1, 3, 4}, 4)-sphere with $p_1 = p_3 = 2$. Also, a zigzag in a 1-self-hedrite is necessarily self-intersecting, a phenomenon which we will encounter again in Chap. 5.

A *railroad* in a self-dual {2, 3, 4}-sphere is a sequence of 4-gons bounded by two zigzags. The *length of the railroad* is the number of 4-gons occurring consecutively. If a 4-gon occurs twice, i.e., if the railroad is self-intersecting, then it is counted

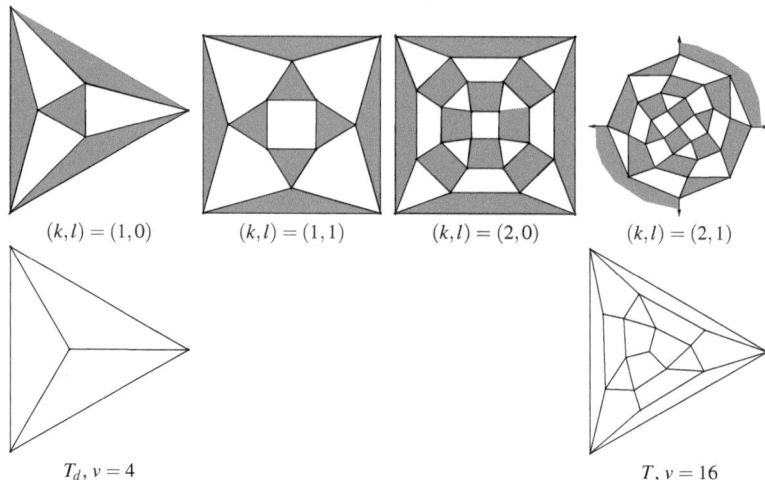

$(k,l) = (1,0)$ $(k,l) = (1,1)$ $(k,l) = (2,0)$ $(k,l) = (2,1)$

$T_d, v = 4$ $T, v = 16$

Fig. 4.18 First examples of 8-hedrites of symmetry O_h or O, expressed as $GC_{k,l}(Octahedron)$, and the corresponding self-dual {3, 4}-spheres $Med^{-1}(GC_{k,l}(Octahedron))$ with $k + l$ odd

twice. Figure 4.16 gives a section of an example. Such railroad becomes an ordinary railroad in the medial graph of length twice the one of the original railroad. A self-dual {2, 3, 4}-sphere is tight if and only if it has no railroad. See in Fig. 4.17 an example of a railroad in a 4-self hedrite.

Similarly, any 2-self-hedrite is *pure*, i.e., has only simple zigzags. We conjecture (and it is checked for $v \leq 40$) that any self-intersecting zigzag in a 3- or 4-self-hedrite has signature (e_1, e_2) with $e_1 = 0$.

For each $v \geq 3$, there is one 2-self-hedrite A_v with $\mathbf{z} = (2^{v-1}, 2(v - 1))$. Its symmetry is C_{2v} for odd and C_{2h} for even v. In fact, A_3 is the 2-nd one in Fig. 4.19. The medials of A_v form the series $J_{4,2m}$ of only 2-connected 4-hedrites from Table 4.1.

Our enumeration method for i-self-hedrites with $i \geq 2$ is to consider all $2i$-hedrites G', determine for them the graphs G_1, G_2 such that $G' = Med(G_1) = Med(G_2)$ and keep the ones that have G_1 isomorphic to G_2. The obtained plane graph, if it exists, we denote by $Med^{-1}(G') = G_1 \simeq G_2$. Using the enumeration of $2i$-hedrites, we can derive the i-self-hedrites, see Table 4.3.

Theorem 4.9 (i) *The symmetry groups of 2-self-hedrites are* $C_2, C_{2v}, C_{2h}, D_2, D_{2h}$.
(ii) *The symmetry groups of 3-self-hedrites are* C_1, C_s, C_2 C_{2v}.
(iii) *The symmetry groups of 4-self-hedrites are* $C_1, C_i, C_s, C_2, C_{2h}, C_{2v}, C_3, C_{3v}$,
       $C_4, C_{4v}, S_4, D_2, D_{2d}, D_{2h}, T, T_d$.
(iv) *Minimal representatives for each symmetry group (i)–(iii) above are given in*
       Figs. 4.19, 4.20, *and* 4.21.

It is known [DuDe04] that all 8-hedrites of symmetry O or O_h are of the form $GC_{k,l}(Octahedron)$ for some integers $0 \leq l \leq k$. See Theorem 6.2 and Fig. 4.18.

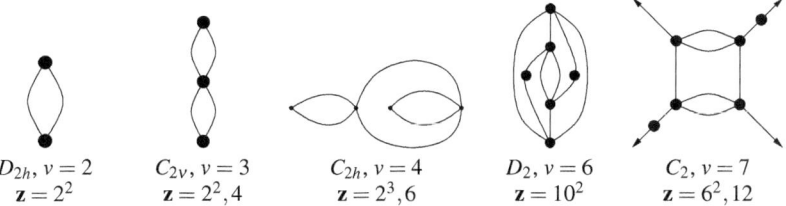

Fig. 4.19 Minimal representatives for the symmetries of 2-self-hedrites

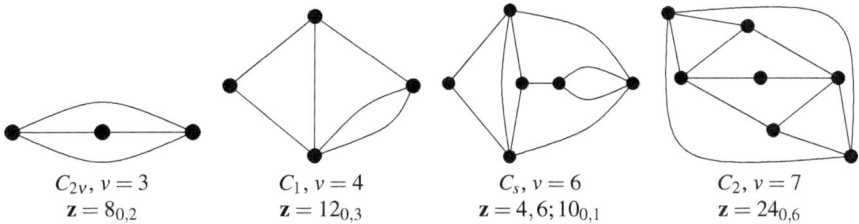

Fig. 4.20 Minimal representatives for the symmetries of 3-self-hedrites

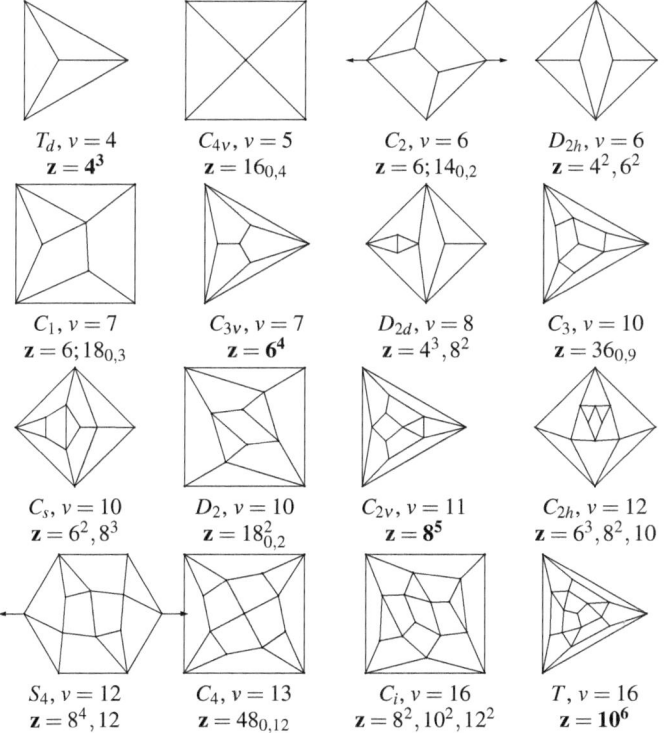

Fig. 4.21 Minimal representatives for the symmetries of 4-self-hedrites; all tight and pure among them are 4 with boldfaced z: 1-st, 6-th, 11-th, and 16-th

Table 4.3 Number of i-self-hedrites, $2 \leq i \leq 4$, with $v \leq 40$

v	2	3	4	v	2	3	4	v	2	3	4	v	2	3	4
2	1	0	0	12	4	29	24	22	10	90	191	32	9	239	584
3	1	1	0	13	6	30	33	23	7	119	198	33	9	256	631
4	2	1	1	14	5	42	40	24	7	131	234	34	14	232	748
5	2	4	1	15	5	47	48	25	10	124	276	35	10	290	760
6	3	6	2	16	8	48	69	26	10	162	304	36	14	308	857
7	3	7	4	17	5	64	73	27	8	170	332	37	16	286	956
8	3	11	6	18	6	72	92	28	12	158	407	38	11	342	1002
9	3	16	8	19	8	70	114	29	10	190	421	39	11	359	1070
10	5	16	15	20	6	89	130	30	9	210	476	40	16	332	1239
11	4	26	16	21	8	104	148	31	14	202	550				

Theorem 4.10 *(i) All 2-self-hedrites of symmetry D_2 or D_{2h} are of the form $Med^{-1}(GC_{k,l}(Bundle_4))$ with $k + l$ odd.*
(ii) All 4-self-hedrites of symmetry T or T_d are of the form $Med^{-1}(GC_{k,l}(Octahedron))$ with $k + l$ odd.

Proof (ii) If G is a 4-self-hedrite of symmetry T or T_d, then its medial $G' = Med(G)$ is an 8-hedrite of symmetry O or O_h. So, $G' = GC_{k,l}(Octahedron)$ for some (k, l). The automorphism group of the plane graph G' is transitive on triangles; so, we only need to determine when the triangles are not all in $\mathscr{F}_1(G')$ or $\mathscr{F}_2(G')$ (see Sect. 1.3). Clearly, this corresponds to $k + l$ odd.

(i) The proof is similar and uses the facts that 4-hedrites of symmetry D_4, D_{4h} are of the form $GC_{k,l}(Bundle_4)$ for some (k, l). □

If G is a 4-self-hedrite with simple zigzags, then $Med(G)$ is an 8-hedrite with simple central circuits. By Sect. 4.3, pure 8-hedrites G' are obtained by taking the ones of Fig. 4.11 and splitting each central circuit C_i into m_i parallel central circuits. Then we have to determine for which $m = (m_i)$ the triangles are in two parts $\mathscr{F}_1(G')$ and $\mathscr{F}_2(G')$ which are equivalent under an automorphism of G'.

This requires a detailed analysis of the automorphism and a search of the necessary relations between m_i and parity conditions. The details are very cumbersome but we can get a classification of the pure 4-self-hedrites. In particular, 1, 3, 5, 7-th tight pure 8-hedrites in Fig. 4.11 are the medial graphs of 1, 6, 11, 16-th 4-self-hedrites in Fig. 4.21, respectively. They are all pure tight 4-self-hedrites, together with a 8-vertex D_{2h} and 12-vertex C_{2v} 4-self-hedrites, for which 4-th and 6-th 8-hedrites in Fig. 4.11 are the medial graphs, respectively.

Theorem 4.11 *All pure tight i-self-hedrites with $i = 2, 3, 4$ are either*

(i) any 2-self-hedrites with even v and $\mathbf{z} = (2v - 2)^2$, or
(ii) the 3-self-hedrite given in Fig. 4.22, or
(iii) 6 following 4-self-hedrites: 1-st, 6-th, 11-th, 16-th on Fig. 4.21 and two 4-self-hedrites from Fig. 4.23.

Fig. 4.22 Unique tight pure 3-self-hedrite; its medial is 6-hedrite $8 - 5$, projection of the link 8_6^3

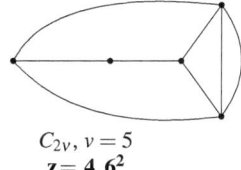

$C_{2v}, v = 5$
$\mathbf{z} = 4, 6^2$

Fig. 4.23 Two pure tight
4-self-hedrites; their medial
graphs are 4-th and 6-th
8-hedrites in Fig.4.11

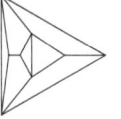 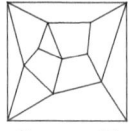

$$D_{2h}, v = 8 \qquad C_{2v}, v = 12$$
$$\mathbf{z} = \mathbf{6}^2, \mathbf{8}^2 \qquad \mathbf{z} = \mathbf{8}^3, \mathbf{10}^2$$

Proof The proof is obtained simply by remarking that the medial of a pure tight
i-self-hedrite is a pure tight $2i$-hedrite which are themselves classified.   □

References

[BrHaHe03] Brinkmann, G., Harmuth, T., Heidemeier, O.: The construction of cubic and quartic
planar maps with prescribed face degrees. Discrete Appl. Math. 128–2(3), 541–554
(2003)

[DeSt03] Deza, M., Shtogrin, M.: Octahedrites, Special Issue "Polyhedra in Science and Art",
Symmetry: Culture and Science. The Quarterly of the International Society for the
Interdisciplinary Study of Symmetry, 11/1-4, 27–64 (2003)

[DeDuSh03] Deza, M., Dutour, M., Shtogrin, M.: 4 -valent plane graphs with 2-, 3- and 4 -gonal
faces. In: Advances in Algebra and Related Topics (in memory of B.H.Neumann;
Proceedings of ICM Satellite Conference on Algebra and Combinatorics, Hong Kong
2002), pp. 73–97. World Scientific Publishing Co. (2003)

[DuDe11] Dutour Sikirić, M., Deza, M.: 4-regular and self-dual analogs of fullerenes, Mathematics and Topology of Fullerenes. In: Ori, O., Graovac, A., Cataldo, F. (eds.) Carbon
Materials, Chemistry and Physics, vol. 4, pp. 103–116. Springer, New York (2011)

[DuDe04] Dutour, M., Deza, M.: Goldberg-Coxeter construction for 3- or 4-valent plane graphs.
Electron. J. Comb. **11–1**, R20 (2004)

[Grün67] Grünbaum, B.: Convex Polytopes. Wiley-Interscience, New York (1967); second
edition, Graduate Texts in Mathematics, vol. 221. Springer, New York (2003)

[Heid98] Heidemeier, O.: Die Erzeugung von 4-regulären, planaren, simplen, zusammenhängenden Graphen mit vorgegebenen Flächentypen. Diplomarbeit,Universität Bielefeld, Fakultät für Wirtschaft und Mathematik(1998)

[Rol76] Rolfsen, D.: Knots and Links, Mathematics Lecture Series 7, Publish or Perish,
Berkeley (1976); second corrected printing: Publish or Perish, Houston (1990)

[Thi] Thistlewaite, M.: Homepage. http://www.math.utk.edu/morwen

Chapter 5
ZC-Circuits of 5- and 6-Regular Spheres

In this chapter, based mainly on [DDS13a] and [DeDu12], we consider ($\{3, 4\}$, 5)-spheres (named *icosahedrites*) and ($\{1, 2, 3\}$, 6)-spheres. Both cases allow to consider zigzags. But in contrast to the 3- and 4-regular cases, the second structure of edges appear: weak zigzags for 5- and central circuits for 6-regular graphs.

The icosahedrite case is specific because the growth rate of their number is exponential; so their enumeration is limited to 32 vertices. For the ($\{1, 2, 3\}$, 6)-spheres, we give two enumeration methods that allow us to reach 100 vertices and correct previous enumeration results of our paper [DeDu12].

In both cases, we consider tightness notions of both circuit notions and the corresponding extremal problem for tight ones and enumeration of those with only simple circuits. For ($\{1, 2, 3\}$, 6)-spheres, we also consider a kind of the Goldberg–Coxeter construction which has rather specific properties.

5.1 Icosahedrites

We call *icosahedrite* any ($\{3, 4\}$, 5)-sphere. All of them, except Icosahedron, have negatively curved faces: 4-gons. They form the simplest nontrivial class of 5-regular bifaced plane graphs. For them it holds that $p_3 = 20 + 2p_4$ and $v = 12 + 2p_4$. Apropos, all eight ($\{1, 2, 3\}$, 5)-spheres are given, with their $((p_1, p_2, p_3); v)$ and the symmetry, in Fig. 5.1.

The only known b-gon-transitive (and the only known with at most two orbits of 3-gons) ($\{3, b\}$, 5)-spheres are Snub Cube, Snub Dodecahedron, and snub $APrism_b$ (including snub $APrism_3$, i.e., Icosahedron). Archimedean ($\{3, b\}$, 5)-plane tilings $(3.3.4.3.4)$, $(3.3.3.4.4)$, $(3.3.3.3.6)$ have $b = 4, 4, 6$, respectively. They are transitive on b-gons and vertices; they also, respectively, have 1, 1, 2 orbits of 3-gons.

All 38 groups of symmetry of icosahedrites are given in Sect. 1.5. They were found in [DDS13a], which gives also the smallest examples for all of them, except C_5, C_{5v}, C_{5h}, S_{10}, O_h and I. In Figs. 5.4 and 5.5 we added "min.", when the icosahedrite is minimal for its group. In fact, an v-vertex icosahedrite has 3-, 4- or 5-fold

© Springer India 2015
M. Deza et al., *Geometric Structure of Chemistry-Relevant Graphs*,
Forum for Interdisciplinary Mathematics 1, DOI 10.1007/978-81-322-2449-5_5

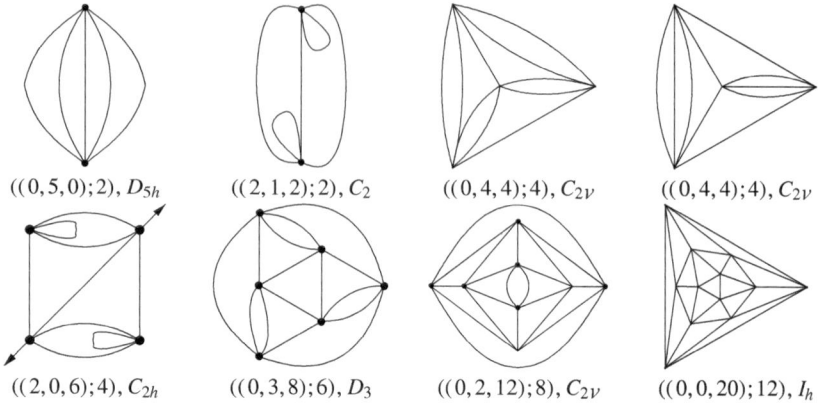

$((0,5,0);2), D_{5h}$ $((2,1,2);2), C_2$ $((0,4,4);4), C_{2v}$ $((0,4,4);4), C_{2v}$

$((2,0,6);4), C_{2h}$ $((0,3,8);6), D_3$ $((0,2,12);8), C_{2v}$ $((0,0,20);12), I_h$

Fig. 5.1 The 5-valent graphs with faces of size at most 3; their $(\mathbf{p}; v)$ and symmetry are given

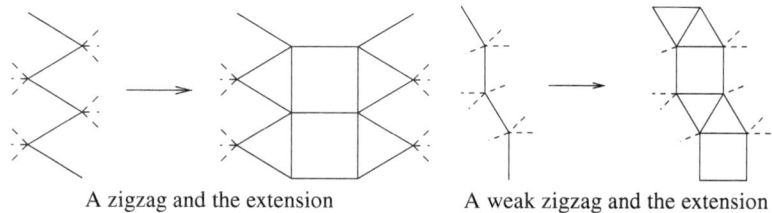

A zigzag and the extension   A weak zigzag and the extension

Fig. 5.2 The local structure of extending an icosahedrite from a zigzag or a weak zigzag

symmetry if it holds $p_4 \equiv 0$ (mod 3), $p_4 \equiv 2$ (mod 4) or $p_4 \equiv 0$ (mod 5), i.e., $v \equiv 0$ (mod 6), $v \equiv 0$ (mod 8) or $v \equiv 2$ (mod 10), respectively.

For a given icosahedrite, a *weak zigzag* WZ is a circuit of edges such that one alternates between the left and right way but never extreme left or right. The usual *zigzag* is a circuit such that one alternates between the extreme left and extreme right way. A zigzag or weak zigzag is called *simple* if any edge of it occurs only once. Clearly, the weak zigzags (as well as the usual zigzags) doubly cover the edge-set.

Given a simple zigzag or a simple weak zigzag, we can duplicate it, creating a *railroad* or *weak railroad*, by adding 4-gons and 3-gons and get another icosahedrite. The local scheme is described in Fig. 5.2. Figure 5.3 describes such schemes. An icosahedrite that does not originate from an extension is called *tight* or *weakly tight*, respectively. These two properties are independent: see Fig. 5.3.

Conjecture 5.1 (*i*) *All icosahedrites with only simple zigzags are 1-st, 4-th or 6-th from Fig. 5.5; note that those 3 graphs have also only simple weak zigzags (Fig. 5.4).*

(*ii*) *There are v-vertex weakly tight icosahedrites with only simple weak zigzags for any $v \neq 20$ divided by 4. The 2-nd and 9-th icosahedrites in Fig. 5.5 could be the cases $m = 1, 2$ of an infinite series of such $16m$-vertex D_{4d} icosahedrites.*

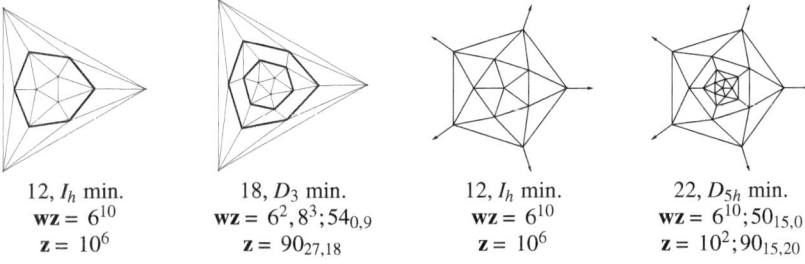

12, I_h min.	18, D_3 min.	12, I_h min.	22, D_{5h} min.
$\mathbf{wz} = 6^{10}$	$\mathbf{wz} = 6^2, 8^3; 54_{0,9}$	$\mathbf{wz} = 6^{10}$	$\mathbf{wz} = 6^{10}; 50_{15,0}$
$\mathbf{z} = 10^6$	$\mathbf{z} = 90_{27,18}$	$\mathbf{z} = 10^6$	$\mathbf{z} = 10^2; 90_{15,20}$

Fig. 5.3 A simple weak zigzag WZ_6 in Icosahedron; the weak railroad extension on WZ_6; a simple zigzag Z_{10} in Icosahedron; the railroad extension on Z_{10}. Among the above four, only 4-th is not tight, while only 2-nd is not weakly tight

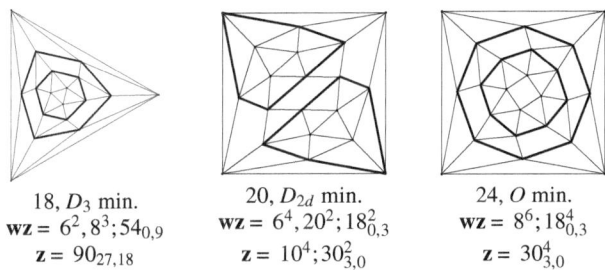

18, D_3 min.	20, D_{2d} min.	24, O min.
$\mathbf{wz} = 6^2, 8^3; 54_{0,9}$	$\mathbf{wz} = 6^4, 20^2; 18_{0,3}^2$	$\mathbf{wz} = 8^6; 18_{0,3}^4$
$\mathbf{z} = 90_{27,18}$	$\mathbf{z} = 10^4; 30_{3,0}^2$	$\mathbf{z} = 30_{3,0}^4$

Fig. 5.4 All the above icosahedrites are tight and have two "parallel" simple weak zigzags, but only on the 1-st and 3-rd ones they form a weak railroad, while 2-nd one is weakly tight

(iii) There is no upper bound on the number of zigzags of tight icosahedrites and the number of or weak zigzags of weakly tight icosahedrites.

A 3-gon surrounded by nine 3-gons or a 4-gon surrounded by twelve 3-gons are bounded by simple weak zigzags of length 6 or 8; denote them by WZ_6 and WZ_8. In fact, they are only weak zigzags, in which every vertex occurs only once.

In Table 5.1 we list the numbers of v-vertex icosahedrites for $v \leq 32$. From this list it appears likely that any icosahedrite is 3-connected.

The z-knotted v-vertex icosahedrites have $\mathbf{z} = 5v$ with signature ($e_1 = \frac{5v}{2} - e_2, e_2$). In fact, $e_2 = 18, 25$ if $v = 18, 24$ and $e_2 \in [24, 42], [23, 45], [24, 50], [27, 55]$ if $v = 26, 28, 30, 32$; also, e_2 is even exactly if $\frac{v}{2}$ is odd.

Theorem 5.1 *A v-vertex ($\{3, 4\}, 5$)-sphere exists if and only if v is even, $v \geq 12$ and $v \neq 14$. Their number N_v grows at least exponentially with v.*

Proof If there is a weak zigzag WZ_6 in an icosahedrite G, then we can insert a *corona* (6-ring of three 4-gons alternated by three pairs of adjacent 3-gons); instead of it and get an icosahedrite with 6 more vertices. Since there are such icosahedrite with $v = 18, 20, 22$ (see Fig. 5.4), we can generate the required spheres. There are always two options when inserting the corona; so N_v grows exponentially. □

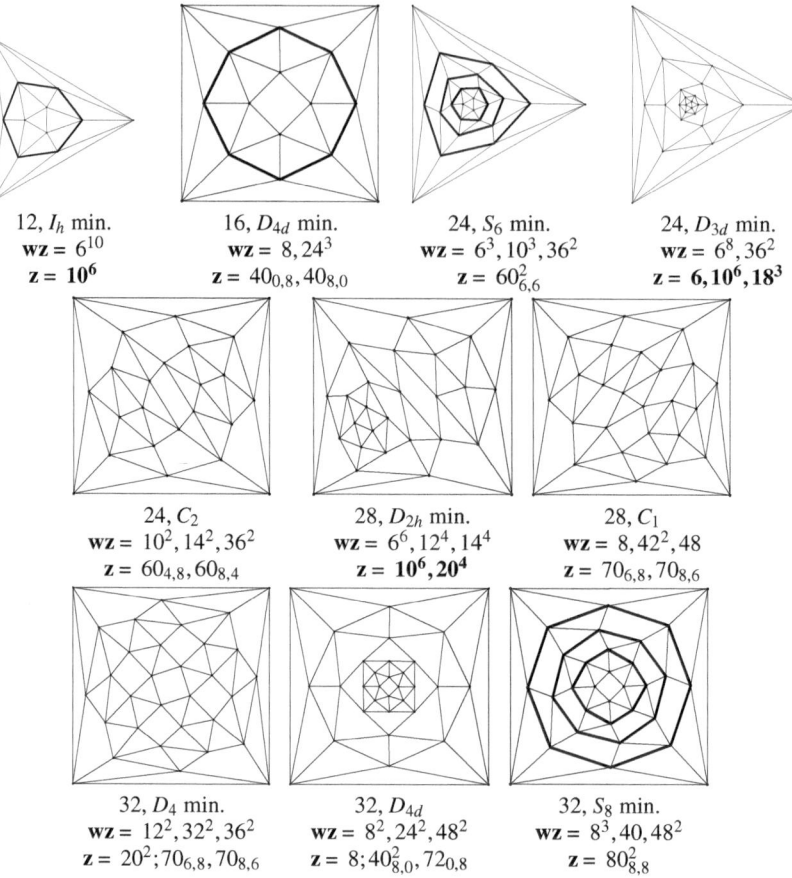

12, I_h min.
wz $= 6^{10}$
z $= 10^6$

16, D_{4d} min.
wz $= 8, 24^3$
z $= 40_{0,8}, 40_{8,0}$

24, S_6 min.
wz $= 6^3, 10^3, 36^2$
z $= 60_{6,6}^2$

24, D_{3d} min.
wz $= 6^8, 36^2$
z $= \mathbf{6, 10^6, 18^3}$

24, C_2
wz $= 10^2, 14^2, 36^2$
z $= 60_{4,8}, 60_{8,4}$

28, D_{2h} min.
wz $= 6^6, 12^4, 14^4$
z $= \mathbf{10^6, 20^4}$

28, C_1
wz $= 8, 42^2, 48$
z $= 70_{6,8}, 70_{8,6}$

32, D_4 min.
wz $= 12^2, 32^2, 36^2$
z $= 20^2; 70_{6,8}, 70_{8,6}$

32, D_{4d}
wz $= 8^2, 24^2, 48^2$
z $= 8; 40_{8,0}^2, 72_{0,8}$

32, S_8 min.
wz $= 8^3, 40, 48^2$
z $= 80_{8,8}^2$

Fig. 5.5 First seven are all icosahedrites with only simple weak zigzags and at most 30 vertices; all known icosahedrites with only simple usual zigzags are 1-st, 4-th and 6-th among them. We also give 3 of 14 icosahedrites with only weak simple zigzags and 32 vertices. All the above graphs are tight and all, but the 3-rd and last one having double weak road are weakly tight

In fullerenes or other ($\{a, b\}, k$)-spheres with nonnegatively curved faces, p_a is fixed and the structure is made up of patches of b-gons. The parameterization of such graphs, generalizing one in [Thur98] and including the case of one complex parameter—Goldberg–Coxeter construction [DuDe04]—are thus built; see Chap. 6. A consequence of this is the polynomial growth of the number of such spheres. This does not happen for icosahedrites and their parameterization, if any, looks elusive (Table 5.2).

Finally, the list of icosahedrites with $v \le 32$ vertices permit to see sharply increasing statistical trends concerning, especially a silent majority of them: those of symmetry C_1 and consisting of 1 or 2 zigzags. Starting from $v = 26$ (see Fig. 5.6), appear sets of *siblings* , i.e., icosahedrites with the same ($\mathbf{z}, \mathbf{wz}$); their number is 1, 4 for $v = 26, 28$.

Table 5.1 The numbers N_v of all and N_v' of z-knotted ones v-vertex icosahedrites with $v \le 32$

v	12	14	16	18	20	22	24	26	28	30	32
N_v	1	0	1	1	5	12	63	246	1,395	7,668	45,460
N_v'	0	0	0	1	0	0	3	45	195	1,359	6,926

Table 5.2 Zigzag structure of the icosahedrites with only simple weak zigzags and 32 vertices; all are tight and all but 1-st one are weakly tight

Nr	G	$\mathbf{z}$	$\mathbf{wz}$
(32,7758)	S_8	$80_{8,8}^2$	$8^3, 40, 48^2$
(32,1113)	C_1	$80_{8,8}^2$	$18, 46, 48^2$
(32,28583)	C_2	$80_{6,10}, 80_{10,6}$	$16^2, 20, 30^2, 48$
(32,15612)	$C_!$	$80_{6,10}, 80_{10,6}$	$8^2, 12, 36, 48^2$
(32,45456)	C_1	$80_{6,10}, 80_{10,6}$	$10^3, 34, 48^2$
(32,33724)	C_1	$80_{6,10}, 80_{10,6}$	$22, 42, 48^2$
(32,24032)	C_2	$80_{4,12}, 80_{12,4}$	$12, 48^2, 52$
(32,18008)	D_{4d}	$8; 40_{8,0}^2, 72_{0,8}$	$8^2, 24^2, 48^2$
(32,6536)	D_4	$20^2; 60_{4,4}^2$	$12^2, 32^2, 36^2$
(32,17004)	C_2	$10^2; 70_{4,11}^2$	$6^4, 16^2, 18^2, 32, 36$
(32,41622)	C_2	$10^2; 22_{0,1}, 118_{16,27}$	$6^5, 20, 26, 42^2$
(32,26043)	C_2	$26_{1,0}, 134_{27,28}$	$20, 22^2, 24^2, 48$
(32,10533)	C_2	$38_{0,4}, 40_{2,2}, 70_{14,0}$	$14^3, 22^2, 24, 42$
(32,887)	C_1	$20^4; 40_{0,2}, 40_{2,0}$	$12^4, 14, 24^2, 50$

Fig. 5.6 Unique set of more than one icosahedrite with at most 26 vertices having the same $(\mathbf{z}, \mathbf{wz})$

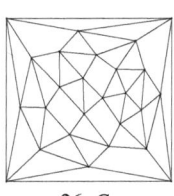

26, C_2
$\mathbf{wz} = 52; 78_{0,13}$
$\mathbf{z} = 130_{39,26}$

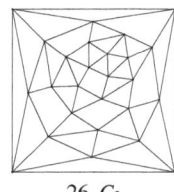

26, C_2
$\mathbf{wz} = 52; 78_{0,13}$
$\mathbf{z} = 130_{39,26}$

Starting from $v = 28$, appear icosahedrites with $\mathbf{wz} = \mathbf{z}$ (case A) or with $\mathbf{wz}$ coming from $\mathbf{z}$ by reversing order in signature of all zigzags (case B); all are C_1. Among the 3 such icosahedrites for $v = 28$, we have two *A-superknots* with $\mathbf{z} = \mathbf{wz} = 140_{37,33}, 140_{45,25}$ and one *B-superknot* with $(\mathbf{z}, \mathbf{wz}) = (140_{41,29}, 140_{29,41})$.

Among 9 cases with $v = 30$, all icosahedrites are A-superknots with $\mathbf{z} = \mathbf{wz} = 150_{i,75-i}$, where $i = 29, 31.33, 35, 37, 41$; in the subcases $i = 37, 41$, they form the sets of 2, 3 siblings. Among 172 cases with $v = 32$, five graphs have 2 zigzags, while others are A_i A-superknots with $\mathbf{z} = \mathbf{wz} = 160_{i,80-i}$ and B_i B-superknots

Table 5.3 The number of A- and B-superknots with 32 vertices

i	27	29	31	33	35	37	39	41	43	45	47	49	51
A_i	1	2	1	5	14	6	12	8	8	11	10	1	1
B_i	1	2	2	1	6	6	14	11	18	14	11	2	1

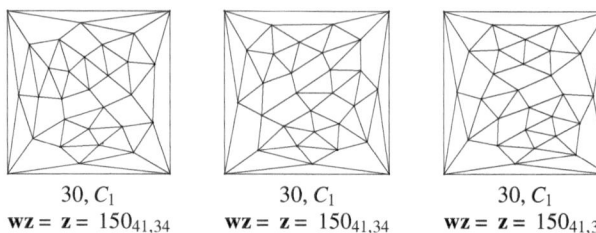

30, C_1 30, C_1 30, C_1

$\mathbf{wz} = \mathbf{z} = 150_{41,34}$ $\mathbf{wz} = \mathbf{z} = 150_{41,34}$ $\mathbf{wz} = \mathbf{z} = 150_{41,34}$

Fig. 5.7 The only sets of more than one icosahedrite with $v \le 30$, having the same $(\mathbf{z}, \mathbf{wz})$ and $\mathbf{z} = \mathbf{wz}$, are above triple and a pair of 30-vertex icosahedrites with $\mathbf{z} = \mathbf{wz} = 150_{37,38}$

with $(\mathbf{z}, \mathbf{wz}) = (160_{i,80-i}, 160_{80-i,i})$, for any odd number on $i \in [27, 51]$. See the numbers A_i, B_i in Table 5.3 and the largest family of 14 A-superknots-siblings in Fig. 5.7.

Conjecture 5.2 *(i) If (α_1, α_2) is a signature of a z- or wz-knotted icosahedrite, then α_1 is odd.*
(ii) An A-superknot exists if and only if v is even and $v \ge 28$, and a B-superknot exists if and only if $v \ge 28$ and v is divided by 4. All of them have symmetry C_1 and their number exponentially increases with increasing v.

5.2 Generation Method of ({1, 2, 3}, 6)-Spheres

If G is a 6-regular plane graph, then by Euler formula it satisfies the equality:

$$\sum_{i \ge 1} p_i(3 - i) = 6. \tag{5.1}$$

So any $(\{1, 2, 3\}, 6)$-sphere has $p_2 = 6 - 2p_1$ and $v = 2 + \frac{1}{2}(p_3 - p_1)$ vertices. Clearly, $(p_1, p_2) = (3, 0), (2, 2), (1, 4), (0, 6)$ are all possibilities.

In [DeDu08] we proved that for any $v \ge 2$ there exist a $(\{2, 3\}, 6)$-sphere with v vertices. In any $(\{2, 3\}, 6)$-sphere, one can collapse its 2-gons into simple edges. By doing so one obtains a graph with vertices of degree at most 6 and with faces of size 3 only. So the dual will be a 3-regular graph with faces of size at most 6.

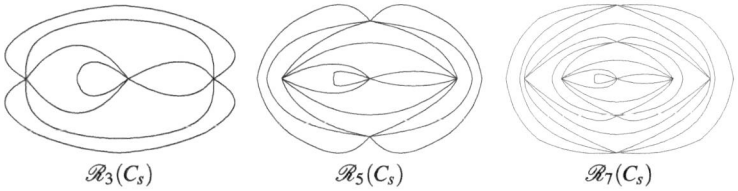

Fig. 5.8   First terms of the series of ({1, 2, 3}, 6)-spheres $\mathscr{R}_{2i+1}$ with $p_1 = 1$

Theorem 5.2 ([DeDu12]) *With the exception of the graphs Bundle$_6$ = $6 \times K_2$ and $3 \times K_3$ (first two in Fig. 5.10), any ({2, 3}, 6)-sphere is obtained from a ({3, 4, 5, 6}, 3)-sphere by adding vertices of degree two and taking the dual.*

Proof Let G be a ({2, 3}, 6)-sphere and let G^* be its dual. Then, by removing from G^* its vertices of degree 2, one gets a 3-regular graph G_1. It can happen that G_1 has no vertices and is reduced to a simple circular edge e. In this case, if one adds six vertices on e and takes the dual, one will get the first exceptional graph with 2 vertices. If G_1 has one face F, which is a 1-gon, then we have to add 5 vertices of degree 2 on the edge e of F. Necessarily, any face adjacent to F has to be a 1-gon, but this is, clearly, impossible. If F is a 2-gon and F is adjacent to at least one 2-gon, then G_1 is reduced to a graph with two vertices and three edges. The corresponding ({2, 3}, 6)-sphere is the second exceptional graph.

Assume that F is adjacent to F_1, F_2 with F_i being a a_i-gon and $a_i \geq 3$. If one of a_i is 3, then the other is 6 and this gives a 1-gon. Thus, the only possibility is $a_1 = a_2 = 4$. This implies that we have a graph with 4 vertices, two 4-gonal and two 2-gonal faces. But consideration of all possibilities rules out this option. So G_1 is a 3-regular plane graph with faces of size within $\{3, 4, 5, 6\}$.   □

The method can be generalized (in Theorem 5.3) to deal with graphs with 1-gons. Note that for most ({3, 4, 5, 6}, 3)-spheres one cannot add those vertices of degree 2, in order to get the required spheres, because whenever we add such a 2-gon, we have two faces of size lower than 6 that are adjacent. Graphs admitting such adjacency are relatively rare in the set of ({3, 4, 5, 6}, 3)-spheres. Some such graphs are the ({5, 6}, 3)-spheres with all twelve 5-gons organized in pairs.

The above theorem gives a method to enumerate ({2, 3}, 6)-spheres. First enumerate the ({3, 4, 5, 6}, 3)-spheres using the program CPF, which is available from [BFDH97] and whose algorithm has been described in [BrHaHe03]. After such enumeration is done, it remains only to add the six vertices of degree 2 in all possibilities. The numbers of graphs are shown as N_0 in Table 5.2 for $v \leq 100$.

Theorem 5.3 ([DeDu12]) *With the exception of the graphs Trifolium T_1, T_2 (first two in Fig. 5.13), and the spheres of the infinite series depicted in Figs. 5.8 and 5.9, any ({1, 2, 3}, 6)-sphere with $p_1 > 0$ is obtained from a ({3, 4, 5, 6}, 3)-sphere by taking the dual and then splitting some edges according to the following two schemes:*

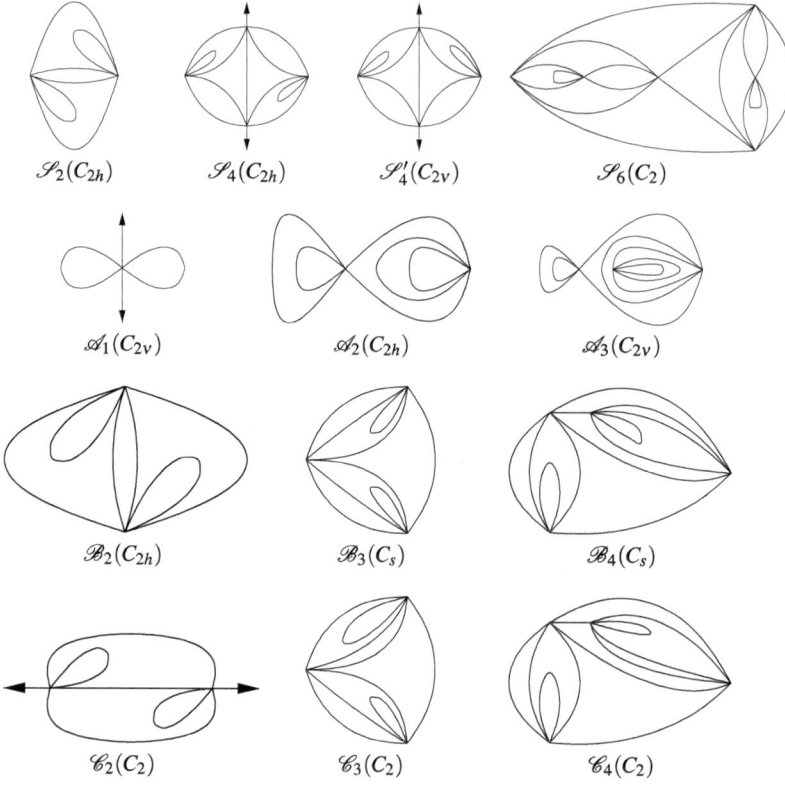

Fig. 5.9 First terms of the series $\mathscr{S}_{2i}, \mathscr{S}'_{4i}, \mathscr{A}_i, \mathscr{B}_{i+1}$ and $\mathscr{C}_{i+1}$ of $(\{1, 2, 3\}, 6)$-spheres with $p_1 = 2$. $\mathscr{S}_{4i+2}$ is of symmetry C_2, $\mathscr{S}_{4i}$ of symmetry C_{2h} and $\mathscr{S}'_{4i}$ of symmetry C_{2v}

Proof Let us take a $(\{1, 2, 3\}, 6)$-sphere G with at least one 1-gon F in its face-set. Clearly, F cannot be adjacent to another 1-gon. If F is adjacent to a 2-gon, then simple considerations yield that G belongs to the infinite series $\mathscr{A}_i$ of Fig. 5.9. So we can assume in the following that all 1-gons, say $F_1, \ldots, F_s$ are adjacent to 3-gons $G_1, \ldots, G_s$. If one of the G_i is adjacent to two 2-gons, then we get the sphere $\mathscr{B}_2$ $(C_{2h}, n = 2)$ depicted in Fig. 5.9. If one of the G_i is adjacent to exactly one 1-gon, then we get the following partial diagram:

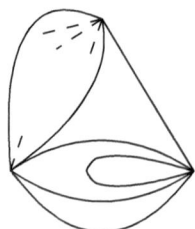

Table 5.4 Number N_i of v-vertex ({1, 2, 3}, 6)-spheres with $p_1 = i \in \{0, 1, 2, 3\}$ and $v \le 100$

v	N_0	N_1	N_2	N_3	v	N_0	N_1	N_2	N_3	v	N_0	N_1	N_2	N_3	v	N_0	N_1	N_2	N_3	v	N_0	N_1	N_2	N_3
1	0	0	1	1	21	86	46	19	1	41	582	187	26	0	61	1864	361	38	1	81	4165	676	63	1
2	1	0	3	0	22	133	33	23	0	42	758	133	53	0	62	2328	316	58	0	82	5239	480	75	0
3	1	1	3	1	23	112	62	16	0	43	679	180	27	1	63	1999	420	56	1	83	4608	766	51	0
4	3	1	6	1	24	165	44	38	0	44	869	172	54	0	64	2572	310	82	1	84	5501	590	124	1
5	2	3	5	0	25	144	57	20	1	45	749	199	43	0	65	2252	450	52	0	85	4980	666	66	0
6	7	2	8	0	26	205	54	27	0	46	1000	149	44	0	66	2718	322	78	0	86	6005	608	79	0
7	5	6	6	1	27	176	75	22	1	47	868	250	30	0	67	2467	434	41	1	87	5145	793	65	0
8	12	5	13	0	28	251	61	37	1	48	1101	182	73	1	68	3035	414	78	0	88	6482	581	113	0
9	10	8	8	1	29	214	95	19	0	49	989	235	35	2	69	2595	496	53	0	89	5652	877	54	0
10	19	6	12	0	30	299	61	40	0	50	1259	194	57	0	70	3348	336	87	0	90	6706	586	122	0
11	16	14	9	0	31	265	96	20	1	51	1076	270	40	0	71	2904	566	44	0	91	6097	820	68	2
12	29	11	18	1	32	360	89	44	0	52	1410	210	62	1	72	3520	415	108	0	92	7330	758	103	0
13	24	17	10	1	33	305	111	28	0	53	1228	313	33	0	73	3161	515	45	1	93	6264	890	69	1
14	42	16	16	0	34	429	80	33	0	54	1535	218	63	0	74	3875	450	69	0	94	7804	633	86	0
15	35	23	15	0	35	375	134	31	0	55	1389	280	45	0	75	3349	552	68	1	95	6907	970	73	0
16	59	18	23	1	36	488	105	51	1	56	1753	278	78	1	76	4210	451	86	1	96	8124	756	143	0
17	48	33	12	0	37	436	133	24	1	57	1485	333	44	1	77	3695	664	59	0	97	7315	909	59	1
18	79	22	22	0	38	581	118	37	0	58	1929	238	54	0	78	4416	466	91	0	98	8821	804	103	0
19	69	36	13	1	39	495	159	32	1	59	1690	388	37	1	79	3995	610	48	1	99	7542	1004	83	0
20	100	34	29	0	40	677	112	60	0	60	2077	283	95	0	80	4918	536	119	0	100	9435	743	133	1

Clearly, such diagram extends to one of the graphs of the infinite series $\mathscr{B}_i$, $\mathscr{C}_i$ depicted in Fig. 5.9.

So we can now assume that the G_i is adjacent to 3-gons only. If one of the 3-gons adjacent to a G_i turns out to be another G_j, then we get the map $\mathscr{C}_2$ from Fig. 5.9. So we assume further that those 3-gons are not of the type G_i.

The faces G_i contains two vertices v_i, v_i' with v_i being contained in F_i. If $v_i = v_i'$, then we get the exceptional sphere Trifolium. So we assume further that $v_i \neq v_i'$. If $v_i = v_j'$ for $i \neq j$, then some easy considerations give the sphere T_2 as the only possibility. So let us assume now that the vertex v_i is contained in a 2-gon. Then we have the following local configuration:

From that point, after enumeration of all possibilities we get the infinite series $\mathscr{R}_{2i+1}$, $\mathscr{S}_{2i}$ and $\mathscr{S}_{4i}'$ of Figs. 5.8 and 5.9. So now we have that all vertices v_i are contained in four 3-gons. This implies that G is obtained from a $(\{3, 4, 5, 6\}, 3)$-sphere by taking the dual and then splitting some edges according to the above-mentioned schemes.   □

Remark 5.1 Another enumeration method uses that for a given v-vertex $(\{1, 2, 3\}, 6)$-sphere G, the truncation $Tr(G)$ is a $6v$-vertex $(\{2, 4, 6\}, 3)$-sphere. The $(\{2, 4, 6\}, 3)$-spheres can be enumerated by remarking that a 2-gon can be collapsed into a single edge and the reduced graph is still a $(\{2, 4, 6\}, 3)$-sphere. Thus, one can enumerate the $(\{2, 4, 6\}, 3)$-spheres by taking the $(\{2\}, 3)$-spheres and the $(\{4, 6\}, 3)$-spheres and adding 2-gons whenever possible. The $(\{4, 6\}, 3)$-spheres themselves can be enumerated by CPF very efficiently. Given a $(\{2, 4, 6\}, 3)$-sphere, testing if it originates as a truncation of a $(\{2, 4, 6\}, 3)$-sphere is very easy (Table 5.4).

It turns out that both methods give the same set of graphs. But the second method is much faster and allows to go up to 100 vertices, though higher number of vertices are certainly within reach. It also does not require a complex discussion of cases and any infinite series. We found several errors in our original article [DeDu12]: the infinite series $\mathscr{S}_{4i}'$ was missing and the symmetries of the series $\mathscr{S}_{2i}$ were wrong.

Obviously, the above theorem gives a method to enumerate the $(\{1, 2, 3\}, 6)$-spheres. The enumeration results for $v \leq 100$ are shown in Table 5.2.

5.3 Symmetry Groups of the $(\{1, 2, 3\}, 6)$-Spheres

Theorem 5.4 ([DeDu12])

(i) *The symmetry groups of a* $(\{2, 3\}, 6)$-*sphere are* C_1, C_s, C_i, C_2, C_{2v}, C_{2h}, S_4, C_3, C_{3v}, C_{3h}, S_6, D_2, D_{2h}, D_{2d}, D_3, D_{3h}, D_{3d}, D_6, D_{6h}, T, T_h, T_d.

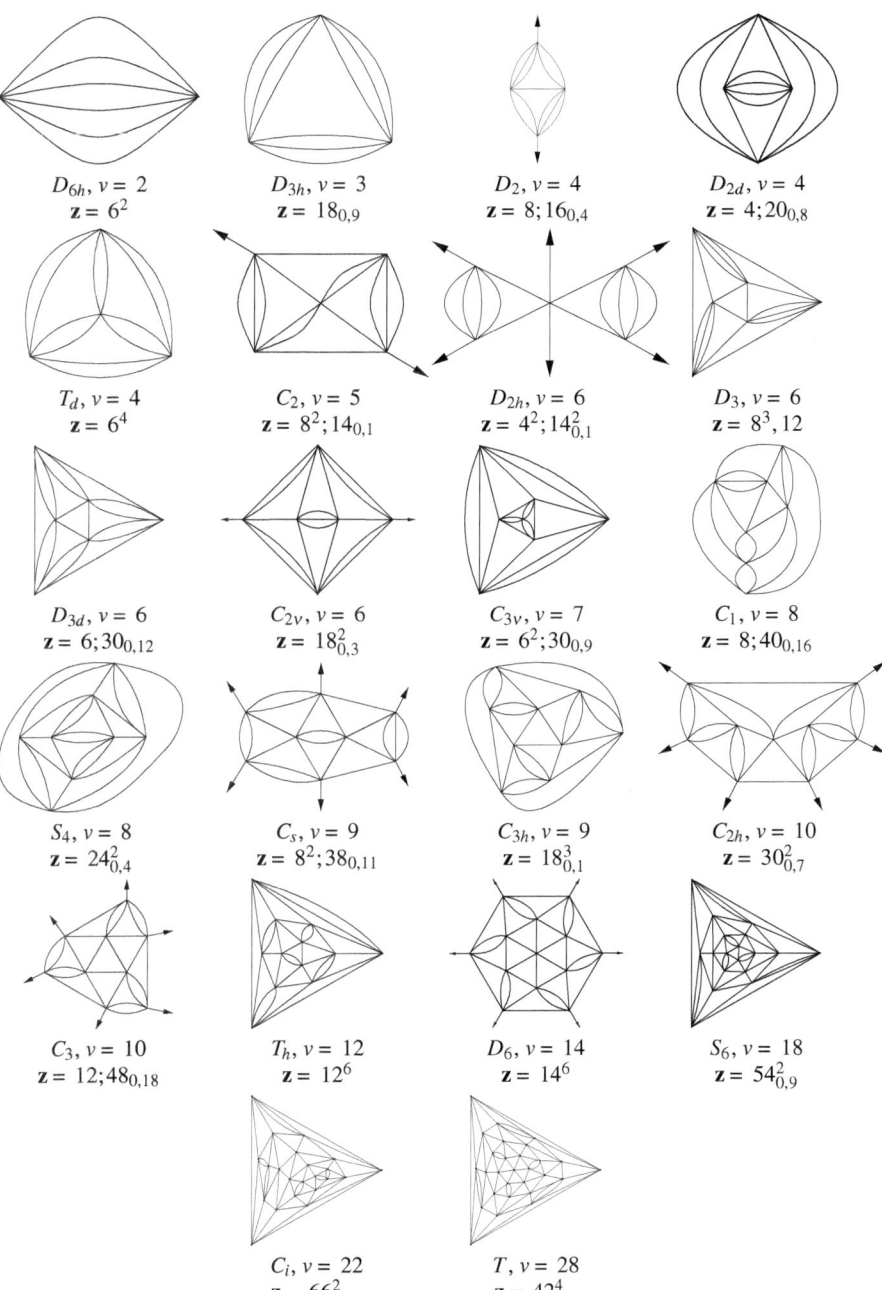

$D_{6h}, v = 2$
$\mathbf{z} = 6^2$

$D_{3h}, v = 3$
$\mathbf{z} = 18_{0,9}$

$D_2, v = 4$
$\mathbf{z} = 8; 16_{0,4}$

$D_{2d}, v = 4$
$\mathbf{z} = 4; 20_{0,8}$

$T_d, v = 4$
$\mathbf{z} = 6^4$

$C_2, v = 5$
$\mathbf{z} = 8^2; 14_{0,1}$

$D_{2h}, v = 6$
$\mathbf{z} = 4^2; 14^2_{0,1}$

$D_3, v = 6$
$\mathbf{z} = 8^3, 12$

$D_{3d}, v = 6$
$\mathbf{z} = 6; 30_{0,12}$

$C_{2v}, v = 6$
$\mathbf{z} = 18^2_{0,3}$

$C_{3v}, v = 7$
$\mathbf{z} = 6^2; 30_{0,9}$

$C_1, v = 8$
$\mathbf{z} = 8; 40_{0,16}$

$S_4, v = 8$
$\mathbf{z} = 24^2_{0,4}$

$C_s, v = 9$
$\mathbf{z} = 8^2; 38_{0,11}$

$C_{3h}, v = 9$
$\mathbf{z} = 18^3_{0,1}$

$C_{2h}, v = 10$
$\mathbf{z} = 30^2_{0,7}$

$C_3, v = 10$
$\mathbf{z} = 12; 48_{0,18}$

$T_h, v = 12$
$\mathbf{z} = 12^6$

$D_6, v = 14$
$\mathbf{z} = 14^6$

$S_6, v = 18$
$\mathbf{z} = 54^2_{0,9}$

$C_i, v = 22$
$\mathbf{z} = 66^2_{0,15}$

$T, v = 28$
$\mathbf{z} = 42^4_{0,3}$

Fig. 5.10 Minimal representatives for the symmetry groups of ({2, 3}, 6)-spheres

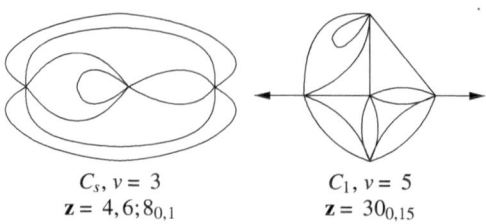

$$C_s, v = 3 \qquad\qquad\qquad C_1, v = 5$$
$$\mathbf{z} = 4, 6; 8_{0,1} \qquad\qquad \mathbf{z} = 30_{0,15}$$

Fig. 5.11 Minimal representatives for the symmetry groups of $(\{1, 2, 3\}, 6)$-spheres with $p_1 = 1$

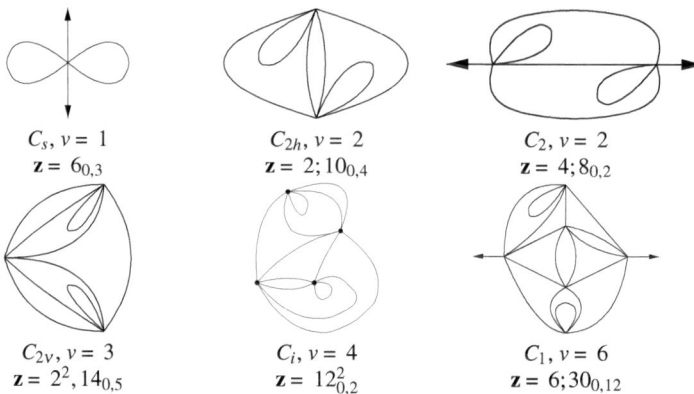

$$C_s, v = 1 \qquad\qquad C_{2h}, v = 2 \qquad\qquad C_2, v = 2$$
$$\mathbf{z} = 6_{0,3} \qquad\qquad \mathbf{z} = 2; 10_{0,4} \qquad\qquad \mathbf{z} = 4; 8_{0,2}$$

$$C_{2v}, v = 3 \qquad\qquad C_i, v = 4 \qquad\qquad C_1, v = 6$$
$$\mathbf{z} = 2^2, 14_{0,5} \qquad\qquad \mathbf{z} = 12_{0,2}^2 \qquad\qquad \mathbf{z} = 6; 30_{0,12}$$

Fig. 5.12 Minimal representatives for the symmetry groups of $(\{1, 2, 3\}, 6)$-spheres with $p_1 = 2$

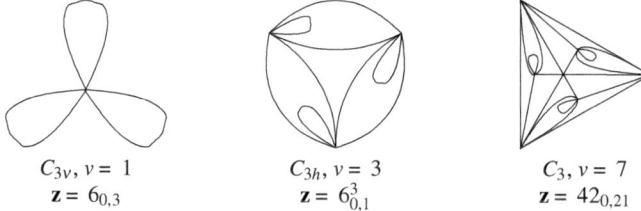

$$C_{3v}, v = 1 \qquad\qquad C_{3h}, v = 3 \qquad\qquad C_3, v = 7$$
$$\mathbf{z} = 6_{0,3} \qquad\qquad \mathbf{z} = 6_{0,1}^3 \qquad\qquad \mathbf{z} = 42_{0,21}$$

Fig. 5.13 Minimal representatives for the symmetry groups of $(\{1, 3\}, 6)$-spheres

- (ii) *The symmetry groups of a $(\{1, 2, 3\}, 6)$-sphere with $p_1 = 1$ are C_1, C_s.*
- (iii) *The symmetry groups of a $(\{1, 2, 3\}, 6)$-sphere with $p_1 = 2$ are $C_1, C_s, C_i, C_2,$ C_{2v}, C_{2h}.*
- (iv) *The symmetry groups of a $(\{1, 3\}, 6)$-sphere are C_3, C_{3v}, C_{3h}.*
- (v)   *The minimal representatives for each symmetry group (i)–(iv) above, are given in Figs. 5.10, 5.11, 5.12 and 5.13, respectively.*

Proof The method is to consider the possible axes of symmetry they are passing through; faces, edges, or vertices. As a consequence, the possibilities for a k-fold axis of symmetry are 2, 3, or 6. The only groups that could occur, besides 22 given in the theorem, are D_{6d}, C_6, C_{6h} or C_{6v}.

If a 6-fold axis occurs, then it necessarily passes through two vertices, say, v_1, and v_2. Around this vertex one can add successive rings of triangles as in the classical structure of the triangular lattice. At some point one gets a 2-gon and thus, by the 6-fold symmetry, six 2-gons. Then, one can continue the structure uniquely and the structure is defined uniquely. This completion is the same as the one around v and it implies the existence of a mapping that inverts v with the transformation inverting v_1 and v_2 and the group are D_6, D_{6h} or D_{6d}. The group D_{6d} is ruled out because 2-fold axis passes through the 2-gons. $\square$

In the next section we will describe explicitly, via new GC-construction, the $(\{2, 3\}, 6)$-spheres, of symmetry D_6, D_{6h} and T, T_h, T_d, as well as $(\{1, 3\}, 6)$-spheres.

All $(\{1, 2, 3\}, 6)$-maps on the projective plane $\mathbb{P}^2$ are antipodal quotients (i.e., with halved p-vector and number of vertices) of $(\{1, 2, 3\}, 6)$-spheres whose groups contain the inversion, i.e., $(\{2, 3\}, 6)$-spheres of symmetry C_i, C_{2h}, D_{2h}, D_{3d}, D_{6h}, S_6, T_h and $(\{1, 2, 3\}, 6)$-spheres with $p_1 = 2$ of symmetry C_{2h}.

5.4 The Goldberg–Coxeter Construction for 6-Regular Graphs

Here we introduce a new Goldberg–Coxeter construction. It takes a 6-regular sphere G_0, two integers k, l and returns two 6-regular spheres G_1, G_2 with $GC_{k,l}(G_0) = \{G_1, G_2\}$. The construction satisfies a multiplicativity property based on the ring of Eisenstein integers. In the case $k = l = 1$ we call the construction *oriented tripling* and we have a more explicit description of it. This GC-construction generalizes the one discussed in Sect. 6.2 for 3- or 4-regular plane graphs and allows to describe explicitly all $(\{1, 3\}, 6)$-spheres, as well as the $(\{2, 3\}, 6)$-spheres of symmetry D_6, D_{6h} and those of symmetry T, T_h, T_d.

If G is a $(\{1, 2, 3\}, 6)$-sphere, then the dual G^* is a plane graph with 6-gonal faces and thus, bipartite. The tessellation $\{6; 3\}$ of Euclidean plane by regular 6-gons is represented in Fig. 5.14. The centers of its 6-gons form $\{3; 6\}$. Also, with vertices of $\{6; 3\}$, they form $\{3; 6\}$, rotated by $90°$ and scaled by $\frac{1}{3}\sqrt{3}$.

We use two vectors $\mathbf{v}_1$, $\mathbf{v}_2$ to represent the points of $\{6; 3\}$. In complex coordinates $\mathbf{v}_1 = 1$ and $\mathbf{v}_2 = w$ with $w = e^{i\pi/3}$. The lattice $L = \mathbb{Z}\mathbf{v}_1 + \mathbb{Z}\mathbf{v}_2$ is called the *Eisenstein ring* . The point A is the origin and the point $B(k, l)$ is $k + lw$. The points in the same bipartite component as A are $L_A = (1 + w)L$, while the points in the same component as $B(1, 0)$ are $L_B = 1 + (1 + w)L$. Both sets L_A and L_B are stable under multiplication. Their union is the *bilattice* of the vertices of $\{6; 3\}$. We will first define GC-construction for $k + lw \in L_B$. Then we will extend it to any $k, l \neq 0$. See Fig. 5.15 for $6 \times K_2$.

Theorem 5.5 *If $z = k + lw \in L_B$ and G_0 is a 6-regular plane graph with v_0 vertices, then one can define a plane graph $G' = GC_z(G_0) = GC_{k,l}(G_0)$ such that it holds:*

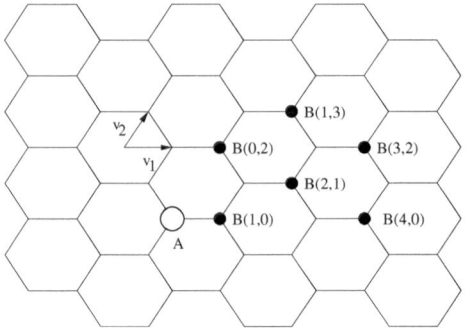

Fig. 5.14 The tiling by hexagons, the point A and some points in the other bipartite component

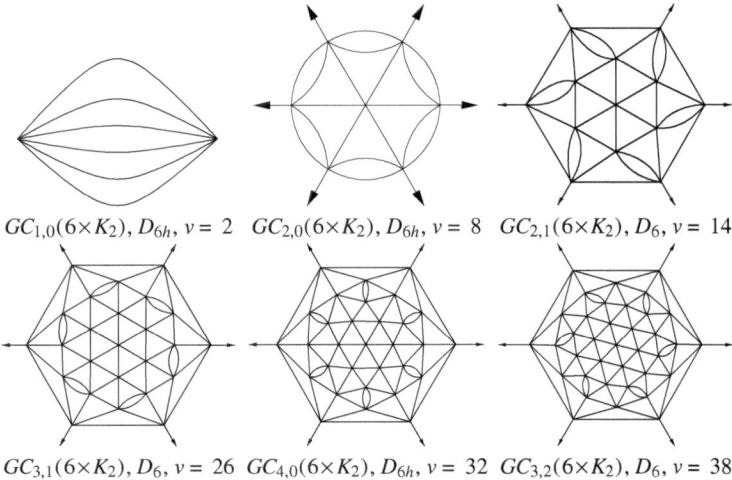

$GC_{1,0}(6\times K_2), D_{6h}, v = 2$ $GC_{2,0}(6\times K_2), D_{6h}, v = 8$ $GC_{2,1}(6\times K_2), D_6, v = 14$

$GC_{3,1}(6\times K_2), D_6, v = 26$ $GC_{4,0}(6\times K_2), D_{6h}, v = 32$ $GC_{3,2}(6\times K_2), D_6, v = 38$

Fig. 5.15 Smallest $(\{2, 3\}, 6)$-spheres of symmetry D_6, D_{6h} in terms of the GC-construction

 (i) G' is a 6-regular plane graph with $v_0(k^2 + kl + l^2)$ vertices.
 (ii) Any face of G_0 corresponds to a face of G' and all new faces of G' are 3-gons.
(iii) G' has all rotational symmetries of G_0 and, if $l = 0$ or $k = 0$, all symmetries.
(iv) $GC_{1,0}(G_0) = G_0$ and $GC_z(G_0) = GC_{zw^2}(G_0)$.
 (v) $GC_z(GC_{z'}(G_0)) = GC_{zz'}(G_0)$.
(vi) $GC_z(G_0) = GC_{\bar{z}}(\overline{G_0})$, where $\overline{G_0}$ differs from G_0 only by a plane symmetry.

Proof The dual G_0^* is a plane graph with all faces being 6-gons. If $z = k + lw \in L_B$, then the point $B(k, l)$ belongs to the same connected component as B.

   The point $c = -w^2 z$ is the center of a 6-gon and we build around it 6 points P_q:

$$P_q = c - w^q c \text{ for } 0 \le q \le 5.$$

These six points form a *master hexagon* that correspond to the original hexagon. Every hexagon of G_0^* can be thus modified and we can arrange them together at the boundary between adjacent hexagons. So, we can obtain another plane graph with 6-gonal faces. By taking the dual one more time, we get $GC_{k,l}(G_0)$. Checking the remaining properties is relatively easy.   □

See in Fig. 5.20 the local structures of the GC-construction $GC_{3,2}$ and $GC_{4,0}$.

Theorem 5.6 *If G is a $(\{2, 3\}, 6)$-sphere of symmetry D_6 or D_{6h}, then $G = GC_{k,l}(6 \times K_2)$ with $k + lw \in L_B$.*

Proof In the dual G^*, the 6-fold axis passes through a 6-gon F and the 2-gons of G correspond to the vertices of degree 6. But the position of these 2-gons define a master hexagon around F and so, we get the structure of a graph $GC_{k,l}$ $(6 \times K_2)$.   □

In Theorem 5.5 we have defined the Goldberg–Coxeter construction $GC_{k,l}$ for $k + lw \in L_B$. Now we want to define it for any $k, l \neq 0$. For that we first introduce the notion of *oriented tripling*, which will correspond to $GC_{1,1}(G)$.

Given a 6-regular plane graph G, for each bipartite class C of its dual G^* we define a graph $Or_C(G)$ with the following properties:

  (i) $Or_C(G)$ is a 6-regular plane graph with 3 times as many vertices.
 (ii) Each vertex of G corresponds to 3 vertices of $Or_C(G)$ and 4 triangular faces.
(iii) Every symmetry of G preserving C also occurs as a symmetry of $Or_C(G)$.

The local configuration of the operation is shown in Fig. 5.16. For every face F of G, we orient the edges of F counterclockwise. So for every bipartite class C of G^*, we get an orientation of the edges of G. Around a vertex V and its six adjacent vertices, there are three vertices U, to which the arc $\{V, U\}$ is oriented from V to U. They are the vertices $1, 3, 5$ in Fig. 5.16.

So if G^* has two inequivalent bipartite components C_1 and C_2, then $Or_{C_1}(G)$ and $Or_{C_2}(G)$ are not necessarily isomorphic and the smallest such example is shown in Fig. 5.17. In Fig. 5.18 we give two examples of the action of oriented tripling when the obtained graph is unique.

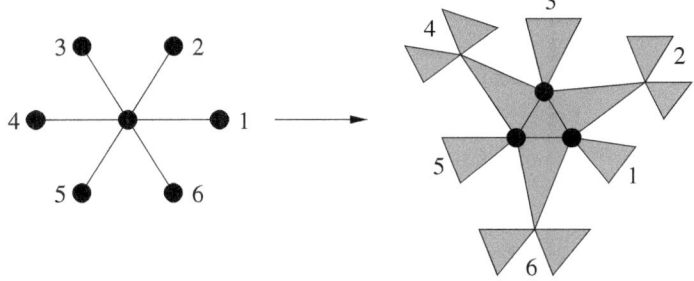

Fig. 5.16 Local configuration around a vertex of the oriented tripling operation

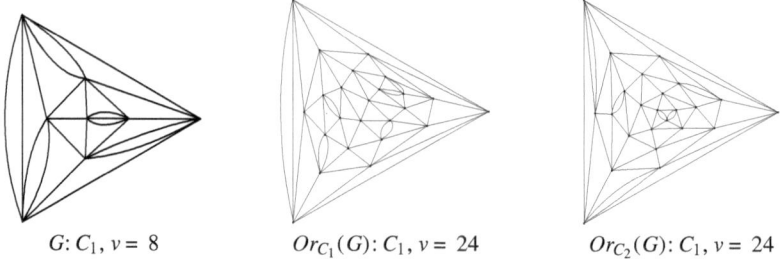

$G: C_1, v = 8$ $Or_{C_1}(G): C_1, v = 24$ $Or_{C_2}(G): C_1, v = 24$

Fig. 5.17 Smallest $(\{2, 3\}, 6)$-sphere G having two nonisomorphic oriented triplings $Or(G)$

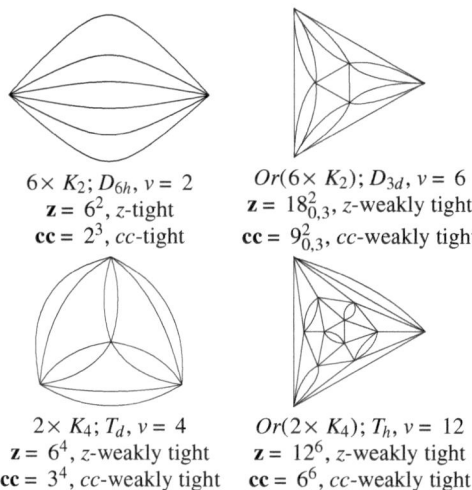

$6 \times K_2; D_{6h}, v = 2$ $Or(6 \times K_2); D_{3d}, v = 6$
$\mathbf{z} = 6^2$, z-tight $\mathbf{z} = 18_{0,3}^2$, z-weakly tight
$\mathbf{cc} = 2^3$, cc-tight $\mathbf{cc} = 9_{0,3}^2$, cc-weakly tight

$2 \times K_4; T_d, v = 4$ $Or(2 \times K_4); T_h, v = 12$
$\mathbf{z} = 6^4$, z-weakly tight $\mathbf{z} = 12^6$, z-weakly tight
$\mathbf{cc} = 3^4$, cc-weakly tight $\mathbf{cc} = 6^6$, cc-weakly tight

Fig. 5.18 Two examples of $(\{2, 3\}, 6)$-spheres with unique oriented tripling and their tightness

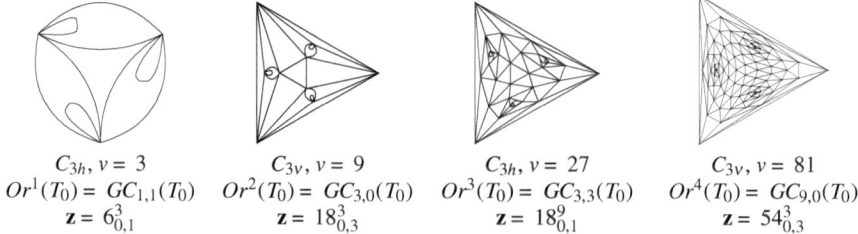

$C_{3h}, v = 3$ $C_{3v}, v = 9$ $C_{3h}, v = 27$ $C_{3v}, v = 81$
$Or^1(T_0) = GC_{1,1}(T_0)$ $Or^2(T_0) = GC_{3,0}(T_0)$ $Or^3(T_0) = GC_{3,3}(T_0)$ $Or^4(T_0) = GC_{9,0}(T_0)$
$\mathbf{z} = 6_{0,1}^3$ $\mathbf{z} = 18_{0,3}^3$ $\mathbf{z} = 18_{0,1}^9$ $\mathbf{z} = 54_{0,3}^3$

Fig. 5.19 The terms of the series of $(\{1, 3\}, 6)$-spheres $Or^i(Trifolium)$ for $i = 1, 2, 3, 4$

For Trifolium T_0, we can define a sequence $Or^i(T_0)$ of graphs as $Or^0(T_0) = T_0$ and $Or^{i+1}(T_0)$; see Fig. 5.19.

We now introduce the Goldberg–Coxeter construction in the general case. For a sphere G, denote by $Tr(G)$ the *truncation* of G, i.e., (see Sect. 1.1) the sphere obtained by replacing every vertex of degree k of G by a "small" k-gonal face.

We also use the following result: if G is a 3-regular sphere with faces of even size, then it is possible to color the faces of G, so that any two adjacent faces have different colors. Such a coloring is unique up to permutation of the colors. If G_0 is a graph with vertices of even degree, then its dual is bipartite and the three colors in $Tr(G_0)$ come from the vertices of G_0 and the two classes of faces in G_0.

Theorem 5.7 *For a 6-regular v_0-vertex plane graph G_0 and two integers k, l with $k, l \neq 0$, we can define two 6-regular spheres G_1, G_2 with $GC_{k,l}(G_0) = \{G_1, G_2\}$. This will satisfy the following properties:*

(i) *$Tr(G_i) = GC_{k,l}(Tr(G_0))$ for $i = 1, 2$ with $GC_{k,l}$ being the Goldberg–Coxeter construction for 3-regular spheres.*

(ii) *G_1 and G_2 are 6-regular plane graphs with $v_0(k^2 + kl + l^2)$ vertices.*

(iii) *Every face of G_0 corresponds to a face of G_1 and G_2 with all new faces of G_1 and G_2 being 3-gons.*

(iv) *$GC_{1,1}(G_0) = \{Or_C(G_0), Or_{C'}(G_0)\}$.*

(v) *If $k + lw \in L_B$, then $G_1 = G_2$.*

(vi) *$GC_{k',l'}(G_1) = GC_{k_p,l_p}(G_0)$ with $k_p + l_p w = (k + lw)(k' + l'w)$.*

Proof Let us take a 3-coloring in white, red, and blue of the faces of $Tr(G_0)$ with white corresponding to the faces coming from vertices of G_0. The 3-regular sphere $GC_{k,l}(Tr(G_0))$ has the faces of G_0 and some 6-gonal faces; so all its faces are of even size and we can find a 3-coloring of them.

One can see directly that all the white faces of $Tr(G_0)$ have the same color in $GC_{k,l}(Tr(G_0))$; we color them white. If $k \equiv l \pm 1$ (mod 3) (this contains the case $k + lw \in L_B$), then the faces of $Tr(G_0)$ coming from faces of G_0 will not be white in $GC_{k,l}(Tr(G_0))$. Thus by shrinking the white faces, we get a graph that is actually the $GC_{k,l}(G_0)$ defined in Theorem 5.5 if $k + lw \in L_B$.

If $k \equiv l$ (mod 3), then all faces of $Tr(G_0)$ correspond in $GC_{k,l}(Tr(G_0))$ to white faces. The remaining 6-gonal faces have color red and blue. This gives two sets of shrinkable faces and thus two possible graphs. All properties follow easily.   □

Theorem 5.8 (i) *Any $k + lw \neq 0$ can be written as $(1 + w)^s(k' + l'w)w^u$ with $s \geq 0$, $u \in \{0, 1\}$ and $k' + l'w \in L_B$.*

(ii) *It holds $GC_{k,l}(G_0) = GC_{k',l'}(Or^s(G_0))$, i.e., $GC_{k,l}(G_0)$ is obtained by applying the oriented tripling s times and then the GC-construction from Theorem 5.5.*

Proof (i) The ring of Eisenstein integers is a unique factorization domain. That is, every $k + lw \neq 0$ can be factorized into the relevant primes. The condition $k \equiv l$ (mod 3) is equivalent to $k + lw$ being divisible by $1 + w$. Thus by repeated application of this, we can write

$$k + lw = (1 + w)^s(k_2 + l_2w) \text{ with } k_2 \equiv l_2 \pm 1 \pmod{3}.$$

If $k_2 \equiv l_2 + 1$ (mod 3), then we are done; otherwise, we divide by w.

(ii) follows from the multiplicativity property (vi) of Theorem 5.7.   □

Theorem 5.9 *Let G be a* $(\{1, 3\}, 6)$*-sphere. The following hold:*

(i) $G = GC_{k,l}(Trifolium)$ *with* $0 \leq l \leq k$ *and has* $k^2 + kl + l^2$ *vertices.*
(ii) *G has symmetry* C_{3v} *if* $l = 0$, C_{3h} *if* $l = k$ *and* C_3*, otherwise.*

Proof (i) In fact, $Tr(G)$ is a $(\{2, 6\}, 3)$-sphere. Either from [GrZa74] or [Thur98], we know that such spheres are obtained as $GC_{k,l}(3 \times K_2)$ with $GC_{k,l}$ denoting here the Goldberg–Coxeter construction for 3-regular graphs. Since the faces of $GC_{k,l}(3 \times K_2)$ are of even size, it is possible to define a 3-coloring of those faces. The 2-gonal faces should not be in all 3 different colorings. This can occur only if $k \equiv l$ (mod 3). So, $k + lw$ can be factorized as $(1 + j)(k' + l'w)$ and we get

$$Tr(G) = GC_{k,l}(3 \times K_2) = GC_{k',l'}(GC_{1,1}(3 \times K_2)) = GC_{k',l'}(Tr(Trifolium)).$$

(ii) The symmetry of $GC_{k,l}(3 \times K_2)$ is D_{3h} if $k = 0$ or $k = l$ and D_3, otherwise. If $k \equiv l$ (mod 3), then all 2-gons of $GC_{k,l}(3 \times K_2)$ are in the same color, say, white. The 3-gonal faces, that are not white, are of two possible colors red and blue. A symmetry of $Tr(G) = GC_{k,l}(3 \times K_2)$ induces a symmetry of G if and only if it preserves all three colors of the coloring. This reduces by a factor of 2 the symmetry group and we get C_3, C_{3h} and C_{3v} as possible groups.   □

Theorem 5.10 *If G is a* $(\{2, 3\}, 6)$*-sphere of symmetry* T, T_d *or* T_h*, then*

(i) $G = GC_{k,l}(2 \times K_4)$ *with* $0 \leq l \leq k$ *and has* $4(k^2 + kl + l^2)$ *vertices.*
(ii) *G has symmetry* T_d *if* $l = 0$, T_h *if* $l = k$ *and* T*, otherwise.*

Proof In fact, $Tr(G)$ is a $(\{4, 6\}, 3)$-sphere with a subgroup T of symmetry. By Theorem 6.2 of [DeDu05], this implies that the symmetry group of $Tr(G)$ is O or O_h. By [DuDe04] Theorem 5.2, $Tr(G)$ is $GC_{k,l}(Cube)$. We need now to determinate which graphs $GC_{k,l}(Cube)$ are of the form $Tr(G)$. For that we need to consider the 3-coloring of the faces. Thus the 4-gonal faces are colored by at most two colors. Hence, $k \equiv l$ (mod 3) and so, $k + lw = (1 + w)(k' + l'w)$, implying

$$Tr(G) = GC_{k,l}(Cube) = GC_{k',l'}(GC_{1,1}(Cube)) = GC_{k',l'}(Tr(2 \times K_4))$$

and so, (i) follows from Theorem 5.7.

   If a $(\{2, 3\}, 6)$-sphere G is of symmetry T_d or T_h, then the symmetry group of $Tr(G)$ is O_h and such spheres are described as $GC_{k,0}(Cube)$ and $GC_{k,k}$ $(Cube)$.   □

Theorem 5.11 *The number of* $(\{1, 2, 3\}, 6)$*-spheres with* $p_1 = i$ *and less than* v *vertices grows as* $O(v^{4-i})$*.*

Proof For a v-vertex $(\{1, 2, 3\}, 6)$-sphere with $p_1 = i$, $Tr(G)$ is a $(\{2, 4, 6\}, 3)$-sphere with $p_2 = i$, $p_4 = 6 - 2i$ and $6v$ vertices. So, the number of faces of size 2 or 4 is $6 - i$. The 3-regular plane graphs, whose faces have size at most 6 and the set of non-6-gonal faces is fixed, are described by the parametrization theory of Thurston

[Thur98]. For example, the $(\{2, 4, 6\}, 3)$-spheres with $p_2 = i$ are described by $4 - i$ Eisenstein integers. Using this theory, [Sah94] obtained some upper bound on the number of fullerenes. The proof applies just as well for our classes of graphs and gives us the required upper bound.   □

5.5 Zigzags and Central Circuits of 6-Regular Graphs

Recall that in zigzags and central circuits of a 6-regular graph, all edges of self-intersection are, by Proposition 1.1 (ii), of type II, i.e., $e_1 = 0$ in all signatures and $\alpha_i = \alpha_i' = 0$ in the Theorem below.

Theorem 5.12 *Let us take a 6-regular plane graph G with z-vector $\ldots, l_i{}^{a_i}_{\alpha_i, \beta_i}, \ldots$ and c-vector $\ldots, k_j{}^{b_j}_{\alpha_j', \beta_j'}, \ldots$. Then the z-vector and c-vector of $GC_{1+4u,0}(G)$ are*

$$\ldots, \{l_i(1 + 3u)\}^{a_i(1+u)}_{\alpha_i, \beta_i}, \ldots, \ldots, \{2k_j(1 + 3u)\}^{2ub_j}_{\alpha_j', \beta_j'}, \ldots$$

and

$$\ldots, \left\{l_i \frac{1 + 3u}{2}\right\}^{ua_i}_{\alpha_i, \beta_i}, \ldots, \ldots, \{k_j(1 + 3u)\}^{b_j(1+2u)}_{\alpha_j', \beta_j'}, \ldots$$

Proof The proof uses the GC-construction previously built. One goes into the dual and subdivides the hexagons. Figure 5.20 shows that any central circuit of G corresponds to $1 + 2u$ central circuits (named B in the figure) and that the zigzags in A on one side correspond to zigzags in G. The result follows similarly for z-vector.   □

Theorem 5.13 (i) *Any $(\{1, 2, 3\}, 6)$-sphere with $p_1 \geq 1$ has at least one self-intersecting central circuit and at least one self-intersecting zigzag.*
(ii) *In any $(\{1, 3\}, 6)$-sphere, all central circuits and zigzags are self-intersecting.*

Proof (i) The self-intersection is evident from Fig. 5.21.
(ii) If a central circuit in a $(\{1, 2, 3\}, 6)$-sphere G is simple, then it splits G into two domains D_1 and D_2. If one denotes $n_{i,j}$ the number of faces of size i into

Fig. 5.20 The local structures of the Goldberg–Coxeter construction $GC_{3,2}$ and $GC_{4,0}$

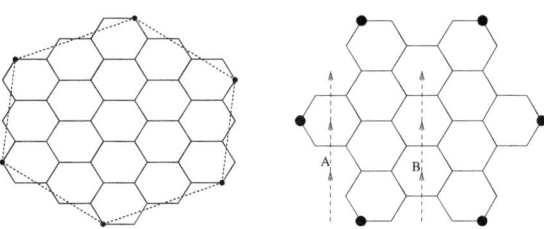

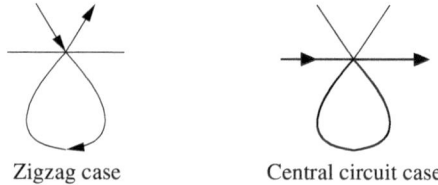

Zigzag case   Central circuit case

Fig. 5.21 Self-intersection induced by a 1-gon

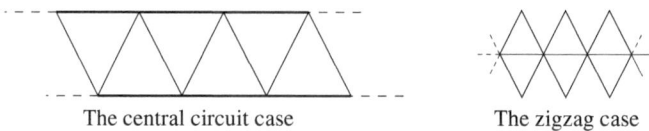

The central circuit case   The zigzag case

Fig. 5.22 A cc-railroad and a z-rairoads bounded by two central circuits, respectively, zigzags

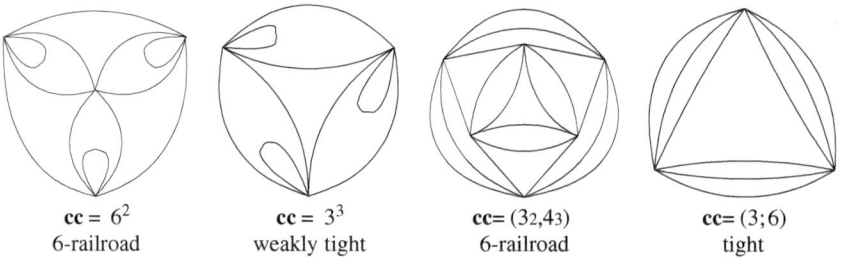

cc = 6^2 cc = 3^3 cc= $(3_2, 4_3)$ cc= $(3; 6)$
6-railroad   weakly tight   6-railroad   tight

Fig. 5.23 Examples of ($\{1, 3\}, 6$)- and ($\{2, 3\}, 6$)-spheres with and without railroads

the domain D_j, then one has obviously $2n_{1,j} + n_{2,j} = 3$. So, if $n_{2,j} = 0$, then
there is no solution. The proof for zigzags is the same.   □

A z-, respectively cc-railroad is the circuit of 3-gons bounded by two parallel
zigzags, respectively, central circuits. See Fig. 5.22 for illustration. Railroads can
self-intersect, and since a 3-gon has 3 edges and 3 vertices, they can also triply
self-intersect (see, for example, the first sphere in Fig. 5.23).

A ($\{1, 2, 3\}, 6$)-sphere is called z-tight if for any zigzag, there is at least one 1-gon
or 2-gon on each of its side of the sphere. It is called z-weakly tight if for any zigzag
there is no zigzag parallel to it. We define the corresponding notions for central
circuits. See Figs. 5.24 and 5.25 for some illustration of those notions.

For a ($\{1, 2, 3\}, 6$)-sphere G, let us call a p-gon with $p = 1, 2$ incident to a zigzag
or central circuit if it shares an edge with it. Call it weakly incident if it is not incident
to it but still prevents the existence of a railroad.

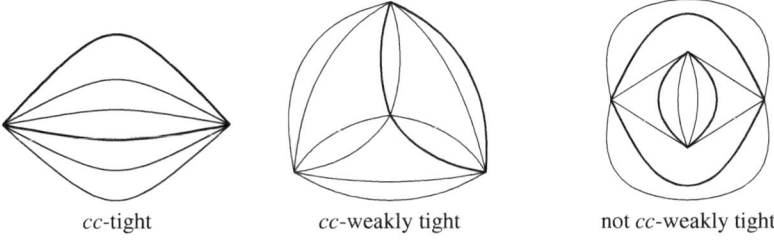

cc-tight   cc-weakly tight   not cc-weakly tight

Fig. 5.24 Illustration of the notions of *cc*-tightness of a ($\{2, 3\}, 6$)-sphere

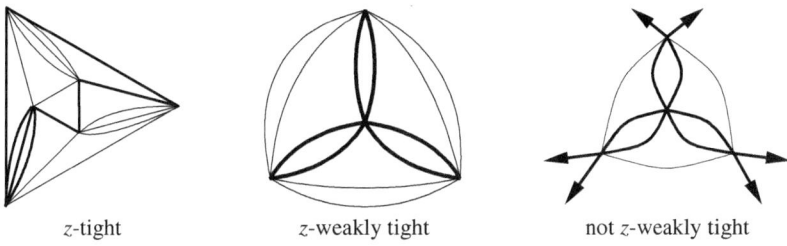

z-tight   z-weakly tight   not z-weakly tight

Fig. 5.25 Illustration of the notions of *z*-tightness of a ($\{2, 3\}, 6$)-sphere

Theorem 5.14 *For a* ($\{1, 2, 3\}, 6$)-*sphere G we have:*

(i) *If G is z-, respectively, cc-tight, then it has at most $\frac{p_1}{2} + p_2$ zigzags, respectively central circuits.*

(ii) *If G is z-, respectively, cc-weakly tight, then it has at most $p_1 + 2p_2$ zigzags, respectively central circuits.*

Proof If *G* is *cc*-tight, then any central circuit *C* has at least one 1- or 2-gon on each side. Since the number of sides is $p_1 + 2p_2$ and there is two sides per central circuit, this gives (i), 6-regular analog of Proposition 1.4. The zigzag case is identical.

If *G* is *cc*-weakly tight, then a *p*-gon ($p = 1, 2$) is incident or weakly incident, to at most *p* central circuits. Since it is weakly tight on each side of central circuits, there is at least one incident or weakly incident central circuit. Thus the maximal number of central circuits is $p_1 + 2p_2$. The proof for zigzags is identical. □

Conjecture 5.3 (i) *Any* ($\{1, 2, 3\}, 6$)-*sphere has the number of vertices v and the number of zigzags of the same parity.*

(ii) *The z-vector of a* ($\{1, 2, 3\}, 6$)-*sphere is the doubling of its cc-vector if and only if v and the number of central circuits are of the same parity.*

(iii) *(implied by (i), (ii)) Any z-knotted v-vertex* ($\{1, 2, 3\}, 6$)-*sphere has odd v, and any cc-knotted v-vertex* ($\{1, 2, 3\}, 6$)-*sphere is z-knotted if and only if v is odd.*

Theorem 5.15 *For a* ($\{1, 3\}, 6$)-*sphere G it holds:*

(i) *Every central circuit corresponds uniquely to a zigzag of doubled length, i.e.,* **z** = 2**cc** *disregarding signatures.*

(ii) G cannot be cc-, or z-tight. If G is cc- or z-weakly tight, then the number of central circuits, zigzags is 1 or 3.

Proof For a central circuit C, denote by $t_1, \dots, t_N$ the triangles on the side of F_2. Clearly, the set of edges of triangles t_i not contained in C, define a zigzag.

By Theorem 5.4, all $(\{1, 3\}, 6)$-spheres have symmetry C_3, C_{3v} or C_{3h}. Hence, they have a 3-fold axis of rotation and hence, the 1-gons belong to a single orbit under the group. The faces of a 6-regular plane graph are partitioned in two classes, say F_1, F_2, since its dual graph is bipartite. Clearly, the 1-gons are all in one partition class, say, F_1. A cc-, z-circuit has two sides, and the faces in those sides all belong to the same partition class. Thus, on one side of any ZC-circuit, there is only 3-gons.

If C is a central circuit in a cc-weakly tight $(\{1, 3\}, 6)$-sphere G, then on the side of F_1 there is a 1-gon and there are at most 3 central circuits. 2 is excluded by the group action.   □

Theorem 5.16 *Table 5.6 for the maximal number of zigzags and central circuits and both notions of weak tightness and tightness hold.*

Proof For $(\{1, 3\}, 6)$-spheres, Theorem 5.15 resolves the question. The existence of specific graphs in Fig. 5.31 shows the lower bounds that are indicated. Theorem 5.14 shows the required upper bounds for z-tightness and cc-tightness.

For the notion of weak tightness, we have to provide something more. Let G be a cc-weakly tight $(\{1, 2, 3\}, 6)$-sphere with central circuits $C_1, \dots, C_l$. It has $p_1 = i$, $p_2 = 6 - 2i$. We obtain $2l$ sides, since every central circuits has two sides. A side S is called *lonely* if it is incident or weakly incident to only one 2-gon.

If a side S is incident to exactly one 2-gon, then Fig. 5.26a shows that there is a side of parallel central circuit that is weakly incident two times to this 2-gon. Moreover, if it is incident exactly two times, then there is another lonely side, see Fig. 5.26b. A similar structure shows up if a side is weakly incident to a 2-gon.

Denote by n_{1a} the number of lonely sides in the first case and n_{1b} the number of lonely sides in the second case. Denote by n_{1c} the number of sides incident or weakly incident to exactly one 1-gon. Also, let n_2 be the number of sides incident to exactly two i-gons (identical or not). Let n_3 be the number of sides incident to at least 3 i-gons (identical or not). Obviously, $l = \frac{1}{2}(n_{1a} + n_{1b} + n_{1c} + n_2 + n_3)$.

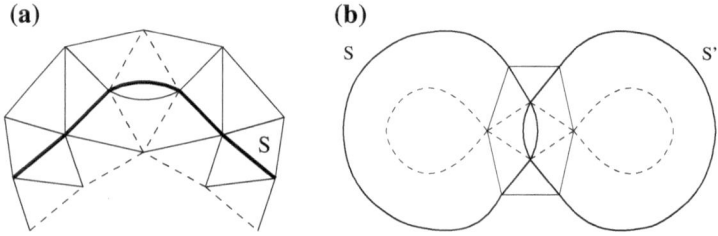

Fig. 5.26 Local structure around a side S incident to a 2-gon. **a** One lonely side S. **b** Two lonely sides S and S'

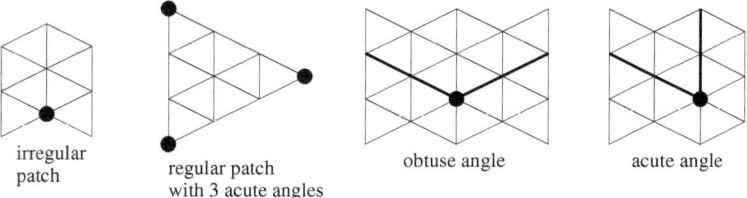

Fig. 5.27 Examples of patches and their angles

In the case (a), a lonely side S is incident to at least three i-gons; so, $n_{1a} \leq n_3$. Clearly, $n_{1c} \leq 2i$. Every 2-gon can be incident to 0, 1 or 2 lonely sides; so, $n_{1a} + \frac{n_{1b}}{2} \leq 6 - 2i$. By an enumeration of incidences we get

$$n_{1a} + n_{1b} + n_{1c} + 2n_2 + 3n_3 \leq s(G) = 2i + 4(6 - 2i) = 24 - 6i.$$

Denote by $\mathscr{P}_i$ the $5D$ polytope defined by these inequalities and $n_{1a}, \ldots, n_3 \geq 0$. We optimize the quantity l over $\mathscr{P}_i$ by using cdd [Fu], which uses exact arithmetic, and found the optimal value to be $9 - 2i$ for $i \leq 3$. The proof for zigzags is identical. $\qquad\square$

In the rest of this section, we give a local Euler formula for central circuits in order to enumerate the $(\{2, 3\}, 6)$-spheres which are cc-weakly tight and with simple central circuits. The method for zigzags is very similar.

Let G be a $(\{2, 3\}, 6)$-sphere. Consider a patch A in G, which is bounded by t arcs, i.e., sections of central circuits (different or coinciding).

We admit also 0-gonal patch A, i.e., just the interior of a simple central circuit. Suppose the patch A is *regular*, i.e., the continuation of any of its bounding arcs (on the central circuit, to which it belongs) lies outside of the patch (see Fig. 5.27). Let $p_2'(A)$ be the number of 2-gonal faces in A.

There are two types of intersections of arcs on the boundary of a regular patch: either intersection in an edge of the boundary, or intersection in a vertex of the boundary. Let us call these types of intersections *obtuse* and *acute*, respectively (see Fig. 5.27); denote by t_{ob} and t_{ac} the respective number of obtuse and acute intersections. Clearly, $t_{ob} + t_{ac} = t$, where t is the number of arcs forming the patch. The following formula can easily be verified:

$$6 - t_{ob} - 2t_{ac} = 2p_2'(A). \qquad (5.2)$$

Theorem 5.17 (cf. Theorem 3.7) *The intersection of every two simple central circuits, respectively zigzags, of a $(\{2, 3\}, 6)$-sphere, if nonempty, has one of the following forms (and so, its size is 2, 4 or 6):*

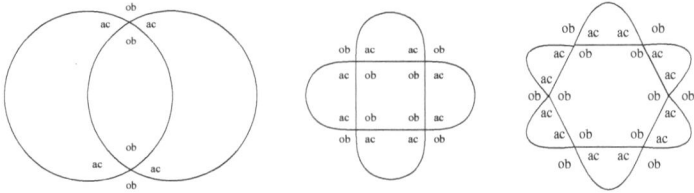

Proof Let us consider the central circuit case, the zigzag case being identical. Let H be the graph, whose vertices are edges of intersection between simple central circuits C and C', with two vertices being adjacent if they are linked by a path belonging to one of C, C'. Then H is a plane 4-regular graph and C, C' define two central circuits in H. Since C and C' are simple, the faces of H are t-gons with even t.

Applying formula (5.2) to a t-gonal face F of H, we obtain that the number $p_2'(F)$ of 2-gons in F satisfies $6 - t_{ob} - 2t_{ac} = 2p_2'(F)$. So the numbers t_{ob} and t_{ac} are even, since $t = t_{ob} + t_{ac}$. Also, $6 - t_{ob} - 2t_{ac} \geq 0$. So, $t \leq 6$.

We obtain the following five possibilities for the faces of H: 2-gons with two acute angles, 2-gons with two obtuse angles, 4-gons with four obtuse angles, 4-gons with two acute and two obtuse angles, 6-gons with six obtuse angles.

Take an edge e of a 6-gon in H and consider the sequence (possibly, empty) of adjacent 4-gons of H emanating from this edge. This sequence will stop at a 2-gon or a 6-gon; the case-by-case analysis of angles yields that this sequence has to stop at a 2-gon (see Fig. 5.28a).

Consider this construction for an edge of a 2-gon in H. If the angles are both obtuse, then the construction is identical and the sequence will terminate at a 2-gon or a 6-gon. If the angles are both acute, then cases (b), (c) of Fig. 5.28 are possible.

In the first case, all 4-gons contain two obtuse angles and two acute angles; so, the sequence of 4-gons finishes with an edge of two obtuse angles. In the second case, there is a 4-gon, whose angles are all obtuse; this 4-gon is unique in the sequence and its position is arbitrary. Every pair of opposite edges of a 4-gon belongs to a sequence of 4-gons considered above. So all angles of a 4-gon are the same, i.e., obtuse. So, all possibilities of intersections are the three cases of the theorem.   □

Theorem 5.18 *The weakly tight* $(\{2, 3\}, 6)$*-spheres, having only simple zigzags (respectively, simple central circuits) are the ones of Figs. 5.29 and 5.30.*

Proof Let us consider first the central circuit case. By Theorem 5.17, every two simple central circuits intersect in at most six vertices. If a $(\{2, 3\}, 6)$-sphere has t central circuits, this gives an upper bound of $6\frac{t(t-1)}{2}$ on the number of vertices of

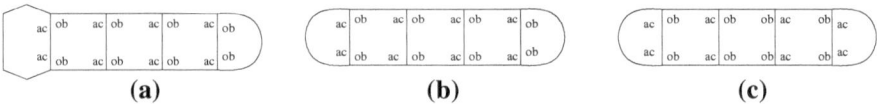

Fig. 5.28 Three cases for sequence of 4-gons

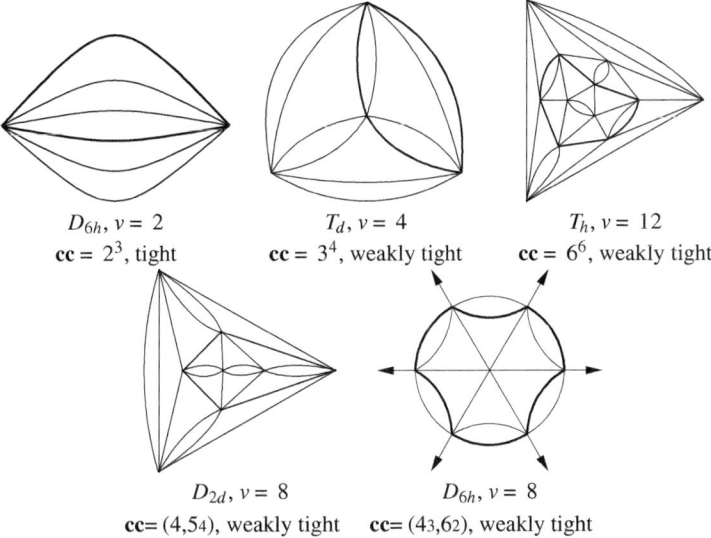

Fig. 5.29 The cc-weakly tight ($\{2, 3\}$, 6)-spheres with simple central circuits

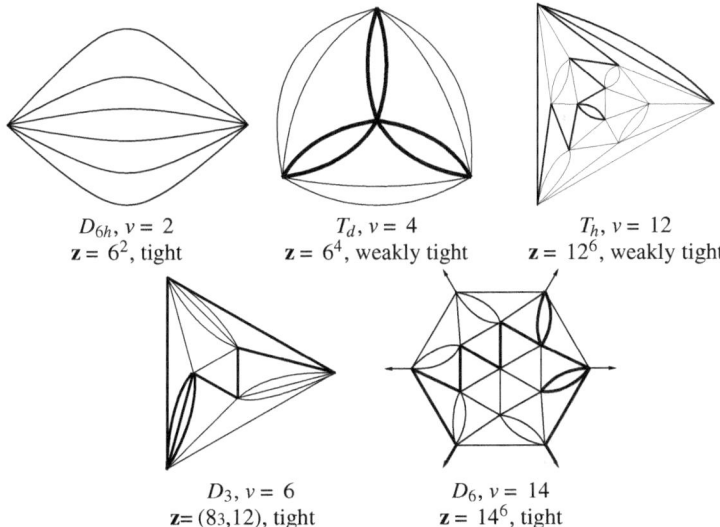

Fig. 5.30 The z-weakly tight ($\{2, 3\}$, 6)-spheres with simple zigzags; all are minimal representatives for their group of symmetry

intersection. Since any vertex can be the intersection of only 3 central circuits we get the upper bound of $t(t-1)$ on the number of vertices. If one uses the upper bound of Table 5.5 on t for weakly tight $(\{2, 3\}, 6)$-spheres, then one gets $t \leq 9$ and the upper bound 72 on the number of vertices, which is too large for the enumeration done in Table 5.5. If one looks at the proof of Theorem 5.16, then one sees that a lonely side implies a self-intersection of a parallel central circuit. So, there are no lonely sides in $(\{2, 3\}, 6)$-spheres with only simple central circuits. This gives the upper bound $t \leq 6$ on the number of central circuits and then 30 on the number of vertices (Fig. 5.31).

For zigzags we have the upper bound $3t(t-1)$ on the number of edges and this gives the same upper bound of $t(t-1)$ on the number of vertices. The enumeration result shown in Figs. 5.29 and 5.30 follow from the determination results of Sect. 5.2.   □

An interesting problem is to determine all $(\{2, 3\}, 6)$-spheres with simple zigzags and/or central circuits. By Theorem 5.12, their number is infinite.

A $(\{1, 2, 3\}, 6)$-sphere is called *z-pure* or *cc-pure* if all zigzags (or central circuits) are simple. It is called *z-knotted* or *cc-knotted* if it has only one zigzag or central circuit. They correspond to plane curves with only triple self-intersection points. Conjecture 5.3 implies that such *z*-knotted sphere has odd number of vertices v.

For $v \leq 54$, there are, besides Trifolium, 7 *cc*-knotted (equivalently, *z*-knotted) $(\{1, 3\}, 6)$-spheres: ones with $v = k^2 + kl + l^2 = 7, 13, 19, 31, 37, 43, 49$, i.e., with $(k, l) = (2, 1), (3, 1), (3, 2), (5, 1), (4, 3), (6, 1), (5, 3)$, but not (among others with odd v and of symmetry C_3) with $(k, l) = (4, 1), (5, 2)$, i.e., for $v = 21, 39$; perhaps, v should be not divisible by both, 2 and 3, for such knotted spheres (Fig. 5.32). From this, we derived the following two conjectures:

Conjecture 5.4 (i) *The symmetries of* $(\{1, 2, 3\}, 6)$-spheres with odd v (i.e., not only for even v), are: for $p_1 = 0$: $C_1, C_2, C_{2v}, C_3, C_{3h}, C_{3v}, C_s, D_{2h}, D_3$ or D_{3h}; for $p_1 = 1$: all (C_1, C_s); for $p_1 = 2$: all but C_i $(C_1, C_2, C_s, C_{2v}, C_{2h})$; for $p_1 = 3$: all (C_3, C_{3v}, C_{3h}).
   (ii) *The symmetries of a z- or cc-knotted* $(\{1, 2, 3\}, 6)$-spheres, except $3 \times K_3$ D_{3h} and Trifolium C_{3v}, are: C_1, C_2, C_3, D_2, D_3 for $p_1 = 0$; C_1 for $p_1 = 1$; C_1, C_2 for $p_1 = 2$; C_3 for $p_1 = 3$; see Table 5.2.
   (iii) *The* $(\{1, 2, 3\}, 6)$-spheres of symmetry T_d, T_h, D_{6h} are z- and cc-pure.
   (iv) *The cc-pure* $(\{2, 3\}, 6)$-spheres have symmetry $T_d, T_h, D_{6h}, D_{3d}, D_{2d}, D_{2h}, D_3, C_{2h}$ or C_{3v}.
   (v) *The* $(\{2, 3\}, 6)$-spheres of symmetry T_d have $v = 4x^2$, $\mathbf{cc} = (3x)^{4x}$, $\mathbf{z} = (6x)^{4x}$ and those of symmetry T_h have $v = 12x^2$, $\mathbf{cc} = (6x)^{6x}$, $\mathbf{z} = (12x)^{6x}$.

Conjecture 5.5 Let $f_i(v)$ denote the maximal number of central circuits in a v-vertex $(\{1, 2, 3\}, 6)$-sphere with $p_1 = i$. We conjecture:

   (i) $f_2(v) = v + 1$. It is realized exactly by the series (one for each $v \geq 1$) having symmetry C_{2h} and $\mathbf{cc} = 1^v, (2v)_{0,v}$.
   (ii) $f_1(v) = \frac{v-1}{2} + 1, \frac{v-1}{2} + 2$ for $v \equiv 3, 1 \pmod 4$ and $\lfloor \frac{v-1}{3} \rfloor + 2$ for even v.

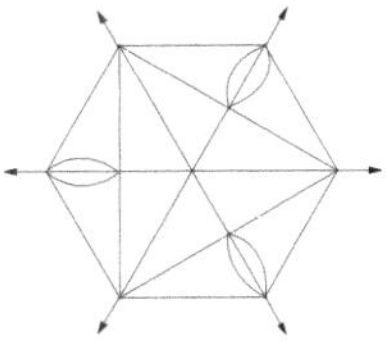

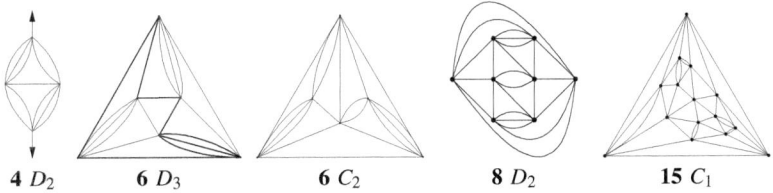

$p_1 = 0, D_{3h}, v = 11$
cc-tight
$5^3, 6^3_{0,1}$

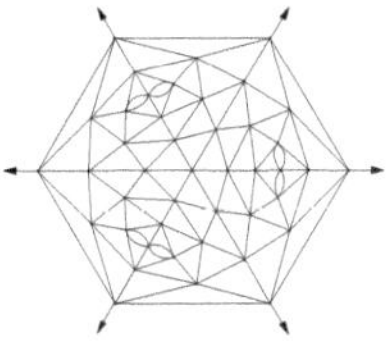

$p_1 = 0, D_{3h}, v = 44$
cc-weakly tight
$10^3, 11^3_{0,1}, 22^3_{0,3}$

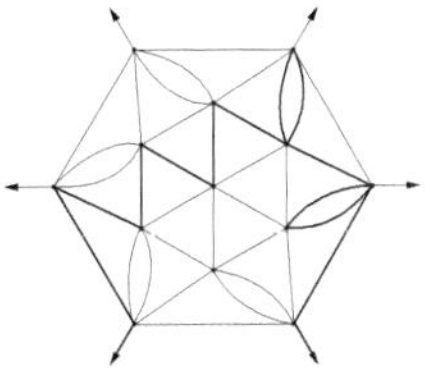

$p_1 = 0, D_6, v = 14$
z-tight
14^6

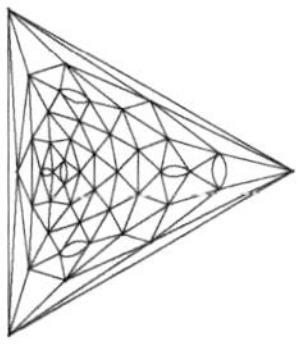

$p_1 = 0, C_{2h}, v = 44$
z-weakly tight
$24^4, 30^2_{0,1}, 54^2_{0,5}$

$p_1 = 1, C_s, v = 4$
cc-tight
$3, 4_{0,1}, 5_{0,1}$

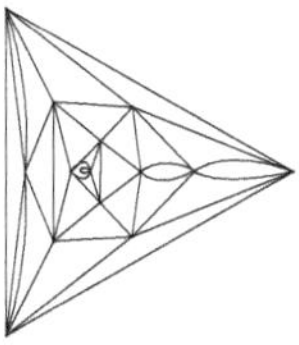

$p_1 = 1, C_s, v = 14$
cc-weakly tight
$5, 11_{0,1}, 12_{0,3}, 7^2_{0,1}$

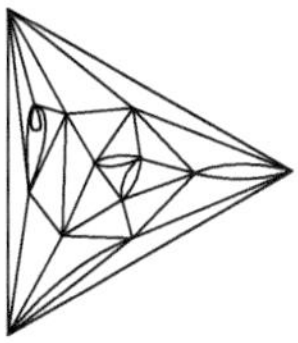

$p_1 = 1, C_1, v = 13$
z-tight
$16_{0,1}, 20_{0,1}, 43_{0,9}$

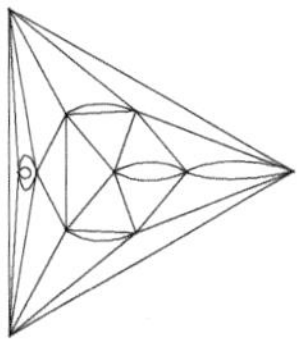

$p_1 = 1, C_s, v = 11$
z-weakly tight
$10, 12, 14^2_{0,1}, 16_{0,1}$

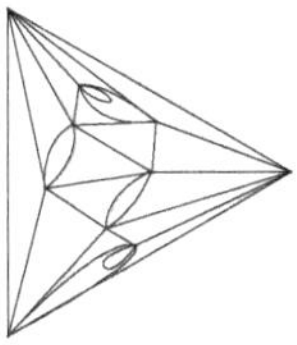

$p_1 = 2, C_2, v = 10$
cc-tight
$8^2_{0,2}, 14_{0,6}$

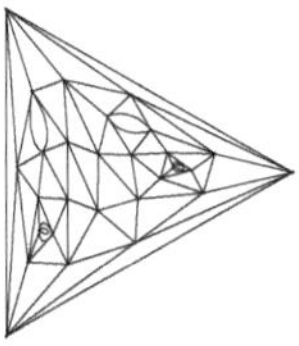

$p_1 = 2, C_2, v = 28$
cc-weakly tight
$14_{0,1}, 14^3_{0,2}, 28_{0,9}$

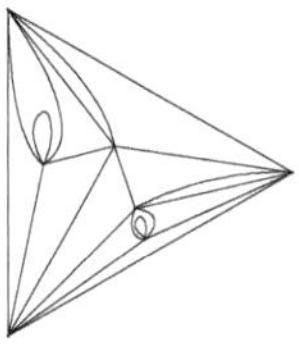

$p_1 = 2, C_2, v = 7$
z-tight
$14_{0,1}, 14^2_{0,2}$

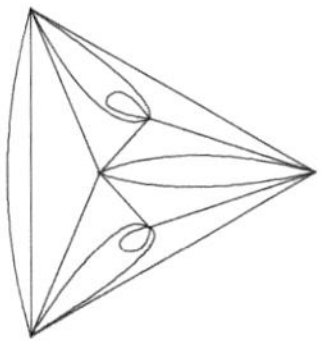

$p_1 = 2, C_{2v}, v = 6$
z-weakly tight
$6^2, 12^2_{0,2}$

Fig. 5.31 The smallest weakly tight and tight ($\{1, 2, 3\}$, 6)-spheres with the maximal known number of zigzags and central circuits

4 D_2 **6** D_3 **6** C_2 **8** D_2 **15** C_1

Fig. 5.32 Small cc-knotted ($\{2, 3\}$, 6)-spheres; see Table 5.5

(iii) $f_0(v) = \frac{v}{2} + 1, \frac{v}{2} + 2$ for $v \equiv 0, 2 \pmod{4}$. For odd v, f_0 is $\lfloor \frac{v}{3} \rfloor + 3$ if $v \equiv 2, 4, 6 \pmod{9}$ and $\lfloor \frac{v}{3} \rfloor + 1$, otherwise.

Table 5.5 Numbers N_i' and N_i'' of z- and cc-knotted v-vertex $(\{1, 2, 3\}, 6)$-spheres with $p_1 = i$

v	N_0', N_0''	N_1', N_1''	N_2', N_2''	N_3', N_3''	v	N_0', N_0''	N_1', N_1''	N_2', N_2''	N_3', N_3''	v	N_0', N_0''	N_1', N_1''	N_2', N_2''	N_3', N_3''
1	0,0	0,0	1,0	1,1	35	110,7	40,10	15,3	0,0	69	701,43	151,26	29,6	0,0
2	0,0	0,0	0,1	0,0	36	0,130	0,24	0,17	0,0	70	0,712	0,109	0,32	0,0
3	1,0	0,0	2,1	0,0	37	144,12	45,7	12,2	1,1	71	729,68	150,27	22,3	0,0
4	0,1	0,0	0,2	0,0	38	0,110	0,31	0,14	0,0	72	0,822	0,94	0,30	0,0
5	0,0	1,0	3,1	0,0	39	150,9	50,14	13,4	0,0	73	804,61	147,16	24,4	1,1
6	0,2	0,1	0,2	0,0	40	0,153	0,38	0,14	0,0	74	0,829	0,108	0,24	0,0
7	2,0	2,0	2,1	1,1	41	151,13	50,13	15,2	0,0	75	883,77	180,36	37,11	0,0
8	0,2	0,2	0,3	0,0	42	0,188	0,36	0,15	0,0	76	0,951	0,116	0,24	0,0
9	5,0	2,1	5,1	0,0	43	200,14	53,7	16,4	1,1	77	839,63	176,34	27,6	0,0
10	0,4	0,2	0,4	0,0	44	0,204	0,39	0,19	0,0	78	0,1002	0,123	0,27	0,0
11	7,0	4,2	5,1	0,0	45	203,16	69,19	20,4	0,0	79	956,76	178,26	21,4	1,1
12	0,11	0,2	0,4	0,0	46	0,235	0,38	0,15	0,0	80	0,1025	0,113	0,34	0,0
13	9,0	5,1	5,1	1,1	47	204,11	73,11	14,3	1,1	81	1060,74	204,37	34,6	0,0
14	0,9	0,6	0,6	0,0	48	0,300	0,44	0,17	0,0	82	0,1106	0,111	0,32	0,0
15	12,1	8,0	6,2	0,0	49	253,18	71,8	21,4	1,1	83	1108,92	225,62	25,5	0,0
16	0,19	0,5	0,6	0,0	50	0,284	0,43	0,21	0,0	84	0,1259	0,131	0,33	0,0
17	9,1	13,1	6,2	0,0	51	292,16	75,13	22,6	0,0	85	1148,87	214,32	38,6	0,0
18	0,23	0,5	0,8	0,0	52	0,335	0,46	0,23	0,0	86	0,1207	0,144	0,35	0,0
19	21,2	12,2	6,2	1,1	53	322,26	82,12	18,4	0,0	87	1345,100	224,48	24,9	0,0
20	0,24	0,8	0,8	0,0	54	0,387	0,63	0,24	0,0	88	0,1371	0,135	0,35	0,0
21	27,3	12,1	10,2	0,0	55	389,27	94,8	17,5	0,0	89	1347,104	279,71	34,5	0,0
22	0,29	0,10	0,9	0,0	56	0,371	0,61	0,23	0,0	90	0,1437	0,165	0,37	0,0
23	33,3	14,4	7,2	0,0	57	419,24	93,17	19,5	0,0	91	1484,122	227,39	36,6	2,2

(continued)

Table 5.5 (continued)

v	N'_0, N''_0	N'_1, N''_1	N'_2, N''_2	N'_3, N''_3	v	N'_0, N''_0	N'_1, N''_1	N'_2, N''_2	N'_3, N''_3	v	N'_0, N''_0	N'_1, N''_1	N'_2, N''_2	N'_3, N''_3
24	0,47	0,11	0,10	0,0	58	0,409	0,80	0,21	0,0	92	0,1524	0,167	0,33	0,0
25	42,3	19,4	12,3	0,0	59	457,39	101,24	18,4	0,0	93	1609,109	246,51	31,8	0,0
26	0,45	0,15	0,8	0,0	60	0,499	0,63	0,25	0,0	94	0,1609	0,180	0,41	0,0
27	65,2	21,6	12,4	0,0	61	507,32	101,11	18,3	1,1	95	1696,149	288,74	30,6	0,0
28	0,74	0,15	0,12	0,0	62	0,511	0,68	0,25	0,0	96	0,1772	0,177	0,39	0,0
29	60,3	32,6	10,2	0,0	63	569,35	119,29	26,6	0,0	97	1797,142	276,43	35,6	1,1
30	0,79	0,21	0,13	0,0	64	0,594	0,78	0,31	0,0	98	0,1750	0,200	0,39	0,0
31	80,3	31,5	9,1	1,1	65	572,46	137,31	29,5	1,1	99	1921,127	293,64	42,12	0,0
32	0,84	0,21	0,12	0,0	66	0,599	0,80	0,21	0,0	100	0,1935	0,182	0,41	0,0
33	99,3	28,5	18,4	1,1	67	670,49	125,29	23,5	1,1					
34	0,106	0,26	0,11	0,0	68	0,649	0,84	0,25	0,0					

Table 5.6 The maximal number of zigzags and central circuits for both notions of tightness of $(\{1, 2, 3\}, 6)$-spheres; bold numbers are definite answer, while intervals give the possible range

	z-tight	z-w. tight	cc-tight	cc-w. tight
$(\{2, 3\}, 6)$-spheres, $p_1 = 0$	**6**	**9**	**6**	$[8, 9]$
$(\{1, 2, 3\}, 6)$-spheres, $p_1 = 1$	$[3, 4]$	$[5, 7]$	$[3, 4]$	$[5, 7]$
$(\{1, 2, 3\}, 6)$-spheres, $p_1 = 2$	**3**	**5**	**3**	$[4, 5]$
$(\{1, 3\}, 6)$-spheres, $p_1 = 3$	**0**	**3**	**0**	**3**

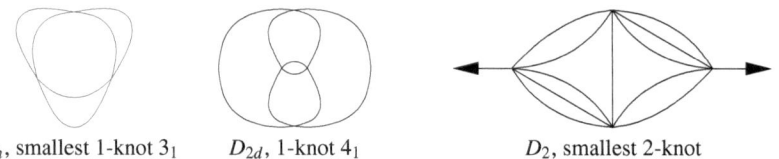

D_{3h}, smallest 1-knot 3_1 D_{2d}, 1-knot 4_1 D_2, smallest 2-knot

Fig. 5.33 Smallest 1- and 2-knots

Remark 5.2 Call a t-*knot* any $2(t+1)$-regular connected plane Eulerian graph without loops, consisting of a unique central circuit. Any 0-knot is a polygon and any 1-knot is a (projection of) usual knot. So the smallest 0- and 1-knots have 2 and 3 vertices. But for any $t \geq 2$. the smallest t-knot is the following: 4-vertex $(\{2, 3\}, 2t+2)$-sphere (D_2) with $\mathbf{p} = (p_2 = 4t - 2, p_3 = 4)$ and $\mathbf{cc} = 4(t + 1)$: it is 4_1 if $t = 1$, and for $t \geq 2$, it comes from $(t - 1)$-figure by adding 4-ring of 2-gons, see Fig. 5.33. The smallest m-regular plane graph without loops, consisting of a unique zigzag, is $Bundle_m = m \times K_2$ (D_{mh}) with $\mathbf{z} = 2m_{m,0}$ for odd and $2m_{0,m}$ for even m. Permitting 1-gons, m-*rose* $m \times K_1$ is the smallest such graph for $\mathbf{z}$ and, if m is odd, for $\mathbf{cc}$.

References

[BFDH97] Brinkmann, G., Delgado Friedrichs, O., Dress, A., Harmuth, T.: CaGe—a virtual environment for studying some special classes of large molecules. MATCH: Commun. Math. Comput. Chem. **36**, 233–237 (1997). http://www.mathematik.uni-bielefeld.de/CaGe/Archive/

[BrHaHe03] Brinkmann, G., Harmuth, T., Heidemeier, O.: The construction of cubic and quartic planar maps with prescribed face degrees. Discrete Appl. Math. **128**–2(3), 541–554 (2003)

[DuDe04] Dutour, M., Deza, M.: Goldberg-Coxeter construction for 3- or 4-valent plane graphs. Electron. J. Comb. **11**–1, R20 (2004)

[DeDu05] Deza, M., Dutour, M.: Zigzag structure of simple two-faced polyhedra. Comb. Probab. Comput. **14**, 31–57 (2005)

[DeDu08] Deza, M., Dutour Sikirić, M.: Geometry of Chemical Graphs: polycycles and two-faced maps, Encyclopedia of Mathematics and its Applications, vol. 119. Cambridge University Press, Cambridge (2008)

[DeDu12] Deza, M., Dutour Sikirić, M.: Zigzag and central circuit structure of $(1, 2, 3, 6)$-spheres. Taiwanese J. Math. **16–3**, 913–940 (2012)

[DDS13a] Deza, M., Dutour Sikirić, M., Shtogrin, M.: Fullerene-like spheres with faces of negative curvature. In: Diudea, M.V., Nagy, C.L. (eds.) Diamond D5 and Related Nanostructures Carbon Materials: Chemistry and Physics, pp. 251–274, vol. 6. Springer, New York (2013)

[Fu] Fukuda, K.: The cdd program. http://www.ifor.math.ethz.ch/fukuda/cdd_home/cdd.html

[GrZa74] Grünbaum, B., Zaks, J.: The existence of certain planar maps. Discrete Math. **10**, 93–115 (1974)

[Sah94] Sah, C.H.: A generalized leapfrog for fullerene structures. Fullerene Sci. Tech. **2–4**, 445–458 (1994)

[Thur98] Thurston, W.P.: Shapes of polyhedra and triangulations of the sphere. In: Rivin, J. Rourke, C., Series, C. (eds.) Geometry and Topology Monographs 1, The Epstein Birthday Schrift, pp. 511–549. Geometry and Topology Publishing, Coventry (1998)

Chapter 6
Goldberg–Coxeter Construction and Parametrization

In this chapter, we consider parametrization and, especially, one with 1 complex parameter, i.e., the *Goldberg–Coxeter construction* $GC_{k,l}(G_0)$ (a generalization of a simplicial subdivision of Dodecahedron considered in [Gold37] and [Cox71]), producing a plane graph from any 3- or 4-regular plane graph G_0 for integer parameters $k, l \geq 0$. See the main features of GC-construction in Table 6.1.

Some classes of such graphs with maximal symmetry can be described as $GC_{k,l}(G_0)$ (see Sect. 6.3). Table 6.2 present the considered graphs $GC_{k,l}(G_0)$. In this Table, r denotes the number of ZC-circuits in $GC_{k,l}(G_0)$. The case $(k, l) = (1, 0)$ corresponds to the initial graph G_0. The columns 1–4 give, respectively, the class of graphs, degree k, p-vector, and all realizable symmetry groups for the graphs $GC_{k,l}(G_0)$. The case $k = l = 1$ corresponds to the medial graph for 4- and to leapfrog graph for 3-regular case; see Fig. 6.1.

Other classes of graphs are more numerous/complicated, and so more complex parameters are needed. We expose the formalism behind such parameterizations by first considering several cases where just two complex parameters are needed. In the case of maps on spheres, [Thur98] has developped a theory of complex parameterization that relates it to complex hyperbolic manifolds. We also consider the 4-valent case and the parameterization of maps on surfaces. Finally, we give conjecture about the signature of the Hermitian form giving the number of triangles in term of the complex parameters.

6.1 The Complex Number Rings $\mathbb{Z}[\omega]$ and $\mathbb{Z}[i]$

The *Square lattice* is denoted by $\mathbb{Z}^2$. The *Root lattice* A_2 (or *Hexagonal lattice*) is defined by $A_2 = \{x \in \mathbb{Z}^3 : x_0 + x_1 + x_2 = 0\}$.

The ring $\mathbb{Z}[\omega]$, where $\omega = e^{\frac{2\pi}{6}i} = \frac{1}{2}(1 + i\sqrt{3})$, of *Eisenstein integers* consists of the complex numbers $z = k + l\omega$ with $k, l \in \mathbb{Z}$ (see also [HaWr96], where ω is replaced by ρ). The norm of such z is denoted by $N(z) = z\bar{z} = k^2 + kl + l^2$, and we

© Springer India 2015

M. Deza et al., *Geometric Structure of Chemistry-Relevant Graphs*,
Forum for Interdisciplinary Mathematics 1, DOI 10.1007/978-81-322-2449-5_6

Table 6.1 Main features of Goldberg–Coxeter construction

	3-regular graph G_0	4-regular graph G_0	6-regular graph G_0
Lattice	Root lattice $A_2 = \{3; 6\}$	Square lattice $\mathbb{Z}^2 = \{4, 4\}$	Bilattice $\{6; 3\}$
Ring	Eisenstein integers $\mathbb{Z}[\omega]$	Gaussian integers $\mathbb{Z}[i]$	Eisenstein integers $\mathbb{Z}[\omega]$
$t(k, l)$	$k^2 + kl + l^2$	$k^2 + l^2$	$k^2 + kl + l^2$
Euler formula	$\sum_i (6 - i) p_i = 12$	$\sum_i (4 - i) p_i = 8$	$\sum_i (3 - i) p_i = 6$
Flat faces	6-gons	4-gons	3-gons
ZC-circuits	Zigzags	Central circuits	Both
Case $k = l = 1$	Leapfrog graph	Medial graph	Oriented tripling

Table 6.2 Main series of q-regular graphs ($q = 3, 4$) G_0, for which we consider $GC_{k,l}(G_0)$

Class	q	p-vector	Groups	$(k, l) = (1, 0)$	$(k, l) = (1, 1)$	r if I	r if II
2_v	3	$p_2 = 3, p_6$	All D_3, D_{3h}	$Bundle_3$	Tr.Triangle	3	1
3_v	3	$p_3 = 4, p_6$	All T, T_d	Tetrah	Tr.Tetrahed	3	3
4_v	3	$p_4 = 6, p_6$	All O, O_h	Cube	Tr.Octahed	6	4
5_v	3	$p_5 = 12,$ p_6	All I, I_h	Dodecah	Tr.Icosahed	6, 10, 15	
$\mathscr{GP}_m$	3	$p_4 = m,$ $p_m = 2, p_6$ $(m \neq 2, 4)$	All D_m, D_{mh}	$Prism_m$	$Tr.Prism_m^*$	Conj. 7.6	
$\mathscr{GA}_m$	4	$p_3 = 2m,$ $p_m = 2, p_4$ $(m \neq 2, 3)$	Some $D_m,$ D_{md}	$APrism_m$	$Med(APrism_m)$	Conj. 7.7	
8-hed	4	$p_3 = 8, p_4$	All O, O_h	Octahed	Cuboctahed	4	3, 6
4-hed	4	$p_2 = 4, p_4$	All D_4, D_{4h}	$Foil_2$	$Foil_4$	2	2
5-hed	4	$p_2 = 3,$ $p_3 = 2, p_4$	All D_3, D_{3h}	Trefoil 3_1	$Med(3_1) = 6_1^3$	3	1
6-hed	4	$p_2 = 2,$ $p_3 = 4, p_4$	Some D_2, D_{2d}	4_1	$Med(4_1) =$ 8_{14}^2	2, 4	1, 3
7-hed	4	$p_2 = 1,$ $p_3 = 6, p_4$	Some C_2, C_{2v}	7_6^2	$Med(7_6^2)$	3, 5, 7	1, 2, 3, 5

will use the notation $t(k, l) = k^2 + kl + l^2$. If one identifies $x = (x_1, x_2, x_3) \in A_2$ with the Eisenstein integer $z = x_1 + x_2\omega$, then it holds $2N(z) = \|x\|^2$.

One has $\mathbb{Z}^2 = \mathbb{Z}[i]$, where the ring $\mathbb{Z}[i]$ consists of the complex numbers $z = k + li$ with $k, l \in \mathbb{Z}$. The norm of such z is denoted by $N(z) = z\bar{z} = k^2 + l^2$ and we will use the notation $t(k, l) = k^2 + l^2$. The rings $\mathbb{Z}[\omega]$ and $\mathbb{Z}[i]$ are unique factorization rings.

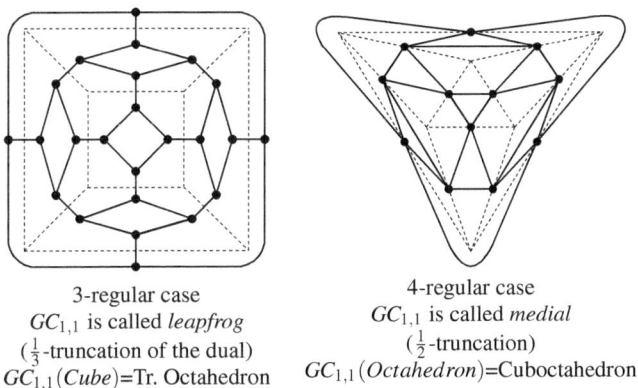

3-regular case	4-regular case
$GC_{1,1}$ is called *leapfrog*	$GC_{1,1}$ is called *medial*
($\frac{1}{3}$-truncation of the dual)	($\frac{1}{2}$-truncation)
$GC_{1,1}(Cube)$=Tr. Octahedron	$GC_{1,1}(Octahedron)$=Cuboctahedron

Fig. 6.1   The case $(k, l) = (1, 1)$ of Goldberg–Coxeter construction

Two Eisenstein or two Gaussian integers z and z' are called *Associated numbers* if the quotient $\frac{z}{z'}$ is an Eisenstein unit (i.e., $1, \omega, \omega^2, \omega^3 = -1, \omega^4 = -\omega, \omega^5 = -\omega^2$) or a Gaussian unit (i.e., $1, i, i^2 = -1, i^3 = -i$). They are called *C-associated numbers* if one of the quotients $\frac{z}{z'}, \frac{\bar{z}}{z'}$ is a such unit. Every Eisenstein or Gaussian integer is associated (or C-associated) to $k + l\omega$ or $k + li$ with $k, l \geq 0$ (or $0 \leq l \leq k$).

The lattices A_2 and $\mathbb{Z}^2$ correspond to regular partitions $\{3; 6\}$ and $\{4; 4\}$ of the plane into regular triangles and squares, respectively. Their skeletons are infinite graphs. Besides having a ring and a lattice structure, their vertex-sets are also metric spaces with the shortest path metrics called (in Computer Graphics and Robot Vision) the *Hexagonal distance* and *4-distance*. (The 4-distance is, in fact, l_1-metric on $\mathbb{Z}^2$.) For $k, l \geq 0$, this distance between 0 and $k + l\omega$ (or $k + li$) is $k + l$.

Thurston [Thur98] developed a global theory of parameter space for sphere triangulations with degree of vertices at most 6. Clearly, the duals of our plane graphs a_v are covered by it. Let s denote the number of vertices of degree less than 6; such vertices reflect positive curvature of the triangulation of the sphere $\mathbb{S}^2$. Thurston has built a parameter space with $s - 2$ degrees of freedom (complex numbers).

It suffices to give relative positions of these vertices. At most $s - 1$ vectors will do, since one position can be taken 0. But once $s - 1$ vertices are specified, the last one is constrained. Then the number of s-parametrized spheres with at most v triangles is $O(v^s)$ by direct integration. The number of such v-triangle spheres is $O(v^{s-1})$ if $s > 1$, by a *tauberian theorem*, i.e., one giving conditions for a series summable by some method to be summable in the usual sense. In fact, the number of triangles is expressed as a non-degenerate Hermitian form $q = q(z_1, \ldots, z_s)$ of signature $(1, s - 1)$. Let H_s be the cone $\{z \in \mathbb{C}^s : q(z) > 0\}$ and M be the discrete linear group preserving q; then the quotient $H_s / (\mathbb{R}_{>0} \times M)$ is of finite covolume.

If we restrict ourselves to some particular symmetries of plane graphs, then it restricts the number of parameters needed for a characterization. General fullerenes have 10 degrees of freedom, while those with symmetry I or I_h have just one degree of freedom. In [FoCrSt87] the fullerenes with symmetry D_5, D_6, T were described by two complex parameters, i.e., by 4 integer parameters.

We believe, that the hypothesis on degree of vertices (in dual terms, that the graph has no q-gonal faces with $q > 6$, i.e., negatively curved ones) in [Thur98] is unnecessary to his theory of parameter space. Also, we think that his theory can be extended to the case of quadrangulations instead of triangulations.

We focus mainly on the classes of plane graphs, which can be parametrized by *one* complex parameter. The GC-construction, defined below, fully describes them.

Remark 1 (i) A natural number $n = \prod_i p_i^{\alpha_i}$ admits a representation $n = k^2 + l^2$ or $n = k^2 + kl + l^2$ if and only if any α_i is even, whenever $p_i \equiv 3 \pmod 4$ (Fermat Theorem) or, respectively, $p_i \equiv 2 \pmod 3$ (see, for example, [CoGu96] and [Con03]).

(ii) One can have $t(k, l) = t(k', l')$ with corresponding complex numbers z, z' not being C-associated. First cases with $gcd(k, l) = gcd(k', l') = 1$ are $6^2 + 30 + 5^2 = 9^2 + 9 + 1^2$ and $8^2 + 1^2 = 7^2 + 4^2$. Also, $7^2 = 5^2 + 15 + 3^2$ and $5^2 = 4^2 + 3^2$.

6.2 The GC-Construction for 3- and 4-Regular Graphs

First, consider the 3-regular plane graphs. The dual graph G_0^* of every such graph G_0 is a *Triangulation*, i.e., a plane graph whose faces are triangles. The GC-construction with integer parameters k and l consists of subdividing every triangle of this triangulation into another set of faces according to Fig. 6.2, which is defined by k and l. The obtained faces, if they are not triangles, can be glued together coherently with

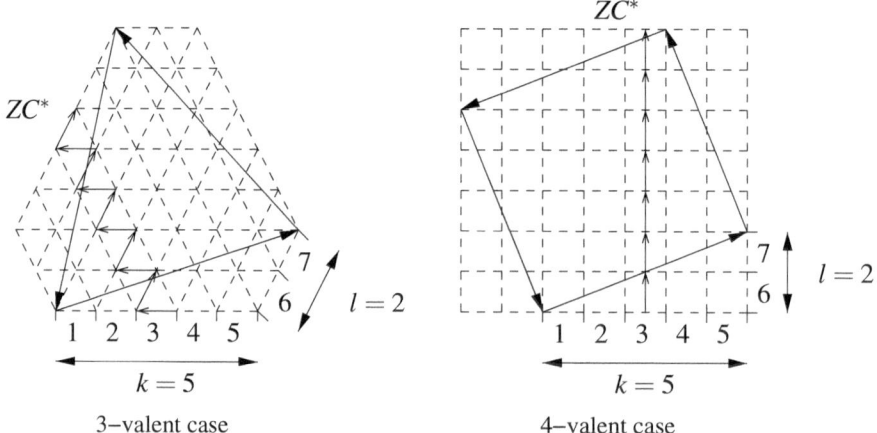

Fig. 6.2 The master polygon and an oriented ZC-circuit for parameters $k = 5, l = 2$

other non-triangular faces (coming from the subdivision of neighboring triangles), in order to form triangles; so, we end up with a new triangulation.

The triangle of Fig. 6.2 has area $\mathscr{A}(k^2+kl+l^2)$ if $\mathscr{A}$ is the area of a small triangle. By transforming every triangle of the initial triangulation in such way and gluing them, one obtains another triangulation, which we identify with a (dual) 3-regular plane graph and denote by $GC_{k,l}(G_0)$. The number of vertices of $GC_{k,l}(G_0)$ (if the initial graph G_0 has v vertices) is $vt(k,l)$ with $t(k,l) = k^2 + kl + l^2$.

For a 4-regular plane graph G_0, the duality operation transforms it into a quadrangulation and this initial quadrangulation is subdivided according to Fig. 6.2, which is also defined by two integer parameters k, l. After merging the obtained non-square faces, one gets another quadrangulation and the duality operation yields graph $GC_{k,l}(G_0)$ having $vt(k,l)$ vertices with $t(k,l) = k^2 + l^2$.

In both, 3- or 4-regular case, the faces of G_0 correspond to some faces of $GC_{k,l}(G_0)$; see Figs. 6.6 and 7.5. If $t(k,l) > 1$, then those faces are not adjacent. See Fig. 6.7 for some example of $GC_{1,1}$ and $GC_{2,1}$ operations.

We illustrate one example in more details in order to show how the construction works. We compute $GC_{3,2}(Octahedron)$ by first taking the dual of Octahedron as in Fig. 6.3a. The resulting squares need to be paved in another way and then put together in Fig. 6.3b. Finally, we take the dual and get the sought graph. In Fig. 6.3c we also show the underlying Cube.

The family $GC_{k,l}(Dodecahedron)$ consists of all 5_v with symmetry I_h or I (see [Gold37, Cox71] and Theorem 6.2). There is large body of literature, where such *Icosahedral fullerenes* appear as Fuller-inspired *Geodesic domes* (in Architecture) and *Virus capsides* (protein coats of virions, cf. [CaKl62]); see, for a survey, [Cox71] and [DDG98]. The GC-construction is also used in numerical analysis for obtaining good triangulations of the sphere; see, for example, [Slo99, ScSw95].

Remind (see Sect. 1.4) that a q-gonal face of a 3- or 4-regular graph is called *of positive, zero, negative curvature* if $q < 6, q = 0, q > 6$ (or $q < 4, q = 4, q > 4$), according to the following Euler formulas for 3- or 4-regular plane graphs:

$$\sum_{i \geq 1}(6-i)p_i = 12 \quad \text{or} \quad \sum_{i \geq 1}(4-i)p_i = 8, \quad \text{respectively.}$$

Proposition 6.1 *Let G_0 be a 3- or 4-regular plane graph and denote the graph $GC_{k,l}(G_0)$ also by $GC_z(G_0)$, where $z = k + l\omega$ or $z = k + li$ if G_0 ia a 3- or 4-regular graph, respectively. The following hold:*

(i) $GC_z(GC_{z'}(G_0)) = GC_{zz'}(G_0)$.

(ii) *If z and z' are two associated Eisenstein or Gaussian integers, then $GC_z(G_0) = GC_{z'}(G_0)$.*

(iii) $GC_{\bar{z}}(G_0) = GC_z(\overline{G_0})$, *where $\overline{G_0}$ denotes the plane graph, which differ from G_0 only by a plane symmetry; if $\overline{G_0} = G_0$ (i.e., $Rot(G_0) \neq Aut(G_0)$) and z, z' are two C-associated Eisenstein or Gaussian integers, then $GC_z(G_0) = GC_{z'}(G_0)$.*

(iv) *If G_0 has no flat faces and if $GC_{k,l}(G_0) = GC_{k',l'}(G_0)$ with $0 \leq l \leq k$ and $0 \leq l' \leq k'$, then $(k,l) = (k',l')$.*

(a)   **(b)**

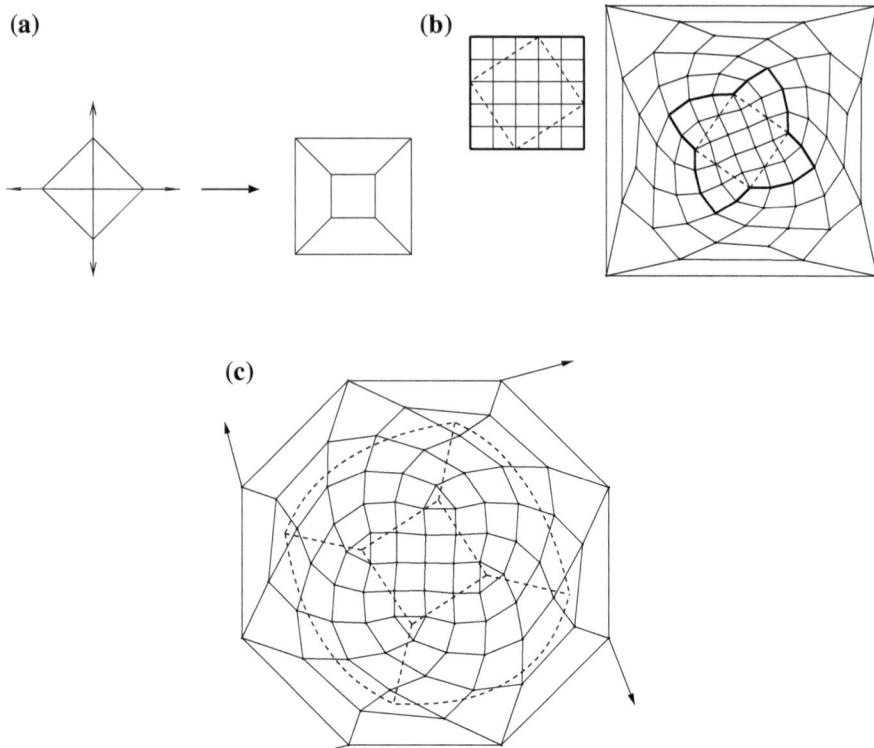

(c)

Fig. 6.3 The three steps of constructing $GC_{3,2}(Octahedron)$. **a** Taking the dual. **b** Transforming the squares. **c** Taking the dual again

Proof (i) follows from the basic construction depicted in Fig. 6.2, which is extended globally. (ii) also follows from it. Let G_0 be a 3-regular plane graph with $GC_{k,l}(G_0) = GC_{k',l'}(G_0)$. The equality for the numbers of vertices implies $t(k, l) = t(k', l')$. The minimum distance between two non-flat faces in $GC_{k,l}(G_0)$ is $k + l$; hence, $k + l = k' + l'$. If one writes $s = k' - k$, then $k' = k + s$ and $l' = l - s$. The equality $t(k, l) = t(k', l')$ yields $s(k - l) + s^2 = 0$ and so, (k', l') is (k, l) or (l, k). Only first case is possible. The 4-regular case can be treated similarly.   □

The above Proposition implies that we can consider only the case $0 \leq l \leq k$ in computations, since all considered graphs have a symmetry plane.

If $l = 0$, then $GC_{k,l}(G_0)$ is called *k-inflation* of G_0; see *t*-inflation in Sect. 4.1. See Fig. 6.4 for the *k*-inflation of cube and octahedron. For $k = 2$, $l = 0$, it is called *Chamfering* (or *Quadrupling*) of G_0, because Goldberg in a foundational paper [Gold34] called the fullerene $C_{80}(I_h) = GC_{2,0}(\text{Dodecahedron})$ *Chamfered Dodecahedron*; see Figs. 6.5 and 6.6. Another case, interesting for Chemistry, is *Capra*, i.e., $GC_{2,1}(G_0)$ (see [Diu03]). All symmetries are preserved if $l = 0$ or $l = k$, while only rotational symmetries are preserved if $0 < l < k$.

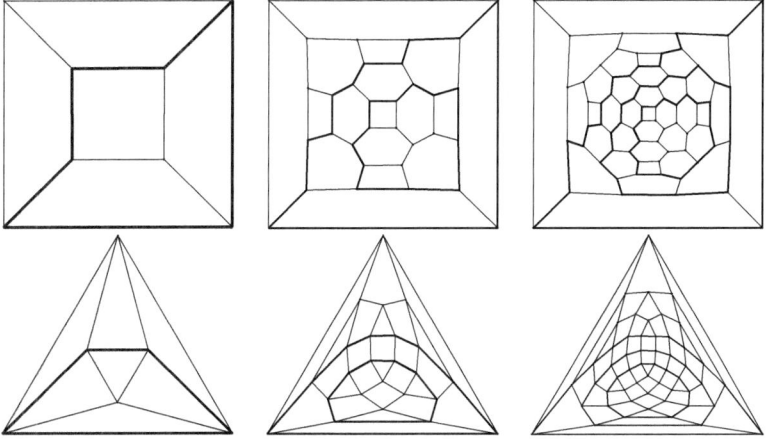

Fig. 6.4 The k-inflation $GC_{k,0}$ for $k = 1, 2, 3$ and G_0 being cube or octahedron

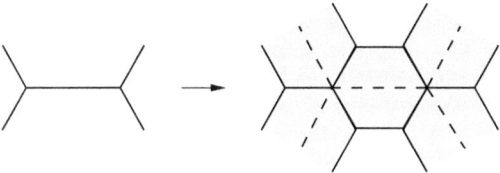

Fig. 6.5 Chamfering seen locally; see also [Fran15]

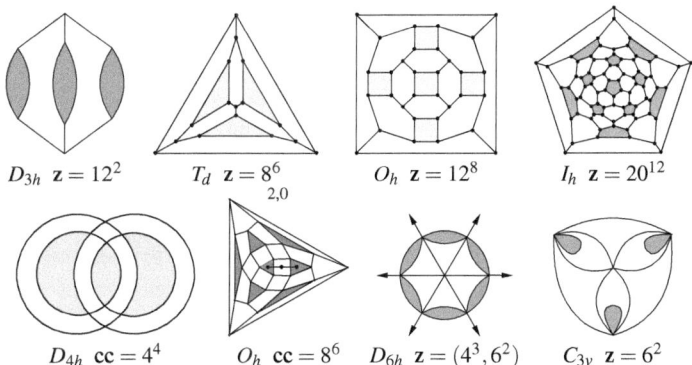

Fig. 6.6 Chamfering $GC_{2,0}(G_0)$ for G_0 being Bundle$_3$, Tetrahedron, Cube, Dodecahedron, Bundle$_4$, Octahedron, Bundle$_6$ and Trifolium

The condition $k \equiv l \pmod 3$ means, that the Eisenstein integer $k + l\omega$ is factorizable by $1 + \omega$, i.e., by the complex number corresponding to the leapfrog operation, $GC_{1,1}$. The condition $k \equiv l \pmod 2$ means, that the Gaussian integer $k + li$ is factorizable by $1 + i$, i.e., by the complex number corresponding to the medial operation, $GC_{1,1}$. Note that $k \equiv l \pmod 2$ is equivalent to $t(k, l) = (k - l)^2 + 2kl$ being even and $k \equiv l \pmod 3$ is equivalent to $t(k, l) = (k - l)^2 + 3kl$ being divisible by 3. It holds $GC_{2k^2,0}(G_0) = GC_{k,k}(GC_{k,k}(G_0))$ for a 4-regular graph G_0, since $(k + ki)^2 = 2k^2i$.

The Goldberg–Coxeter construction for 3- or 4-regular plane graphs can be seen, in algebraic terms, as the scalar multiplication by Eisenstein or Gaussian integers in the parameter space (see [Sah94]). More precisely, $GC_{k,l}$ corresponds to multiplication by complex number $k + l\omega$ or $k + li$ in the 3- or 4-regular case, respectively.

The Goldberg–Coxeter construction can be also defined, similarly, for maps on orientable surfaces. While the notions of medial, leapfrog, and k-inflation go over for non-orientable surfaces, the Goldberg–Coxeter construction is not defined on them.

6.3 Classes of Graphs

Theorem 6.1 (given implicitly in [GrZa74]) *Every graph 2_v is $GC_{k,l}(Bundle_3)$; its symmetry group is D_{3h} if $l = 0$, k and D_3, otherwise.*

[Gold37] proves (iv) of Theorem below, while (i), (ii) are only indicated there.

Theorem 6.2 *(i) Any graph 3_v with symmetry T or T_d is $GC_{k,l}(Tetrahedron)$,*
(ii) any graph 4_v with symmetry O or O_h is $GC_{k,l}(Cube)$,
(iii) any graph 4_v with symmetry D_6 or D_{6h} is $GC_{k,l}(Prism_6)$,
(iv) any graph 5_v with symmetry I or I_h is $GC_{k,l}(Dodecahedron)$,
(v) any 4-hedrite with symmetry D_4 or D_{4h} is $GC_{k,l}(Foil_2)$,
(vi) any 5-hedrite of symmetry D_3 or D_{3h} is $GC_{k,l}(Trefoil)$,
(vii) any 8-hedrite of symmetry O or O_h is $GC_{k,l}(Octahedron)$.

Proof Take a graph 3_v of symmetry T or T_d. Given a face F, the size of its orbit (under the action of the group T) is 4 if F lies on an axis of rotation of order 3, 6 if F lies on an axis of rotation of order 2, or 12 if F is in general position. This implies that all four triangles are on axis of order 3. Take a triangle, say, T_1; after adding p rings of hexagons, one finds a triangle and so, three triangles, say, T_2, T_3, and T_4. The position of triangle T_2 relatively to T_1 defines the Eisenstein integer, corresponding to this graph. One can see easily, that this graph is $GC_{k,l}(Tetrahedron)$.

Take a graph 4_v of symmetry O or O_h. One 4-fold symmetry axis goes through a square, say, sq_1. After adding p rings of hexagons around sq_1, one finds a square and so, by symmetry, four squares, say, sq_2, sq_3, sq_4, and sq_5. The position of the square sq_2 relatively to sq_1 defines an Eisenstein integer $z = k + l\omega$. The graph can be completed in an unique way and this proves, that it is $GC_{k,l}(Cube)$.

Take a graph 5_v of symmetry I or I_h. Any 5-fold axis must go though two pentagons. Since the group I contains six 5-fold axises, this means that every pentagon belongs to one 5-fold axis. Take a pentagon, say, P_1; after adding p rings of hexagons around P_1, one finds five pentagons, in cyclic order, say, P_2, P_3, P_4, P_5, P_6. The position of pentagon P_2 relatively to P_1 defines an Eisenstein integer $k + l\omega$, which is equal to the position of P_3 relatively to P_2 and to the position of P_1 relatively to P_3. The figure formed by P_1, P_2, P_3 is reproduced all over the graph, thanks to the six 5-fold axes. So, the Eisenstein integer defines entirely the graph.

Take a 4-hedrite G with symmetry D_4 or D_{4h}. The 4-fold axis must go through two vertices, say, v_1, v_2 or two 4-gonal faces, say, sq_1, sq_2. After adding p rings of squares around v_1 or sq_1, one finds a 2-gon and so, by symmetry, four 2-gons, say, $\Delta_i, 1 \le i \le 4$. The position of Δ_2, relatively to Δ_1, determines a Gaussian integer $k + li$, such that $G = GC_{k,l}(Foil_2)$.

In an 8-hedrite of symmetry O or O_h, any 3-fold axis go through two triangles. Since there are four 3-fold axes of symmetry, any triangle contains a 3-fold axis of symmetry. The proof is then similar to the case of 5_n with symmetry I or I_h.

The proofs of (iii) and (vi) are special cases of (i) and (ii) in Proposition 6.2. □

For other classes of graphs, the description requires several complex parameters. For them, it is not possible to obtain a description in terms of Goldberg–Coxeter construction of basic graphs, even of a finite number of such graphs.

Proposition 6.2 *(i) Let $\mathscr{GP}_m$ (for $m \ne 2, 4$) denote the class of 3-regular plane graphs with $p_m = 2$, $p_4 = m$ and p_6 6-gonal faces. Every such graph, having a m-fold axis, comes as $GC_{k,l}(Prism_m)$ and has symmetry group D_m or D_{mh}.*

(ii) Let $\mathscr{GF}_m$ (for $m \ne 2, 3$) denote the class of 4-regular plane graphs with $p_m = 2$, $p_2 = $ m-gons and p_4 4-gonal faces. Every such graph, having a m-fold axis, comes as $GC_{k,l}(Foil_m)$ and has symmetry group D_m or D_{mh}.

Proof In a graph $\mathscr{GP}_m$ with an m-fold axis, this axis goes through two m-gonal faces, say, F_1 and F_2. After adding p rings of 6-gons around F_1, one finds a square, say, sq_1 and so, by symmetry, m squares, say, $sq_1, \ldots, sq_m$. The position of sq_1 relatively to F_1 defines an Eisenstein integer $k + l\omega$, such that the graph is $GC_{k,l}(Prism_m)$.

In a graph $\mathscr{GF}_m$, the m-fold axis must go through the two m-gonal faces, say, F_1 and F_2. After adding p rings of squares around F_1, one finds a 2-gon and so, by symmetry, m 2-gons, say, $D_1, \ldots, D_m$. The position of D_1 relatively to F_1 defines a Gaussian integer $k + li$. Once the position of the digons D_i is found, the graph is uniquely determined and so, it is $GC_{k,l}(Foil_m)$. □

6.4 Triangulations of Oriented Maps

Here we consider such triangulations, having a specified set S of vertices of degree $\ne$ 6, with all other vertices being of degree 6. Such map can be described by specifying the relative positions between elements of S using Eisenstein integers.

In this section and followings, a $(\{v_1, \ldots, v_m\}, k)$-*map* denotes a map on an oriented surface with faces of size k, v_i vertices of degree i, i.e., dual of our previous maps. The map is called *sphere* (genus $g = 0$), *torus* ($g = 1$) or a general *oriented map of genus* g. For example, a $(\{v_5 = 2, v_7 = 2, v_6\}, 3)$-torus denotes a triangulation with 2 vertices of degree 5, 7 and an unspecified number of vertices of degree 6. We will be mostly concerned with the case $k = 3$; Euler relation for them reads as

$$\sum_{j \geq 3} v_j(6 - j) = 12(1 - g). \tag{6.1}$$

Formula (6.1) can be interpreted as a Gauss-Bonnet formula and $6 - j$ as the curvature of a vertex of degree j. A triangulation is said to be of *positive curvature* if all vertices have nonnegative curvature. This implies that the possible vertex-degrees belong to $\{3, 4, 5, 6\}$ and the v-vector satisfies $3v_3 + 2v_4 + v_5 = 12$ with v_6 unspecified. All 19 possibilities for (v_3, v_4, v_5) are given below (see Proposition 1.5):

$(0, 0, 12)$ $(0, 1, 10)$ $(0, 2, 8)$ $(0, 3, 6)$ $(0, 4, 4)$ $(0, 5, 2)$ $(0, 6, 0)$
$(1, 0, 9)$ $(1, 1, 7)$ $(1, 2, 5)$ $(1, 3, 3)$ $(1, 4, 1)$ $(2, 0, 6)$ $(2, 1, 4)$
$(2, 2, 2)$ $(2, 3, 0)$ $(3, 0, 3)$ $(3, 1, 1)$ $(4, 0, 0)$

The symmetry groups of fullerenes and other plane graphs of positive curvature were determined in [FoMa95, DeDu05, FoCr97, DDF09, DeDuSh03, DeDu12].

For a given group G of symmetry of a map, $Rot(G)$ denotes the subgroup of index 1 or 2 of G formed by the orientation preserving transformations. The *class of a group* G is the set of groups G' having $Rot(G') = G$. In Table 1.3, we give the possible groups of spheres of positive curvature by their class $Rot(G)$. For any class, the number of vertices of positive curvature is finite and the number of vertices of degree 6 is unspecified.

Since there is essentially only one 6-regular plane triangulation, the positions of the vertices of positive curvature allow to define the map. We want to encode the positions by complex Eisenstein numbers $z \in \mathbb{Z}[\omega]$ with $\omega = e^{i\pi/3}$. The case of 1 parameter corresponds to the Goldberg–Coxeter construction [DuDe04]. In Sect. 6.5, we describe the simple case of two parameters. In Sect. 6.6 we first explain the general theory of complex parameterization of $(\mathscr{V}, 3)$-maps on oriented surfaces. Then we explain Thurston's theory [Thur98] which gives stronger results for the case of spheres of positive curvature. Finally, we explain the extension to $(\mathscr{V}, 4)$-maps, self-dual spheres, and $(\mathscr{V}, 6)$-spheres.

Applications to zigzags are considered in Sect. 7.7. A very basic application of parameterization is for generating maps efficiently, provided that the number of parameters is not too large. Another application considered in [DuFo11] is for eigenvalue estimation, where it was proved that for any interval $[a, b] \subset [-3, 3]$ there is a finite number of graphs of positive curvature having no eigenvalue in I.

We emphasized Eisenstein parameter description but it is, of course, possible to consider descriptions by integral parameters. This is done in [Gra05, Gra04] for fullerenes 5_v and this allows to write parameterizations for each group, not just rotation subgroup. Another real parameter descriptions, by so-called *dihedral angles*, is developed in [Ri94], but it is more suited for describing manifolds than graphs.

Agregating the points groups as in Sect. 1.5, i.e., $\mathbf{T} = \{T, T_d, T_h\}$, $\mathbf{O} = \{O, O_h\}$, $\mathbf{I} = \{I, I_h\}$, $\mathbf{D_m} = \{D_m, D_{mh}, D_{md}\}$ and (for $m \geq 1$) $\mathbf{C_m} = \{C_m, C_{mv}, C_{mh}, S_{2m}\}$, the number of complex parameters by groups is as follows (see Table 1.3):

- $(\{5, 6\}, 3)$-spheres: $\mathbf{C_1}(10)$, $\mathbf{C_2}(6)$, $\mathbf{C_3}$ and $\mathbf{D_2}$ (4), $\mathbf{D_3}(3)$, $\mathbf{D_5}$, $\mathbf{D_6}$ and $\mathbf{T}$ (2), $\mathbf{I}$ (1);
- $(\{4, 6\}, 3)$-spheres: $\mathbf{C_1}$ (4), $\mathbf{C_2} \setminus \{S_4\}$ (3), $\mathbf{D_2}$ and $\mathbf{D_3}$ (2), $\mathbf{D_6} \setminus \{D_{6d}\}$ and $\mathbf{O}$ (1);
- $(\{2, 3\}, 6)$-spheres: $\mathbf{C_1}$ (4), $\mathbf{C_2}$ and $\mathbf{C_3}$ (3), $\mathbf{D_2}$ and $\mathbf{D_3}$ (2), $\mathbf{D_6} \setminus \{D_{6d}\}$ and $\mathbf{T}(1)$;
- $(\{3, 4\}, 4)$-spheres: $\mathbf{C_1}$ (6), $\mathbf{C_2}$ (4), $\mathbf{D_2}$ (3), $\mathbf{D_3}$ and $\mathbf{D_4}$ (2), $\mathbf{O}$ (1);
- $(\{2, 4\}, 4)$-spheres: $\mathbf{D_2}$ (2), $\mathbf{D_4} \setminus \{D_{4d}\}$ (1);
- $(\{3, 6\}, 3)$-spheres: $\mathbf{D_2}(2)$, $\mathbf{T} \setminus \{T_h\}$ (1);
- $(\{2, 6\}, 3)$-spheres: $\mathbf{D_3} \setminus \{D_{3d}\}$ (1);
- $(\{1, 3\}, 6)$-spheres: $\mathbf{C_3} \setminus \{S_6\}$ (1).

Thurston [Thur98] (see also [Sah94]) implies: $(\{a, b\}, k)$-spheres have $p_a - 2$ parameters and the number of v-vertex ones is $O(v^{m-1})$, where $m = p_a - 2 > 1$.

6.5 Two Parameters Constructions

We consider first how one can describe some maps with two parameters. Out first case is the $(\{v_5 = 12, v_6\}, 3)$-spheres of symmetry D_5.

The 5-fold axis of such a sphere has to pass through a vertex of degree 5. There are 5 vertices of degree 5 around it; so, by 5-fold symmetry, 1 complex parameter is needed to describe them. Around those 5 vertices, there are 5 more vertices; so, one more parameter is needed and then the last vertex is uniquely defined.

If one applies the following operations to parameters

- Operation 1: $(z_1, z_2) \mapsto (z_1, z_1 + z_2)$
- Operation 2: $(z_1, z_2) \mapsto (z_1 + \omega^2 z_2, z_1 - z_2)$
- Operation 3: $(z_1, z_2) \mapsto (z_1, z_2)\omega^r$

then one obtains the same sphere as a result (Fig. 6.8). The group generated by those operations is named *Monodromy group*. The number of triangles of S is expressed as $q(z_1, z_2) = 10\{z_1\overline{z_1} + \frac{z_1\overline{z_2} - \overline{z_1}z_2}{\omega - \overline{\omega}}\}$. So, for a given pair (z_1, z_2), a sphere may not exist, for example, if $q(z_1, z_2) < 0$ (Fig. 6.7).

Of course, what has been done for fullerenes of class D_5, applies just as well for fullerenes of class D_6 and similar simple description are possible for the remaining two parameter cases of Table 1.3; see Fig. 6.9 for two such cases.

For the $(\{v_3 = 4, v_6\}, 3)$-spheres a very explicit two parameter description is given in Fig. 6.10. Geometrically, this corresponds to the fact that any such sphere is obtained as the quotient of a $(\{v_6\}, 3)$-torus by a group of order 2 leaving invariant exactly 4 vertices. Clearly, the monodromy group is $PSL(2, \mathbb{Z})$ and the number of triangles is expressed as $\frac{4}{\omega - \overline{\omega}}(z_1\overline{z_2} - \overline{z_1}z_2)$. This description was used in [JoSa09] to compute the eigenvalues of dual $(\{v_3 = 4, v_6\}, 3)$-spheres.

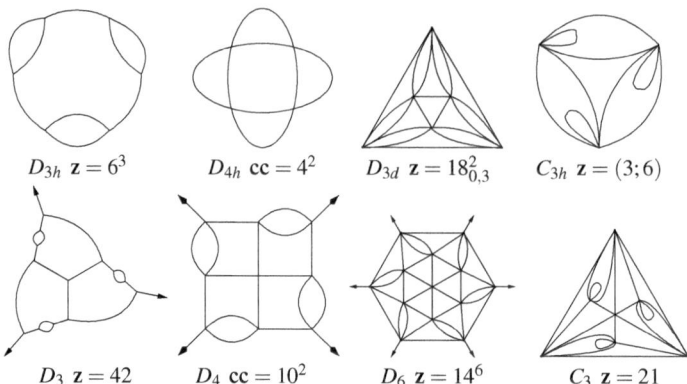

D_{3h} $\mathbf{z} = 6^3$ D_{4h} $\mathbf{cc} = 4^2$ D_{3d} $\mathbf{z} = 18^2_{0,3}$ C_{3h} $\mathbf{z} = (3; 6)$

D_3 $\mathbf{z} = 42$ D_4 $\mathbf{cc} = 10^2$ D_6 $\mathbf{z} = 14^6$ C_3 $\mathbf{z} = 21$

Fig. 6.7 The parameterization of fullerenes of symmetry D_5, D_{5d} or D_{5h} in term of $(z_1, z_2) \in \mathbb{Z}[\omega]^2$ and two parameter operations

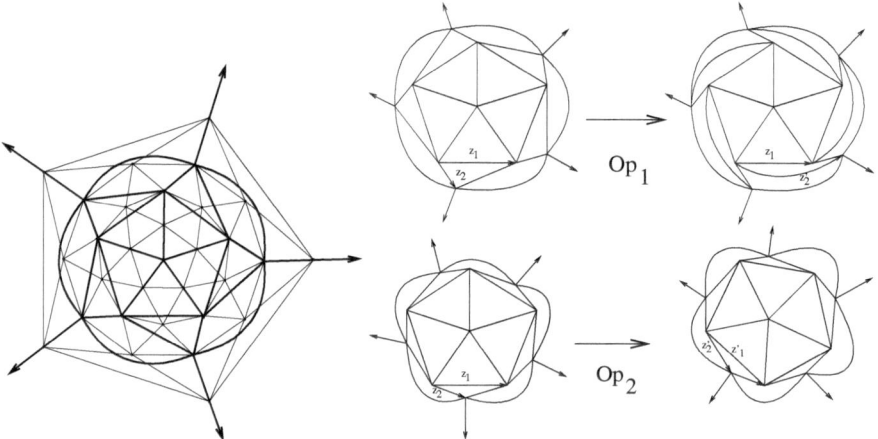

Fig. 6.8 $GC_{1,1}(G_0)$ and $GC_{2,1}(G_0)$ for G_0 being Bundle_3, Bundle_4, Bundle_6 and Trifolium

6.6 General Case of Parameterization of Maps on Oriented Surfaces

The general Eberhard problem is: for a given $g \geq 0$ and vector $(v_i)_{3 \leq i \leq m, i \neq 6}$ satisfying $\sum_{i=3}^{m}(6 - i)v_i = 12(1 - g)$, to determine the set $P(v, g)$ of values of v_6 for which there exist an oriented $(\{v, v_6\}, 3)$-map of genus g. It is proved in [Jen86], that $P(v, g)$ is empty only in the case $g = 1$ and $\mathbf{v} = \{v_5 = 1, v_7 = 1\}$ and the exact determination of $P(v, g)$ is an active subject of research.

Thus for a given v-vector, it is interesting to consider how one can parametrize the oriented $(\{\mathbf{v}, v_6\}, 3)$-maps of genus g. Let us denote by $\mathcal{M}$ such a map, $\widetilde{\mathcal{M}}$ its

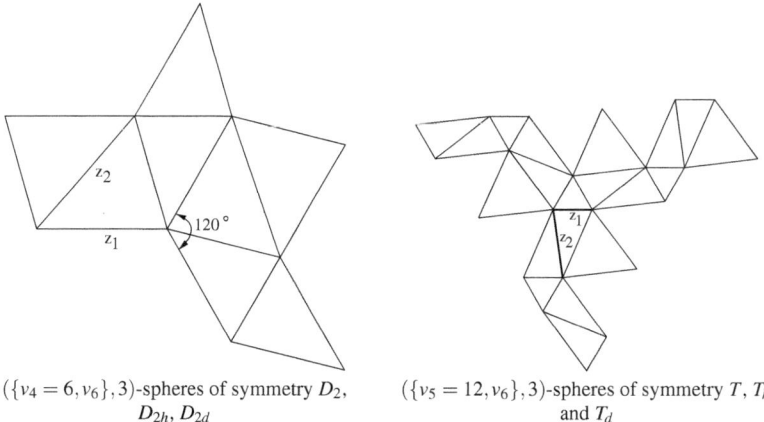

($\{v_4 = 6, v_6\}$, 3)-spheres of symmetry D_2,   ($\{v_5 = 12, v_6\}$, 3)-spheres of symmetry T, T_h
D_{2h}, D_{2d} and T_d

Fig. 6.9 Two parameters description of two classes of spheres with positive curvature

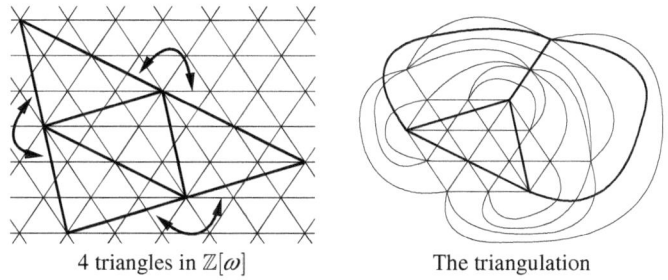

4 triangles in $\mathbb{Z}[\omega]$ The triangulation

Fig. 6.10 Description of ($\{v_3 = 4, v_6\}$, 3)-spheres

universal cover, Γ its fundamental group and S the set of vertices of degree different from 6. By adding edges one by one, we can build a triangulation T on $\mathcal{M}$ having S as vertex-set. Note that the vertex-degrees of $\mathbf{v}$ in T are a priori not related to the vertex-degrees of $\mathbf{v}$ in $\mathcal{M}$. Naturally, many triangulations are possible and they are mapped on the universal cover $\widetilde{\mathcal{M}}$ to Γ-invariant triangulations. We will see below that one can build a parameterization by Eisenstein integers from a triangulation.

Let us a take a triangulation T and encode it combinatorially. A *directed edge* is an edge from a vertex to another vertex. Every edge e is composed of a directed edge $\overrightarrow{e}$ and its *reversal* $r(\overrightarrow{e})$. The *Next operator* n maps a directed edge to the next one in clockwise order around the vertex V in which it is contained. A triangulation T is described by the operators n and r acting on the set of directed edges $DE(T)$. In particular the vertices, edges, and faces of T correspond to the orbits of n, r and nr. For a given vertex V of T, we denote by $\tilde{V}$ the corresponding vertex in the corresponding ($\{\mathbf{v}, v_6\}$, 3)-map. Edges and faces of T will have no direct analogs but homology classes will be mapped to homology classes.

We associate an Eisenstein integer $z_{\vec{e}} \in \mathbb{Z}[\omega]$ to any directed edge $\vec{e}$ of T. For any face $F = \{\vec{e_1}, \vec{e_2}, \vec{e_3}\}$, we impose the relation

$$z_{\vec{e_1}} + z_{\vec{e_2}} + z_{\vec{e_3}} = 0.$$

If we have an edge $e = \{\vec{e_1}, \vec{e_2}\}$ formed by two directed edges, then we impose the consistency relation

$$\omega^{\alpha(\vec{e_1})} z_{\vec{e_1}} + \omega^{\alpha(\vec{e_2})} z_{\vec{e_2}} = 0$$

with $\alpha(\vec{e})$ being a function defined on the directed edges with integer values.

For any vertex V containing the directed edges $(\vec{e_i})_{1 \leq i \leq m}$, we have the relation

$$6 - \deg(\tilde{V}) \equiv \sum_{i=1}^{m} \alpha(\vec{e_i}) - \alpha(r(\vec{e_i})) \pmod{6} \qquad (6.2)$$

where $\deg(\tilde{V})$ is the degree of $\tilde{V}$. Let $D(V) = \deg(\tilde{V})$. If the triangulation T is of genus g, then the first homology group $H_1(T)$ is $\mathbb{Z}^{2g}$. For any cycle C composed of directed edges $\{\vec{e_1}, \ldots, \vec{e_m}\}$ with $\vec{e}_{i+1} = r(nr)^{\pm 1}(\vec{e_i})$, we define the cycle sum

$$I(C, \alpha) = \sum_{j} \alpha(\vec{e_j}) - \alpha(r(\vec{e_j})). \qquad (6.3)$$

This sum depends on the chosen element of the homology class and defines how the orientation is shifted after one moves along C. If one adds 1 to $\alpha(\vec{e})$ and $\alpha(r(\vec{e}))$, then the resulting map $M(T, D, I, \alpha, z)$ does not change. The same happens if one adds 1 to $\alpha(\vec{e})$ for $\vec{e}$ in a face F of the triangulation. In fact, if we choose $2g$ basic cycles C_i of T, then any two functions α, α' satisfying Eq. (6.2) and $I(C, \alpha) \equiv I(C, \alpha') \pmod{6}$ differ by repeated application of above two operations. Eq. (6.2) and the cycle sums I allow to find a corresponding function α if it exists. So, the data of T, D, and I determine the class of maps that one can obtain. In Fig. 6.11 we give an example of a parameterization for $(\{v_3 = 1, v_9 = 1, v_6\}, 3)$-torus.

Theorem 6.3 *The parameter spaces of oriented* $(\{v_1, \ldots, v_m\}, 3)$-*maps of genus* g *have dimension* $\sum_{3 \leq i \neq 6} v_i - 1 + 2g$, *if all faces have gonality divisible by 6, and* $\sum_{3 \leq i \neq 6} v_i - 2 + 2g$, *otherwise.*

Proof Let us take such a map and build a triangulation T on it. Let us write $M = \sum_{3 \leq i \neq 6} v_i$. We then construct a spanning tree of $M - 1$ edges on the set of vertices of degree different from 6. Since the map is of genus g, we have a basis of $2g$ cycles of the $H_1(G)$. We add $2g$ edges to the spanning tree and the remaining edges define a tree in the dual map. Once the position of the $M - 1 + 2g$ edges is defined, we have defined the triangulation uniquely because all other edges can be assigned iteratively. If one of the vertices has a degree not divisible by 6, then its position is defined uniquely once all its neighbors are known and so the dimension decreases by

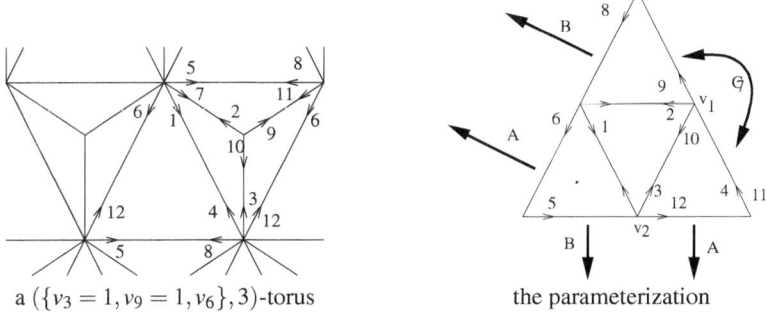

a ($\{v_3 = 1, v_9 = 1, v_6\}$, 3)-torus   the parameterization

Fig. 6.11 A ($\{v_3 = 1, v_9 = 1, v_6\}$, 3)-torus and its parameterization. The identifications along A, B and C yields the equalities $\omega z_{12} + z_6 = 0$, $\omega z_5 + z_8 = 0$ and $z_9 - z_{11} = 0$. We have $\deg(\tilde{v}_1) = 3$ and $\deg(\tilde{v}_2) = 9$

1 in that case. No other relation exists, since one can perturb the remaining parameter and still obtain some corresponding maps. □

Let us denote by $m(v, g)$ the *dimension* in the above theorem.

For a given parameterization (T, D, I), we denote by $q_T(z)$ the number of triangles of the obtained triangulation. The number of triangles contained in a face defined by $f = \{\overrightarrow{e_1}, \overrightarrow{e_2}, \overrightarrow{e_3}\}$ is

$$q_f(z) = \frac{1}{\omega - \overline{\omega}}(z_{\overrightarrow{e_1}}\overline{z_{\overrightarrow{e_2}}} - \overline{z_{\overrightarrow{e_1}}}z_{\overrightarrow{e_2}}).$$

The function q_T counting the number of triangles is thus $q_T = \sum q_f$ and it is an Hermitian form. From this, one can deduce that for a fixed $\mathbf{v}$, the number of oriented ($\{\mathbf{v}, v_6\}$, 3)-maps of genus g with at most n triangles grows like $O(n^{m(v,g)})$. Note that [Sah94] gives more precise estimate $O(n^9)$ for the number of dual fullerenes with exactly n triangles. But we cannot have such a statement in the general case, because, for example, ($\{v_3 = 4, v_6\}$, 3)-spheres exist only if n is divisible by 4.

Conjecture 6.1 If all vertices have degree divisible by 6, then the signature of q_T is

$$(n_+, n_-, n_0) = \left(g + \sum_i v_i Fr\left(\frac{6-i}{6}\right), g, m(v, g) - n_+ - n_-\right),$$

otherwise, denoting again $x - \lfloor x \rfloor$ by $Fr(x)$, the signature is

$$(n_+, n_-, n_0) = \left(g - 1 + \sum_i v_i Fr\left(\frac{6-i}{6}\right), m(v, g) - n_+, 0\right).$$

This conjecture is proved in the case $g = 0$ in [Mos87] and has been checked for the regular maps of genus at most 15 from [CoDo01].

For a given triangulation T of an oriented $(\{v, v_6\}, 3)$-map, we choose m independent parameters $z_1, \ldots, z_m \in \mathbb{Z}[\omega]$. The condition of existence of the map $M(T, D, I, \alpha, z)$ is that $q_f(z) > 0$ for all faces f of T and this defines the *Realizability space*. The *limit realizability space* is the same space with $\mathbb{Z}[\omega]$ replaced by $\mathbb{C}$; it defines a set S in the cone $q_T > 0$. Approaching through a generic point of the boundary of S, which is not in the boundary of $q_T > 0$, one can rearrange the triangulation and get another triangulation T'. The degree of vertex V in T may change, but the degree of $\tilde{V}$ remain the same. Similarly, the cycles C are mapped to cycles C' and $I(C', \alpha') = I(C, \alpha)$. So, D and I are intrinsic to the class of triangulations obtained by rearrangements. Moreover, the parameter set (z'_i) of (T', D, I, α') can be expressed linearly in term of the parameter set (z_i) of (T, D, I, α).

For two quadruples (T, D, I, α) and (T', D, I, α') of parameter sets (z_i), (z'_i), we say that they are *equivalent* if there is a map from T to T' preserving D and I such that (z'_i) can be expressed linearly from (z_i). Such an equivalence preserves edge lengths, triangle areas, and can be extended to adjacent triangulations of T and, correspondingly, T'. Note that some nontrivial equivalence can have $T = T'$; this is the case for the two parameter description considered in Sect. 6.5, for which one triangulation suffices. The group of such transformation is the *monodromy group* and is a subgroup of $GL_m(\mathbb{Z}[\omega])$ leaving invariant the form q_T. It may be that any two triples (T, D, I), (T', D, I) are related by a sequence of such transformations but we have no proof of it and we do not know a counter-example.

6.7 Thurston's Theory for Maps of Positive Curvature

Fix a triple (v_3, v_4, v_5) among the 19 triples of possible curvature. By Theorem 6.3 the number of complex parameters needed to describe $(\{v_3, v_4, v_5, v_6\}, 3)$-spheres is

$$m = m(\{v_3, v_4, v_5\}, 0) = v_3 + v_4 + v_5 - 2$$

The monodromy group is denoted by $M(\{v_3, v_4, v_5\}, 3, 0)$ and it preserves the form q, which is of signature $(1, m - 1, 0)$. This class of monodromy groups was defined and enumerated in [DeMo86, Mos86, Thur98] and, in particular, they are discrete groups. The 19 possible (v_3, v_4, v_5) cases are part of the 94 cases determined there and the form q is the intersection form on $H^1(\mathbb{S}^2 - V, L)$ with V being a set of $m + 2$ points and L a line bundle on $\mathbb{S}^2 - V$.

In [Thur98], it is proved that if $z \in \mathbb{Z}[\omega]^m$ and $q(z) > 0$, then there exists $f \in M(\{v_3, v_4, v_5\}, 3, 0)$ such that $f(z)$ is realizable as a $(\{v_3, v_4, v_5, v_6\}, 3)$-sphere. Thus $\mathbb{H}^m \cap \mathbb{Z}[\omega]^m$ up to the action of the monodromy group is a parameter space for the $(\{v_3, v_4, v_5, v_6\}, 3)$-spheres. As a consequence, the quotient

$$\mathbb{H}^m / (\mathbb{R}_{>0} \times M_3(\{v_3, v_4, v_5\}, 3, 0))$$

is of finite covolume because the number of ($\{v_3, v_4, v_5, v_6\}$, 3)-spheres is finite for any fixed number of triangles.

A manifold can have a compact quotient $\tilde{X}$ in which case it is *cocompact*; otherwise, it has *ends* [HuRa96], which are, in a sense, directions of "non-compactness." In [Thur98, p. 533], a characterization of the ends of the quotients is given, which gives a fortiori a characterization of the cocompact ones.

In our case, an end corresponds to a partition of the vertices of nonzero curvature into two sets S_i for $i = 1, 2$ each having v_q^i vertices of degree q and satisfying to $3v_3^i + 2v_4^i + v_5^i = 6$. It is easy to find such a partition for each of the 19 possible triples (v_3, v_4, v_5). For example, in the fullerene case, we can put 6 pentagons on one part, 6 on the other and there are $\binom{12}{6} = 924$ possibilities. As a consequence, none of those corresponding to ($\{v_3, v_4, v_5, v_6\}$, 3)-spheres are compact. Geometrically, those are dual nanotubes (see Sect. 2.4), i.e., we have two caps C_i with (v_3^i, v_4^i, v_5^i) vertices separated by a number of rings of vertices of degree 6; see Fig. 6.12.

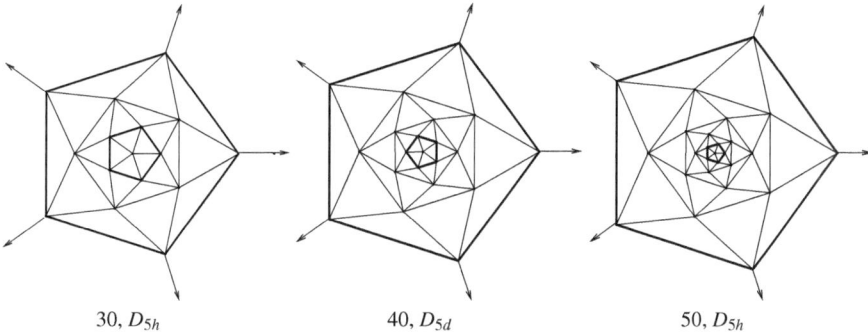

30, D_{5h} 40, D_{5d} 50, D_{5h}

Fig. 6.12 3 dual fullerenes of symmetry D_5 or more; the 1st is the dual of the 1st in Fig. 2.3

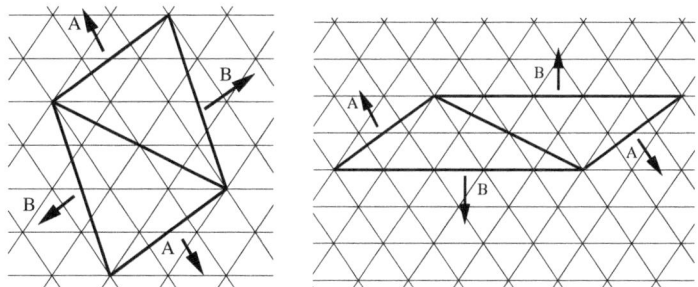

Fig. 6.13 Two representations of a ($\{6\}$, 3)-torus

6.8 Extensions and Other Cases of Parameter Descriptions

A relatively simple extension is to consider vertices of degree 1 or 2. For example, the ($\{v_2 = 3, v_6\}$, 3)-spheres are obtained by applying the Goldberg–Coxeter construction to the sphere reduced to a cycle of length 3 [GrZa74].

One example, that does not fit exactly into the above scheme, is the description of ($\{v_6\}$, 3)-torus by two parameters. It is done by writing down a parallelogram on the plane and identifying the sides. Since all vertices are equivalent in the torus, we have to choose vertices of degree 6 used in the construction. In Fig. 6.13 we show two equivalent representation of a ($\{v_6\}$, 3)-torus. One of them has one horizontal line in the fundamental domain. It is always possible to have such a representation and this is the basis of a 3 integral parameters construction [Ne85].

GC-construction and complex parameterizations for ($\{v_1, v_2, v_3\}$, 6)-spheres are derived in [DeDu12]; see Sect. 5.3.

The theory extends easily for quadrangulations. We are still dealing with triples (T, D, I), but the ring of Eisenstein integers is replaced by the ring of Gaussian integers. For example, for ($\{v_3 = 8, v_4\}$, 4)-spheres, the number of Gaussian integer parameters is 6 and the number for the classes (O, O_h), (D_4, D_{4d}, D_{4h}), (D_3, D_{3d}, D_{3h}), (D_2, D_{2d}, D_{2h}), $(C_2, C_{2h}, C_{2v}, S_4)$, (C_1, C_s, C_i), respectively is 1, 2, 2, 3, 4, and 6. As a byproduct of this parameterization, we also get a method for parametrizing self-dual plane graphs with faces of size 3 or 4, see [DuDe11] for details.

Finally, note that all of the above can be adapted in order to get parameter description of families of maps having a specific symmetry group G, provided that G contains only orientation preserving maps. This is because the symmetry conditions are translated into linear equalities in the parameters.

References

[CaKl62] Caspar, D., Klug, A.: Physical principles in the construction of regular viruses. Cold Spring Harb. Symp. Quant. Biol. **27**, 1–24 (1962)

[Cox71] Coxeter, H.S.M.: Virus macromolecules and geodesic domes. In: Butcher, J.C. (ed.) A Spectrum of Mathematics, pp. 98–107. Oxford University Press/Auckland University Press,Oxford, U.K./Auckland New-Zealand (1971)

[CoGu96] Conway, J., Guy, R.: The Book of Numbers. Copernicus, New York (1996)

[Con03] Conway, J.: Personal Communication (2003)

[CoDo01] Conder, M., Dobcsányi, P.: Determination of all regular maps of small genus. Combin. Theory Ser. B **81**, 224–242 (2001)

[DDG98] Deza, A., Deza, M., Grishukhin, V.P.: Embeddings of fullerenes and coordination polyhedra into half-cubes. Discrete Math. **192**, 41–80 (1998)

[Fran15] del Río-Francos, M.: Chamfering operation on k-orbit maps. Ars Math. Contemp. **8**, 507–524 (2015)

[DeDu12] Deza, M., Dutour Sikirić, M.: Zigzag and central circuit structure of $(1, 2, 3, 6)$ - spheres. Taiwanese J. Math. **16–3**, 913–940 (2012)

[DeMo86] Deligne, P., Mostow, G.D.: Monodromy of hypergeometric functions and nonlattice integral monodromy. Publ. Math. Inst. Hautes Études Sci. **63**, 5–89 (1986)

[DeDu05]   Deza, M., Dutour, M.: Zigzag structure of simple two-faced polyhedra. Comb. Probab. Comput. **14**, 31–57 (2005)

[DDF09]   Deza, M., Dutour Sikirić, M., Fowler, P.W.: The symmetries of cubic polyhedral graphs with face size no larger than 6, MATCH-Commun. Math. Co. **61**, 589–602 (2009)

[DeDuSh03]   Deza, M., Dutour, M., Shtogrin, M.: 4-valent plane graphs with 2-, 3- and 4 -gonal faces. In: Advances in Algebra and Related Topics (in memory of B.H. Neumann; Proceedings of ICM Satellite Conference on Algebra and Combinatorics, Hong Kong 2002), World Scientific Publishing Co, pp. 73–97 (2003)

[Diu03]   Diudea, M.: Capra - a leapfrog related map transformation, Studia Universitatis Babes-Bolyai, Chemia, XLVIII-2 (2003)

[DuDe11]   Dutour Sikirić, M., Deza, M.: 4 -regular and self-dual analogs of fullerenes, mathematics and topology of fullerenes, pp. 103–116. In: Ori, O., Graovac, A., Cataldo, F. (eds.) CarbonMaterials, Chemistry and Physics, vol. 4, Springer Verlag (2011)

[DuDe04]   Dutour, M., Deza, M.: Goldberg-Coxeter construction for 3- or 4-valent plane graphs. Electron. J. Comb. **11–1**, R20 (2004)

[DuFo11]   Dutour Sikirić, M., Fowler, P.W.: Cubic ramapolyhedra with face size no larger than 6. J. Math. Chem. **49**, 843–858 (2011)

[FoMa95]   Fowler, P.W., Manolopoulos, D.E.: An Atlas of Fullerenes. Clarendon Press, Oxford (1995)

[FoCrSt87]   Fowler, P.W., Cremona, J.E., Steer, J.I.: Systematics of bonding in non-icosahedral carbon clusters. Theor. Chim. Acta **73**, 1–26 (1987)

[FoCr97]   Fowler, P.W., Cremona, J.E.: Fullerenes containing fused triples of pentagonal rings. J. Chem. Soc., Faraday Trans. **93**(13), 2255–2262 (1997)

[Gold37]   Goldberg, M.: A class of multisymmetric polyhedra. Tohoku Math. J. **43**, 104–108 (1937)

[Gold34]   Goldberg, M.: The isoperimetric problem for polyhedra. Tohoku Math. J. **40**, 226–236 (1934)

[GrZa74]   Grünbaum, B., Zaks, J.: The existence of certain planar maps. Discrete Math. **10**, 93–115 (1974)

[Gra05]   Graver, J.E.: Catalog of all fullerenes with ten or more symmetries, graphs and discovery, 167–188 DIMACS Series Discrete Mathematics Theoretical Computer Science, 69, American Mathematical Society (2005)

[Gra04]   Graver, J.E.: Encoding fullerenes and geodesic domes. SIAM J. Discrete Math. **17**, 596–614 (2004)

[HaWr96]   Hardy, G.H., Wright, E.M.: An Introduction to the Theory of Numbers, 4th edn. Clarendon Press, Oxford (1996)

[HuRa96]   Hughes, B., Ranicki, A.: Ends of complexes, cambridge tracts in mathematics, 123, pp. xxvi+353. Cambridge University Press, Cambridge (1996)

[Jen86]   Jendrol, S.: On face vectors of trivalent maps. Math. Slovaca **36**, 367–386 (1986)

[JoSa09]   John, P.E., Sachs, H.: Spectra of toroidal graphs. Discrete Math. **309**, 2663–2681 (2009)

[Mos87]   Mostow, G.D.: Braids, hypergeometric functions, and lattices. Bull. Am. Math. Soc. **16**, 225–246 (1987)

[Mos86]   Mostow, G.D.: Generalized Picard lattices arising from half-integral conditions. Publ. Math. Inst. Hautes Études Sci. **63**, 91–106 (1986)

[Ne85]   Negami, S.: Uniqueness and faithfulness of embedding of graphs into surfaces, Doctor thesis, Tokyo Institute of Technology (1985)

[Ri94]   Rivin, I.: Euclidean structures on simplicial surfaces and hyperbolic volume. Ann. Math. **139**, 553–580 (1994)

[Sah94]   Sah, C.H.: A generalized leapfrog for fullerene structures. Fullerene Sci. Technol. **2–4**, 445–458 (1994)

[ScSw95]   Schröder, P., Sweldens, W.: Spherical wavelets: efficiently representing functions on the sphere. In: Proceedings of ACM Siggraph, pp. 161–172 (1995)

[Slo99] Sloane, N.J.A.: Packing planes in four dimensions and other mysteries, algebraic com-
         binatorics and related topics (Yamagata 1997). In: Bannai, E., Harada, M., Ozeki, M.
         (eds.) Yamagata University, Faculty of Science, Department of Mathematics (1999)
[Thur98] Thurston, W.P.: Shapes of polyhedra and triangulations of the sphere. In: Rivin,
         J., Rourke, C., Series, C. (eds.) Geometry and Topology Monographs 1, The Ep-
         stein Birthday Schrift, pp. 511–549. Geometry and Topology Publications, Coventry
         (1998)

Chapter 7
ZC-Circuits of Goldberg–Coxeter Construction

In this chapter, ZC-(zigzag or central circuit) structure will be mainly described using groups. But first, Propositions 7.1 and 7.2 below treat the easiest case: ZC-structure of the k-inflation $GC_{k,0}(G_0)$ of G_0 in terms of ZC-structure of G_0; see example in Fig. 7.5.

We study the ZC-structure of the resulting graph $GC_{k,l}(G_0)$ using the algebraic formalism of the *moving group*, the (k, l)-*product* and a finite index subgroup of $SL_2(\mathbb{Z})$, whose elements preserve the above structure. We also study the intersection pattern of ZC-circuits of $GC_{k,l}(G_0)$ and consider its *projections*, obtained by removing all but one zigzags (or central circuits).

As initial graph G_0 for the Goldberg–Coxeter construction, we consider mainly:

(i) 3- and 4-regular skeletons of platonic and semiregular polyhedra (Table 7.1),
(ii) fullerenes and other chemistry-relevant 3-regular graphs,
(iii) 4-regular plane graphs, which are minimal projections for some interesting alternating links; those links are denoted according to Rolfsen's notation [Rol76].

The column "r if I" shows the number (conjectured or proved) r of ZC-circuits in the case $k \equiv l \pmod 3$ (for degree 3) or (for degree 4) $k \equiv l \pmod 2$, while the column "r if II" shows the remaining case.

7.1 Directed Edge Formalism

Given a graph G, denote by $Mov(G)$ the permutation group on the set of *directed edges*, which is generated by two *basic permutations*, called *left L* and *right R*; $Mov(G)$ is called the *moving group* of G. Directed edges are edges of G_0^* with prescribed direction. We will associate to every pair (k, l) of integers an element of this moving group, which we call (k, l)-*product* of basic permutations, and which encodes the lengths of the ZC-circuits of $GC_{k,l}(G_0)$. For $k = l = 1$, this (k, l)-product is, actually, an ordinary product in the group $Mov(G_0)$. Take a ZC-circuit of

© Springer India 2015

M. Deza et al., *Geometric Structure of Chemistry-Relevant Graphs*,
Forum for Interdisciplinary Mathematics 1, DOI 10.1007/978-81-322-2449-5_7

Table 7.1 GC-construction from 3- or 4-regular platonic and semiregular polyhedra; cf. Table 1.1

Graph G_0	$|Mov(G_0)|$	Reference
Tetrahedron	4	Theorem 7.10
Cube	12	Theorems 7.10 and 7.11, Proposition 7.8 and Conjecture 7.10, Table 7.4
Dodecahedron	60	Theorem 7.10, Proposition 7.8 and Conjecture 7.10, Table 7.4
Octahedron	24	Theorems 7.10 and 7.11 and Proposition 7.8, Table 7.4
Cuboctahedron	576	$= GC_{1,1}(Octahedron)$
Icosidodecahedron	7,200	Conjecture 7.4
Trunc. Tetrahedron	12	$= GC_{1,1}(Tetrahedron)$
Trunc. Octahedron	576	$= GC_{1,1}(Cube)$
Trunc. Cube	20,736	Theorem 7.11
Trunc. Icosahedron	648,000	$= GC_{1,1}(Dodecahedron)$
Trunc. Dodecahedron	648,000	Theorem 7.11
Rhombicuboctahedron	165,888	$= GC_{2,0}(Octahedron)$
Rhombicosidodecahedron	51,840,000	$= GC_{1,1}(Icosidodecahedron)$
Trunc. Cuboctahedron	1,327,104	Theorem 7.11
Trunc. Icosidodecahedron	139,968,000,000	Conjecture 7.4
$Prism_m$	$12(\frac{m}{gcd(m,4)})^3$	Conjecture 7.6
$APrism_m$	$\frac{24}{gcd(m,2)}(\frac{m}{gcd(m,3)})^3$	Conjecture 7.7

$GC_{k,l}(G_0)$ and fix an orientation on it. It will cross some edges of G_0^*. For any directed edge $\overrightarrow{e}$ of oriented ZC-circuit, there are exactly two possible successors $L(\overrightarrow{e})$ and $R(\overrightarrow{e})$; it is clear for zigzags in 3-regular graph G_0, but for central circuits in 4-regular one, it will be obtained from algebraic considerations. The $k + l$ successive left and right choices will define the (k, l)-product. In some cases, the knowledge of normal subgroups of $Mov(G_0)$ will allow an exact computation of the z-vector of $GC_{k,l}(G_0)$ in terms of congruences valid for numbers (k, l). On the other hand, Theorem 7.4 gives a characterization of the graphs G for which $Mov(G)$ is an Abelian group.

7.2 *ZC*-Circuits of Inflation Graphs

Proposition 7.1 *Let G_0 be a 3-regular plane graph with zigzags Z_1, ..., Z_p. Choose an orientation on every zigzag.*

Let G' be the k-inflation of G_0. The graph G' has kp zigzags $Z_{i,j}$ with $1 \leq i \leq p$ and $1 \leq j \leq k$; the length of every $Z_{i,j}$ is k times the length of Z_i. The orientation on Z_i induces an orientation on k zigzags $(Z_{i,j})_{1 \leq j \leq k}$.

The intersection between $Z_{i,j}$ and $Z_{i',j'}$ is equal to the intersection between Z_i and $Z_{i'}$, to twice the self-intersection of Z_i, or to the self-intersection of Z_i, respectively, if $i \neq i'$, $i = i'$, and $j \neq j'$, or $i = i'$ and $j = j'$, respectively. In particular, if $\mathbf{z}(G_0) = \ldots, c_v^{n_v}, \ldots; \ldots, d_{v_{\alpha v1}, \alpha v2}^{m_v}, \ldots$, then $\mathbf{z}(G') = \ldots, k c_v^{k n_v}, \ldots; \ldots, k d_{v_{\alpha v1}, \alpha v2}^{k m_v}, \ldots$ and if $\mathbf{Int}(Z_i) = (a_i, b_i); i_1^{p_1}, \ldots, i_q^{p_q}$, then $\mathbf{Int}(Z_{i,j}) = (a_i, b_i); i_1^{k p_1}, \ldots, i_q^{k p_q}, (2a_i + 2b_i)^{k-1}$.

Proof Let us consider the 3-regular case. The z-structure of G_0^* differs from the one of G_0 only by reversal of type I and type II. The local structure of zigzags changes according to the rule, which is exemplified by the picture below for the case $k = 2$.

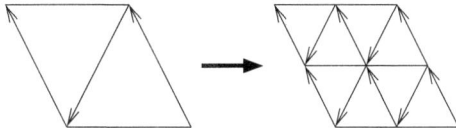

This local picture can be extended to whole graph and we get kp zigzags. The statement about intersections follows easily.   □

The 4-regular case is much more complicate. Take a bipartition $\mathscr{C}_1$, $\mathscr{C}_2$ of the face-set of a 4-regular plane graph G_0. This face-set corresponds to a subset of the face-set of the k-inflation $GC_{k,0}$. The graph $(GC_{k,0}(G_0))^*$ is bipartite also; if k is even, then faces corresponding to $\mathscr{C}_1$ and $\mathscr{C}_2$ in $GC_{k,0}(G_0)$ are in the same part, while if k is odd, then they are in different parts (see Fig. 7.1 for an example). By convention, we take a bipartition $\mathscr{C}_1'$, $\mathscr{C}_2'$ of the face-set of $GC_{k,0}(G_0)$, such that $\mathscr{C}_1'$ contains $\mathscr{C}_1$ (and also $\mathscr{C}_2$, if k is even).

Proposition 7.2 *Let G_0 be a 4-regular plane graph with central circuits C_1, ..., C_p. Choose an orientation on every central circuit*

Let G' be the k-inflation of G_0. Choose the bipartition $\mathscr{C}_1'$, $\mathscr{C}_2'$ of the faces of G' according to the above rule. The graph G' has kp central circuits $C_{i,j}$ with $1 \leq i \leq p$ and $1 \leq j \leq k$; the length of every circuit $C_{i,j}$ is k times the length of C_i.

We define now the orientation on the circuits $C_{i,j}$ in the following way:

- *If $1 \leq j \leq k - 1$, then $C_{i,j}$ is oriented in the opposite way of $C_{i,j+1}$.*
- *If k is odd, then $C_{i,1}$ and $C_{i,k}$ are oriented in the same direction as C_i.*
- *If k is even, then there exists an orientation of all $C_{i,j}$, such that all intersections are of type II.*

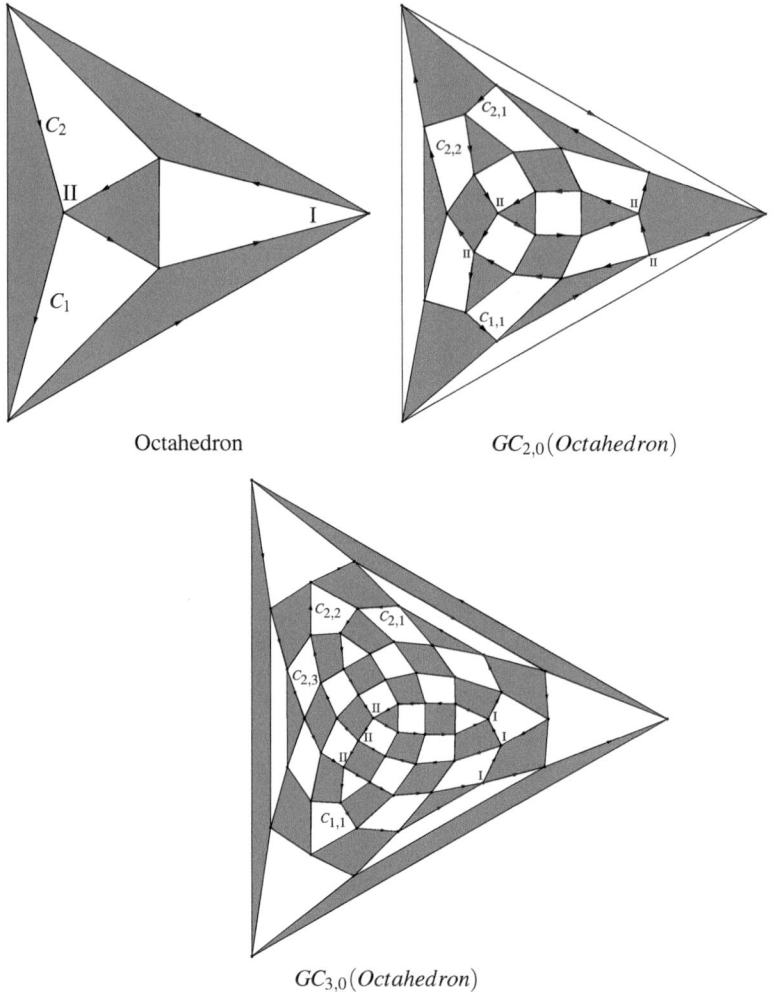

Octahedron $GC_{2,0}(Octahedron)$

$GC_{3,0}(Octahedron)$

Fig. 7.1 Two central circuits C_1, C_2 in octahedron and $C_{1,1}, (C_{2,i})_{1 \leq i \leq k}$ in $GC_{k,0}(Octahedron)$ for $k = 2, 3$

With this orientation one obtains that, if the intersection between C_i and $C_{i'}$ is (α_1, α_2) and $i \neq i'$, then the intersection between $C_{i,j}$ and $C_{i',j'}$ is (α_1, α_2), if k is odd, and $(0, \alpha_1 + \alpha_2)$, if k is even. If the self-intersection of C_i is (α_1, α_2), then the self-intersection of $C_{i,j}$ is (α_1, α_2), $(0, \alpha_1 + \alpha_2)$ if k is odd and even, respectively, while the intersection between $C_{i,j}$ and $C_{i,j'}$ is $(2\alpha_1, 2\alpha_2)$, $(0, 2\alpha_1 + 2\alpha_2)$ if k is odd and even, respectively. In particular, if $\mathbf{cc}(G_0) = \ldots, c_v^{n_v}, \ldots; \ldots, d_{v\alpha_{v1}, \alpha_{v2}}^{m_v}, \ldots$, then $\mathbf{cc}(G') = \ldots, kc_v^{kn_v}, \ldots; \ldots, kd_{v\alpha_{v1}, \alpha_{v2}}^{km_v}, \ldots$ and if $\mathbf{Int}(C_i) = (a_i, b_i); i_1^{p_1}, \ldots, i_q^{p_q}$, then $\mathbf{Int}(C_{i,j}) = I_i; i_1^{kp_1}, \ldots, i_q^{kp_q}, (2a_i + 2b_i)^{k-1}$ with $I_i = (0, a_i + b_i)$ if k is even and $I_{i,i} = (a_i, b_i)$, otherwise.

Proof By definition of the k-inflation in [DeDuSh03], every central circuit C_i of G_0 corresponds to k central circuits of $GC_{k,0}(G_0)$.

If k is odd, then the central circuits $C_{i,1}$ and $C_{i,k}$ have the orientation of C_i; hence, their pairwise intersection is the same. Then the convention of orienting $C_{i,j+1}$ in reverse to $C_{i,j}$, together with the chess-like structure of the bipartition $\mathscr{C}_1', \mathscr{C}_2'$, ensures that the intersection between $C_{i,j}$ and $C_{i',j'}$ is independent of j and j'.

The case of k even is more difficult. Every central circuit C_i corresponds to a set $C_{i,1}, \ldots, C_{i,k}$ of central circuits. By choosing the orientation of $C_{i,1}$, one can assume that it is incident to faces of $\mathscr{C}_1, \mathscr{C}_2$ on the left only. The vertices of the intersection between two (possibly, identical) central circuits $C_{i,1}$ and $C_{i',1}$ belong to faces of $\mathscr{C}_1$ or $\mathscr{C}_2$. By the orientation convention, the intersection between $C_{i,1}$ and $C_{i',1}$ are of type II. By the opposition of orientation between $C_{i,j}$ and $C_{i,j+1}$, the type of vertices of intersection between $C_{i,j}$ and $C_{i',j'}$ is independent of j and j'. In particular, $C_{i,k}$ will also be incident on the left only to faces of $\mathscr{C}_1$ and $\mathscr{C}_2$.

The result on intersection vector follow easily. □

The chosen orientation is necessary for obtaining the above result on intersection vectors; see Fig. 7.1 for an illustration of this point.

7.3 The Moving Group and the (k, l)-product

Given a group Γ acting on a set X, the *stabilizer* (also called *isotropy group*) of an element $x \in X$ is the set of elements $g \in \Gamma$, such that $gx = x$. The action is called *transitive* if for every $x, y \in X$ there exist an element $g \in \Gamma$, such that $gx = y$. The *order of an element* $u \in \Gamma$ is the smallest integer $s > 0$, such that $u^s = Id$. The action is called *free* if the stabilizer of each element of X is trivial.

Lemma 7.1 *If $k, l \geq 0$, then the map*

$$\begin{cases} \phi_{k,l} : \{1, \ldots, k+l\} \to \{1, \ldots, k+l\} \\ \qquad u \mapsto \begin{cases} u+l & \text{if } u \in \{1, \ldots, k\} \\ u-k & \text{if } u \in \{k+1, \ldots, k+l\} \end{cases} \end{cases}$$

is bijective and periodic with period $k + l$; moreover, the successive images of any $x \in \{1, \ldots, k+l\}$ cover entirely the set $\{1, \ldots, k+l\}$ of integers.

Proof If one takes addition modulo $k + l$, then one can write $\phi_{k,l}(u) = u + l$; the lemma follows. □

Let G_0 be a 3- or 4-regular graph. We call *master polygon* a triangle or a square face of G_0^* (see Fig. 6.2). A *directed edge* is an edge of a master polygon with a fixed direction; the set of directed edges is denoted by $\mathscr{DE}$. Given a directed edge $\overrightarrow{e}$, denote its *reverse* (edge with the same vertices, but opposite direction) by $\overleftarrow{e}$.

Any ZC-circuit ZC of $GC_{k,l}(G_0)$, with an orientation, corresponds to a zigzag or a railroad of the dual G_0^*, which we denote ZC^*. If some edges of ZC^* belong to a master polygon, then the orientation of ZC^* determines an entering edge and this entering edge is canonically oriented by ZC^* (see Fig. 6.2).

If $\vec{e}$ is a directed edge and ZC^* go across $\vec{e}$, then the *position p of ZC^**, relatively to $\vec{e}$, is defined as the number of the edge, contained in ZC^*, as numbered in Fig. 6.2; the position of the circuit ZC^*, drawn in Fig. 6.2, is 3. The directed edge, together with its position, determines the circuit ZC^* and its orientation.

Take a circuit ZC^* and a pair $(\vec{e}, p)$ with $\vec{e}$ being a directed edge and p being the position of ZC^*. The directed edge $\vec{e}$ determines a master polygon P, and the next master polygon P' (to which ZC^* belongs) determines a pair $(\vec{e}', p')$. The following equation is a key to all construction that follows:

$$p' = \phi_{k,l}(p) .$$

This equation can be checked on Fig. 7.2 by examining all cases.

The map $(\vec{e}, p) \mapsto (\vec{e}', p')$ is called the *position map* and denoted by $PM(G_0)$. Since the function $\phi_{k,l}$ is $(k+l)$-periodic by Lemma 7.1, one obtains, for any $(\vec{e}, p)$, the relation $PM(G_0)^{k+l}(\vec{e}, p) = (\vec{e}'', p)$ with $\vec{e}'' \in \mathcal{DE}$; let us call *iterated p-position map* and denote by $IPM_p(G_0, k, l)$ the function

$$\begin{cases} IPM_p(G_0, k, l) : \mathcal{DE} \to \mathcal{DE} \\ \qquad\qquad\qquad \vec{e} \mapsto \vec{e}' . \end{cases}$$

Given a circuit ZC^*, let $(\vec{e}, 1)$ be a possible pair of it. Call the *order* of ZC^* and denote by $Ord(ZC)$ the smallest integer s, such that $IPM_1(G_0, k, l)^s \vec{e} = \vec{e}$.

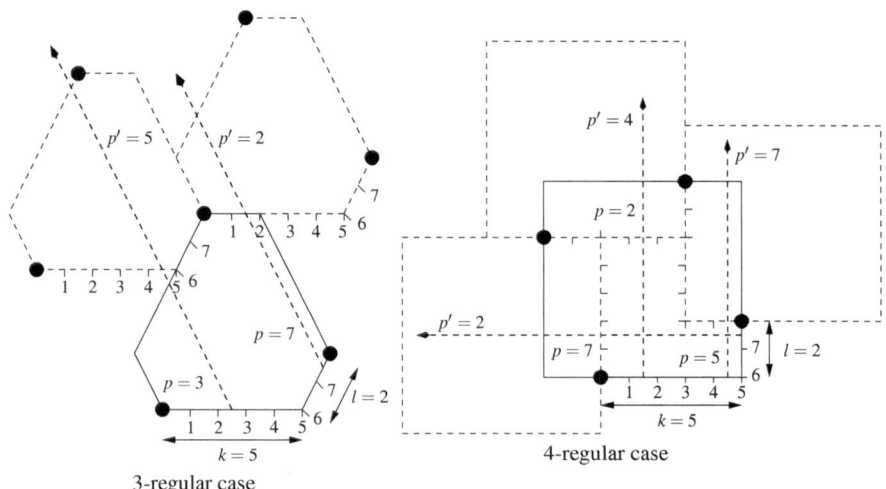

Fig. 7.2 The position map $PM(G_0)$

Theorem 7.1 *If G_0 is a 3- or 4-regular plane graph without faces of curvature zero and $gcd(k, l) = 1$, then $GC_{k,l}(G_0)$ is tight.*

Proof Take a ZC-circuit ZC of $GC_{k,l}(G_0)$. The successive pairs of ZC are denoted by $(\vec{e}_1, p_1), \ldots, (\vec{e}_M, p_M)$ with $M = (k + l)Ord(ZC)$. By the computations done above, $p_{i+1} = \phi_{k,l}(p_i)$.

By Lemma 7.1, there exist i_0 and i_1, such that $p_{i_0} = 1$ and $p_{i_1} = k + l$. First case corresponds to an incidence on the left to a face of nonzero curvature, while the second case corresponds to an incidence on the right.   □

Remark 7.1 If G_0 have a flat face, then, in general, $GC_{k,l}(G_0)$ is not tight if $gcd(k, l) = 1$. We expect that $GC_{k,l}(Prism_6)$ is tight if and only if $gcd(k, l) = 1$.

Definition 7.1 Let G_0 be a 3- or 4-regular plane graph.

(i) In 3-regular case, define two maps L and R, which associate to a given directed edge $\vec{e} \in \mathscr{DE}$ the directed edges $L(\vec{e})$ and $R(\vec{e})$, according to Fig. 7.3.
(ii) In 4-regular case, define the maps g_1, g_2, g_3, which associate to a given directed edge $\vec{e} \in \mathscr{DE}$ the directed edges $g_1(\vec{e}), g_2(\vec{e}), g_3(\vec{e})$, according to Fig. 7.3. Also define $L = g_1$ and $R = g_3 \circ g_2 \circ g_1^{-1}$, where $\circ$ means composition operation.

For a directed edge $\vec{e} \in \mathscr{DE}$ and a position $p \in \{1, \ldots, k + l\}$, it holds:

(i) In 3-regular case, $PM(G_0)(\vec{e}, p) = (\vec{e}', \phi_{k,l}(p))$ with $\vec{e}' = L(\vec{e})$ or $R(\vec{e})$, according to $p \in \{1, \ldots, k\}$ or $\{k + 1, \ldots, k + l\}$.
(ii) In 4-regular case, $PM(G_0)(\vec{e}, p) = (\vec{e}', \phi_{k,l}(p))$ with $\vec{e}' = g_1(\vec{e})$, $g_2(\vec{e})$ or $g_3(\vec{e})$, according to $p \in \{1, \ldots, k - l\}, \{k - l + 1, \ldots, k\}$ or $\{k + 1, \ldots, k + l\}$.

Remind (see Chap. 6) that *directed edges moving group* (in short, *moving group*) $Mov(G_0)$ is the permutation group of the set $\mathscr{DE}$, generated by L and R.

Theorem 7.2 *For any ZC-circuit ZC of $GC_{k,l}(G_0)$ with $gcd(k, l) = 1$, it hold:*
$length(ZC) = 2t(k, l)Ord(ZC)$ *if G_0 is 3-regular and*
$length(ZC) = t(k, l)Ord(ZC)$ *if G_0 is 4-regular.*

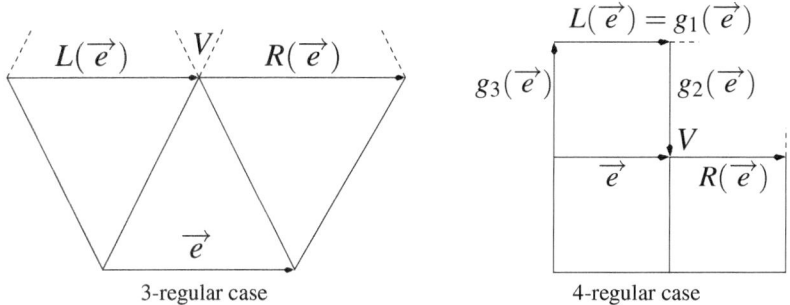

3-regular case   4-regular case

Fig. 7.3 The first and the second maps L, R defined in Definition 7.1

Proof Let us consider the 4-regular case. Given a central circuit C, one can consider the sequence of successive pairs $(\overrightarrow{e_1}, p_1),\ldots,(\overrightarrow{e}_M, p_M)$ with $M = (k+l)Ord(C)$. To every directed edge $\overrightarrow{e_i}$, one can associate a master square, say, SQ_i. Moreover, C can be interpreted as a sequence of squares in the dual graph $(GC_{k,l}(G_0))^*$. So, to every pair $(\overrightarrow{e_i}, p_i)$ one can associate the area A_i of the set of squares in SQ_i between pair $(\overrightarrow{e_i}, p_i)$ and pair $(\overrightarrow{e_{i+1}}, p_{i+1})$. The sets, corresponding to area $A_1,\ldots,A_{k+l}$, can be moved to form a full square of area $t(k,l) = k^2 + l^2$, according to Fig. 7.4. This can be done $Ord(C)$ times. So, the length of C is equal to $t(k,l)Ord(C)$.

In the 3-regular case, the situation is a bit more complicated: for every directed edge $\overrightarrow{e_i}$, we define a master triangle, say, T_i. There is only one triangle $T_{1,i}$, adjacent to T_i and having the directed edge $L(\overrightarrow{e_i})$, and only one triangle $T_{2,i}$, adjacent to T_i and having the directed edge $R(\overrightarrow{e_i})$. The directed edges $L(\overrightarrow{e_i})$ and $R(\overrightarrow{e_i})$ are parallel to the directed edge $\overrightarrow{e_i}$. The area A_i is equal to the area of the set of triangles, which belong to the zigzag going between directed edge $\overrightarrow{e_i}$ and $L(\overrightarrow{e_i})$, $R(\overrightarrow{e_i})$. Those areas can be moved to form a parallelogram (the union of two triangles) of area $2t(k,l)$. So, the length of Z is $2t(k,l)Ord(Z)$.   □

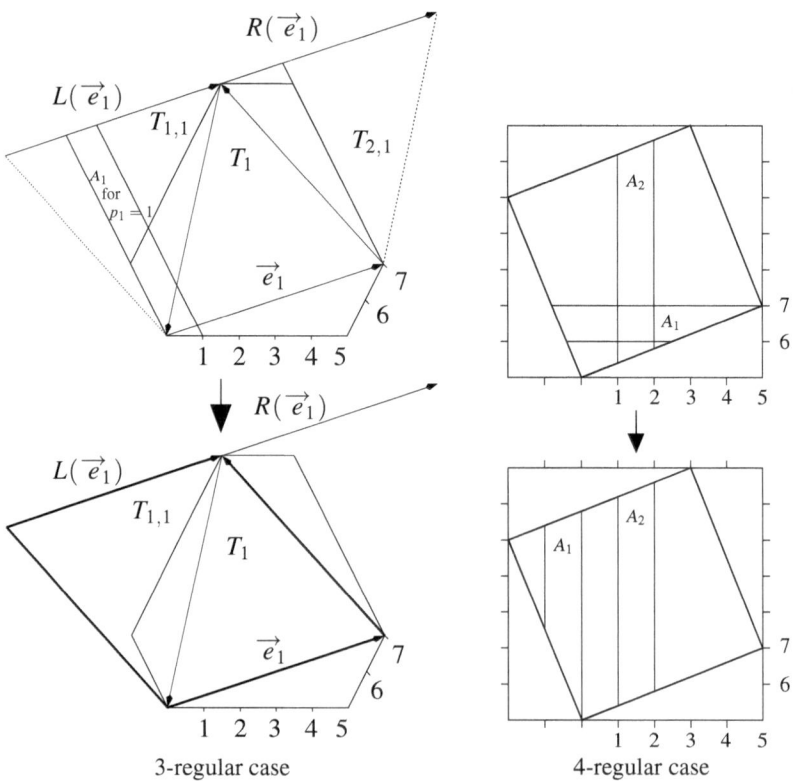

3-regular case   4-regular case

Fig. 7.4 The area covered by ZC-circuits

In this chapter, we call a *partition vector* and denote by $\mathbf{z}$, $\mathbf{cc}$ or, in general, $\mathbf{ZC}$ the vector obtained from z-vectors and cc-vectors by dividing each length by $2t(k, l)$ and $t(k, l)$, respectively (we remove the subscripts specifying self-intersections of different type). In fact, the sum of the components of ZC-vector of any $GC_{k,l}(G_0)$ is the number of edges of G_0.

Theorem 7.3 *Let G_0 be a plane graph; define $s = 6$ or 4 if G_0 is 3- or 4-regular, respectively. The action of $Mov(G_0)$ splits $\mathscr{DE}$ into t orbits of equal size, where t denotes the greatest common divisor of gonalities of all faces of G_0 and of s. The orbit decomposition is as follows:*

(i) If $t = 2$, then for every face F of G_0^*, denote by $DE(F)$ the set of its directed edges having F on the left. G_0^* is bipartite; denote by $\mathscr{F}_1$ and $\mathscr{F}_2$ their corresponding sets of faces. The sets $\mathscr{DE}_i$ of directed edges of faces of $\mathscr{F}_i$ form the orbits of the action of $Mov(G_0)$ on $\mathscr{DE}$.
(ii) If $t = 3$ and G_0 is 3-regular, then there is a tripartition of $\mathscr{DE}$ into 3 orbits O_1, O_2, O_3, such that if $\overrightarrow{e} \in O_i$, then its reverse $\overleftarrow{e}$ is also in O_i.
(iii) In other cases, there is only one orbit.

Proof We will work in G_0^*. If G_0^* is 4-regular, then fix a square, say, sq of G_0^*. Any directed edge of G_0^* can be moved to a directed edge of sq or its reverse. Moreover, if $\overrightarrow{e}$ has sq on its right, then $L^{-1}(\overrightarrow{e})$ has sq on its left. So, any directed edge is equivalent to a directed edge of $DE(sq)$. Hence, there are at most 4 orbits of directed edges. Any directed edge can be moved using L and R, to a directed edge incident to a fixed vertex V. So, in the case of 4 orbits, the minimal degree is at least 4, which is impossible by Euler formula. Therefore, there is 1 or 2 orbits of directed edges.

If G_0 is 3-regular, then fix a triangle, say, Δ of G_0^*. Any directed edge of G_0^* can be moved to a directed edge of $DE(\Delta)$ or to its reverse. Hence, there are at most 6 orbits of directed edges. Any directed edge can be moved using L and R to a directed edge incident to a fixed vertex V. So, in the case of 6 orbits, the minimal degree is at least 6, which is impossible by Euler formula. So, there are 1, 2, or 3 orbits of faces.

If all faces have even gonality, then G_0^* is bipartite and the corresponding bipartition of faces $\mathscr{F}_1 = \{F_1, \ldots\}$, $\mathscr{F}_2 = \{F_1', \ldots\}$ induces a bipartition of $\mathscr{DE}$ by $\{DE(F_1), \ldots\}$ and $\{DE(F_1'), \ldots\}$. So, there are two orbits and, given a directed edge $\overrightarrow{e}$, its reverse belongs to the other orbit. If some faces have odd gonality and G_0 is 4-regular, then there is no such bipartition and so, there is only one orbit. In 3-regular case, if $\overrightarrow{e}$ is in orbit, say, O, then its reverse is also in O and one can identify pairs of opposite directed edges with edges.

Take an edge, say, $e = \{V, V'\}$ in G_0^*, and denote by O the orbit, to which it belongs. We will prove that O contains one third of all edges. By hypothesis, V and V' have degree divisible by 3. By successive iteration of $L \circ R^{-1}$, one gets that at least one third of all edges, incident to V, belong to O. Now, let us prove that for each vertex, exactly one third of all edges belong to O. Let us take a vertex V incident to two edges $e, e' \in O$, which are adjacent; so, at least two third of all edges, which are incident to V, are in O. By hypothesis, there exists a path of edges $e = e_1, \ldots$, $e_N = e'$, such that e_{i+1}, is obtained by application of L or R, the rotation $L \circ R^{-1}$,

$R^{-1} \circ L$, or their inverses. One can assume the path to be of minimal length; this imply that the sequence has no self-intersection. The contradiction arises by application of the Euler formula. So, there are three orbits.   □

Theorem 7.4 *If G_0 is a 3- or 4-regular plane graph, then $Mov(G_0)$ is commutative if and only if the graph G_0 is either a 2_v, a 3_v, or a 4-hedrite.*

Proof In 3-regular case, one can see from Fig. 7.3, that $L \circ R(\overrightarrow{e}) = R \circ L(\overrightarrow{e})$ if and only if **v** has degree 2, 3 or 6. In dual terms, it corresponds to G_0 having 2-, 3-, or 6-gonal faces only. Euler formula $12 = 4p_2 + 3p_3$ for 3-regular plane graphs have solutions $(p_2, p_3) = (3, 0)$ or $(0, 4)$ only.

In 4-regular case, the equality $L \circ R(\overrightarrow{e}) = R \circ L(\overrightarrow{e})$ holds if and only if the vertex V in Fig. 7.3 is of degree 2 or 4. A 4-regular plane graph with all faces being 2- or 4-gons, is exactly a 4-hedrite.   □

Remark 7.2 The problem of generating every graph in the three classes of the above Theorem has been solved: all 2_v come by GC-construction from Bundle$_3$ [GrZa74], while all 3_v (see [GrünMo63]) and all 4-hedrites (see [DeSt03]) are described by two complex parameters.

Given a pair $(k, l) \in \mathbb{Z}^2$, define the *residual group* $Res_{k,l}$ to be the quotient of A_2 or $\mathbb{Z}^2$ (seen as a group) by the sub-group generated by complex numbers $k + l\omega$, $\omega(k + l\omega)$ or, respectively, $k + li, i(k + li)$.

Conjecture 7.1 *(i) $Mov(G_0)$ is isomorphic to a subgroup of $Mov(GC_{k,l}(G_0))$.*
(ii) If the group $Mov(G_0)$ is commutative, then $Mov(GC_{k,l}(G_0))$ is also commutative and $Mov(GC_{k,l}(G_0))/Mov(G_0)$ is isomorphic to $Res_{k,l}$.
(iii) If G_0 is a graph 3_v (respectively, a 4-hedrite), such that $G_0 \neq GC_{k,l}(G_1)$ for G_1 being any other graph 3_v (respectively, any other 4-hedrite), then $|Mov(G_0)| = \frac{v^2}{4}$ (respectively, $|Mov(G_0)| = v^2$).
(iv) A corollary of (iii): all orders of moving groups are the numbers $\frac{v^2}{4t(k,l)}$ (respectively, $\frac{v^2}{t(k,l)}$) with $t(k,l)$ dividing $\frac{v}{4}$, for 3_v (respectively, $\frac{v}{2}$, for 4-hedrites).

Remark 7.3 The order of $Mov(GC_{k,l}(G_0))$ seems to depend on (k, l) in a complicate way and $Mov(G_0)$ is not, in general, a *normal* subgroup of $Mov(GC_{k,l}(G_0))$.

The following definition of (k, l)-*product* can be considered for any group Γ, but here we used it only for the case, when Γ is a moving group of some 3- or 4-regular plane graph G_0. It seems to us, that the majority of notions of this section are new in both, combinatorial and algebraic, contexts. However, an analogous expression of this product itself was proposed in [No87], on the Fisher–Griess Monster group.

Definition 7.2 (the (k, l)-product) Let Γ be a group and g_1, g_2 be two of its elements. Given a pair $(k, l) \in \mathbb{N}^2$ with $gcd(k, l) = 1$, define an element of Γ be their (k, l)-*product* (and denote it by $g_1 \odot_{k,l} g_2$) in the following way:
Define inductively the sequence $(p_0, \dots, p_{k+l})$ by $p_0 = 1$, $p_i = \phi_{k,l}(p_{i-1})$.

Set $S_i = g_1$ if $p_i - p_{i-1} = l$ and $S_i = g_2$ if $p_i - p_{i-1} = -k$; then set

$$g_1 \odot_{k,l} g_2 = S_{k+l} \dots S_2 S_1 .$$

By convention, set $g_1 \odot_{1,0} g_2 = g_1$ and $g_1 \odot_{0,1} g_2 = g_2$.

In the following Theorem, this formalism is used to translate the GC-construction in terms of representation of permutations as product of cycles. For an element $u \in Mov(G_0)$, denote by $\mathbf{ZC}(u)$ the vector $\dots, c_k^{m_k}, \dots$ with multiplicities m_k being the half of the number of cycles of length c_k in the permutation u acting on the set $\mathscr{DE}$. For any subset S of $Mov(G_0)$, denote by $ZC(S)$ the set of all $\mathbf{ZC}(u)$ with $u \in S$.

Theorem 7.5 *Let G_0 be a 3- or 4-regular plane graph. The following hold:*

(i) $IPM_1(G_0, k, l) = L \odot_{k,l} R$ and
(ii) the partition vector $\mathbf{ZC}$ of $GC_{k,l}(G_0)$ is $\mathbf{ZC}(u)$ with $u = IPM_1(G_0, k, l)$.

Proof In 3-regular case, the result follows from the very definition of $IPM_1(G_0, k, l)$. In 4-regular case, the situation is a bit more complicated. Given a sequence of positions $(p_0, p_1, \dots, p_{k+l})$, $S_i = g_1, g_2, g_3$, according to $p_{i-1} \in \{1, \dots, k-l\}$, $\{k-l+1, \dots, k\}, \{k+1, \dots, k+l\}$, and it holds $IPM_1(G_0, k, l) = S_{k+l} \circ \dots \circ S_2 \circ S_1$.

Any multiplication by g_2 is followed by a multiplication by g_3; hence, the relation $IPM_1(G_0, k, l) = g_1 \odot_{k,l} g_3 \circ g_2 \circ g_1^{-1} = L \odot_{k,l} R$ follows:

Take any ZC-circuit ZC and define its *sequence of pairs* as $(\vec{e}_1, p_1), \dots, (\vec{e}_M, p_M)$.

It holds $M = (k+l)Ord(ZC)$ and the values $p_i = 1, k+l$ appear $Ord(ZC)$ times. If one reverses the orientation on ZC, then the sequence of pairs becomes $(\overleftarrow{e}_M, k+l+1-p_M), \dots, (\overleftarrow{e}_1, k+l+1-p_1)$ with $\overleftarrow{e}_i$ being the reverse directed edge of $\vec{e}_i$. So, to every ZC-circuit of length $Ord(ZC)$ correspond two cycles of length $Ord(ZC)$: $(IPM_1(G_0, k, l)^i \vec{e}_1)_{0 \le i \le Ord(ZC)-1}$ and $(IPM_1(G_0, k, l)^i \overleftarrow{e}_M)_{0 \le i \le Ord(ZC)-1}$. $\square$

Remark 7.4 The following hold:

(i) $g_1 \odot_{1,1} g_2 = g_2 g_1$, $g_1 \odot_{k,1} g_2 = g_2 g_1^k$ and $g_1 \odot_{k,k-1} g_2 = (g_2 g_1)^{k-1} g_1$.
(ii) $g_1 \odot_{2q+1,2} g_2 = g_1 (g_1^q g_2)^2$ for any integer q.

The Proposition below gives Euclid algorithm formulas, which can be used to compute $g_1 \odot_{k,l} g_2$ in an efficient way.

Proposition 7.3 *If $(k, l) \in \mathbb{N}^2$ with $gcd(k, l) = 1$, then the following hold:*

(i) If q is an integer, then it holds:

$$\begin{cases} g_1 \odot_{k,l} g_2 = g_1 \odot_{k-ql,l} g_2 g_1^q & \text{if } k - ql \ge 0, \\ g_1 \odot_{k,l} g_2 = g_2^q g_1 \odot_{k,l-qk} g_2 & \text{if } l - qk \ge 0. \end{cases}$$

(ii) $\{g_1 \odot_{k,l} g_2\}^{-1} = g_2^{-1} \odot_{l,k} g_1^{-1}$.

(iii) $g_1 \odot_{k,l} g_2 = g_1^k g_2^l$ if g_1 and g_2 commute.
(iv) $g_1 \odot_{k,l} g_2 \neq Id$ if g_1 and g_2 do not commute.

Proof (i) and (ii) can be obtained by writing down the expressions on both sides and identification. The properties (i) and (ii) allow to compute $g_1 \odot_{k,l} g_2$ by applying the Euclid algorithm to the pair (k, l); at each step of Euclid algorithm, the pair (g_1, g_2) is modified into another pair (g_1', g_2'). It follows from (i) and (ii), that g_1 and g_2 do not commute; so, at any step of the computation, the pair of elements will not commute. So, it is not possible that $g_1 \odot_{k,l} g_2 = Id$, since Id commutes with every element and it yields the commutativity of g_1 and g_2.   □

Corollary 7.1 *If partition vector* **ZC** *of* $GC_{k,l}(G_0)$ *is* 1^p *for one pair* (k, l)*, then* $Mov(G_0)$ *is commutative.*

Proof The partition vector 1^p corresponds to $IPM_1(G_0, k, l) = Id$; Theorem 7.3 (iv) yields that $Mov(G_0)$ is commutative.   □

Remark 7.5 The only known example of graph G_0, such that partition vector of $GC_{k,l}(G_0)$ can be 1^p, is Bundle$_3$.

Proposition 7.4 *Let* $\Gamma = \langle g_1, g_2 \rangle$ *be a group, generated by two elements* g_1 *and* g_2*. Let* n_1 *and* n_2 *be the orders of* $\overline{g_1}$ *and* $\overline{g_2}$ *seen as group elements. Let* K *be a proper normal subgroup of* Γ *and let* $(k, l) \in \mathbb{N}^2$ *with* $gcd(k, l) = 1$*. Then it holds:*

(i) *If* Γ/K *is noncommutative, then* $g_1 \odot_{k,l} g_2 \notin K$.
(ii) *If* Γ/K *is commutative and* $\overline{g_2} = \overline{g_1}^{-1}$ *(with* $\overline{x} = xK$*), then* $g_1 \odot_{k,l} g_2 \in K$ *if and only if* $k - l \equiv 0$ *modulo the index of* K *in* Γ*.*
(iii) *If* Γ/K *is commutative,* $gcd(n_1, n_2) > 1$*, and the map*

$$\Psi : \mathbb{Z}_{n_1} \times \mathbb{Z}_{n_2} \to \Gamma/K$$
$$(k, l) \mapsto \overline{L}^k \overline{R}^l$$

is an isomorphism, then $g_1 \odot_{k,l} g_2 \notin K$ *for every* k, l *with* $gcd(k, l) = 1$*.*

Proof The (k, l)-product goes over to the quotient, i.e., $\overline{g_1 \odot_{k,l} g_2} = \overline{g_1} \odot_{k,l} \overline{g_2}$; so, (i) follows from 7.3 (iii).

If the quotient is commutative, then $\overline{g_1 \odot_{k,l} g_2} = \overline{g_1}^k \overline{g_2}^l = \overline{g_1}^{k-l}$. The quotient is generated by $\overline{g_1}$; so, (ii) follows.

In case (iii), $g_1 \odot_{k,l} g_2 \in K$ if and only if $\overline{g_1}^k \overline{g_2}^l = 1$, i.e., k and l are divisible, respectively, by n_1 and n_2. So, the condition $gcd(k, l) = 1$ implies $k = l = 0$.   □

7.4 The Stabilizer Group

Denote by $\mathscr{P}(G_0)$ the set of all pairs (g_1, g_2) with $g_i \in Mov(G_0)$. Denote by U_{g_1, g_2} the smallest subset of $\mathscr{P}(G_0)$, containing the pair $(g_1, g_2) \in \mathscr{P}(G_0)$, which is stable by the operations $(x, y) \mapsto (x, yx)$ and $(x, y) \mapsto (yx, y)$.

Theorem 7.6 *If G_0 is a 3- or 4-regular plane graph, then it holds:*

(i) *The sequence of subsets $U_{i,L,R}$, defined by $U_{0,L,R} = \{(L, R)\}$ and*

$$U_{n+1,L,R} = \{(u, v), (u, vu), (vu, v) \text{ with } (u, v) \in U_{n,L,R}\},$$

satisfy to $U_{n,L,R} = U_{L,R}$ for n large enough.
(ii) *The set of all possible ZC-vectors of $GC_{k,l}(G_0)$ is the set formed by all partition vectors $\mathbf{ZC}(u)$ and $\mathbf{ZC}(v)$ with $(u, v) \in U_{L,R}$.*

Proof Since G_0 is finite, $Mov(G_0)$ is finite and so, $U_{L,R}$ too. The sequence $(U_{n,L,R})_{n \in \mathbb{N}}$ is increasing and so, by finiteness, there exists an n_0, such that $U_{n_0,L,R} = U_{n_0+1,L,R}$. By construction, the set $U_{n_0,L,R}$ is stable by the operations $(x, y) \mapsto (x, yx), (yx, y)$, which yields (i).

Fix a pair $(k, l) \in \mathbb{N}^2$ with $gcd(k, l) = 1$. By successive applications of Proposition 7.3 (i), one obtains $L \odot_{k,l} R = g_1 \odot_{1,0} g_2$ or $g_1 \odot_{0,1} g_2$ with $(g_1, g_2) \in U_{L,R}$. So, $L \odot_{k,l} R = g_1$ or g_2. Hence, the possible ZC-vectors of $GC_{k,l}(G_0)$ are obtained from g_1 or g_2. On the other hand, if $(g_1, g_2) \in U_{L,R}$. Then, by reversing the process described in (i), that led to (g_1, g_2), one obtains two pairs $(k_i, l_i) \in \mathbb{N}^2$, $gcd(k_i, l_i) = 1$ (with $i = 1$ or 2), such that $L \odot_{k_i,l_i} R = g_i$. □

The *modular group* $SL_2(\mathbb{Z})$ is the group of all 2×2 integral matrices of determinant 1. This group is generated by the matrices $T = \begin{pmatrix} 0 & 1 \\ -1 & 0 \end{pmatrix}$ and $U = \begin{pmatrix} 0 & 1 \\ -1 & -1 \end{pmatrix}$. The group $PSL_2(\mathbb{Z})$ is the quotient of $SL_2(\mathbb{Z})$ by its center $\{I_2, -I_2\}$ with I_2 being the identity matrix. The matrices T and U satisfy to $T^2 = -I_2$ and $U^3 = I_2$.

Lemma 7.2 (i) *The group $PSL_2(\mathbb{Z})$ is isomorphic to the group generated by two elements x, y subject to the relations:*

$$x^2 = Id \text{ and } y^3 = Id .$$

(ii) *The group $SL_2(\mathbb{Z})$ is isomorphic to the group generated by two elements x, y subject to the relations:*

$$x^4 = Id, \quad x^2 y = yx^2 \text{ and } y^3 = Id .$$

Proof (i) is proved in [Ne72]. In order to prove (ii), we will use (i) and the surjection

$$\phi : SL_2(\mathbb{Z}) \to PSL_2(\mathbb{Z})$$
$$M \mapsto \{M, -M\} .$$

Let $W = I_2$ be a word in letters T and U. Write $W = S_1^{n_1} \dots S_N^{n_N}$ with $S_i = T$, U if i is odd, even, respectively. Using the relation $T^4 = I_2$ and $U^3 = I_2$, one can assume that $n_i \in \{1, 2, 3\}$, $n_i \in \{1, 2\}$ if i is odd, even, respectively. Using the relation $T^2 U = U T^2$, one can reduce ourselves to the case of $n_i = 1$ if i odd and greater than 1. Using the morphism ϕ and the property (i), one obtains $m_i = 0$ if i is even. So, the expression can be rewritten as $T^h = I_2$.   □

Definition 7.3 Let Γ be a group.

  (i)   Let $\mathscr{P}(\Gamma)$ be the set of all pairs (g_1, g_2) of elements of Γ.
 (ii)   The *derived group* $D(\Gamma)$ is the group generated by all $uvu^{-1}v^{-1}$ with $u, v \in \Gamma$; it is a normal subgroup of Γ and it is trivial if and only if Γ is commutative.
(iii)   The group $D(\Gamma)$ acts on Γ and $\mathscr{P}(\Gamma)$ in the following way:

$$\begin{cases} Int : D(\Gamma) \times \Gamma \to \Gamma \\ \quad\quad (a, g) \mapsto Int_a(g) = aga^{-1}, \\ Int : D(\Gamma) \times \mathscr{P}(\Gamma) \to \mathscr{P}(\Gamma) \\ \quad\quad (a, (g_1, g_2)) \mapsto Int_a(g_1, g_2) = (Int_a(g_1), Int_a(g_2)) \ . \end{cases}$$

The set of equivalence classes of $\mathscr{P}(\Gamma)$ under this action is denoted by $\mathscr{CP}(\Gamma)$.

The maps Int_a are automorphisms, which are usually called *interior*.

Theorem 7.7 *There exists a group action*

$$\phi : SL_2(\mathbb{Z}) \times \mathscr{CP}(\Gamma) \to \mathscr{CP}(\Gamma)$$
$$(M, c) \mapsto \phi(M)c \ ,$$

such that if $\phi(M)(g_1, g_2) = (h_1, h_2)$, then $g_1 \odot_{(k,l)M} g_2$ is conjugated to $h_1 \odot_{k,l} h_2$, where $(k, l)M = (ak + cl, bk + dl)$ for $M = \begin{pmatrix} a & b \\ c & d \end{pmatrix}$.

Proof Let us define:

$$\phi(T)(g_1, g_2) = (g_2, g_2 g_1^{-1} g_2^{-1}) = Int_{g_2}(g_2, g_1^{-1}) \text{ and}$$
$$\phi(U)(g_1, g_2) = (g_2, g_2 g_1^{-1} g_2^{-2}) = Int_{g_2}(g_2, g_1^{-1} g_2^{-1}) \ .$$

This defines maps from $\mathscr{P}(\Gamma)$ to $\mathscr{P}(\Gamma)$ and so, maps from $\mathscr{CP}(\Gamma)$ to $\mathscr{CP}(\Gamma)$.

If $M \in SL_2(\mathbb{Z})$, then one can find an expression $M = S_1 \ldots S_N$ with $S_i = T$ or U; so, we define:

$$\phi(M) : \mathscr{CP}(\Gamma) \to \mathscr{CP}(\Gamma)$$
$$c \mapsto \phi(M)c = \phi(S_1) \ldots \phi(S_N)c \ .$$

In order to prove, that ϕ is well defined, one needs to prove the independence of $\phi(M)$, over the different expressions of M, in terms of T and U. By standard, but tedious, computations one gets, using the definition of $\phi(T)$ and $\phi(U)$:

$$\begin{cases} \phi(T)^4(g_1, g_2) = \phi(U)^3(g_1, g_2) = Int_{g_2 g_1^{-1} g_2^{-1} g_1}(g_1, g_2), \\ \phi(U)\phi(T)^2(g_1, g_2) = \phi(T)^2\phi(U)(g_1, g_2) \ . \end{cases}$$

The above computations prove the independence of $\phi(M)$ over the different possible expressions of M, since, by Lemma 7.2, all relations, satisfied by T and U, are generated by $T^4 = U^3 = I_2$ and $T^2U = UT^2$. One obtains the relation $\phi(MM') = \phi(M)\phi(M')$ by concatenating two expressions of M and M' in terms of T and U.

One gets $\phi \begin{pmatrix} 1 & 1 \\ 0 & 1 \end{pmatrix} = (g_2g_1, g_2)$ and $\phi \begin{pmatrix} 1 & 0 \\ 1 & 1 \end{pmatrix} = (g_1, g_2g_1)$, which yields the asked relation for the matrices $\begin{pmatrix} 1 & 1 \\ 0 & 1 \end{pmatrix}$ and $\begin{pmatrix} 1 & 0 \\ 1 & 1 \end{pmatrix}$. Since those matrices generate $SL_2(\mathbb{Z})$, the relation is always true. $\square$

Note that the (k, l)-product $g_1 \odot_{k,l} g_2$ is defined for every pair (k, l) with $k \geq 0$, $l \geq 0$ and $gcd(k, l) = 1$. Using the matrices T or U, one can extend it for every pair (k, l) with $gcd(k, l) = 1$, keeping in mind the important fact, that it is defined only up to conjugacy. The obtained extension, still denoted by $g_1 \odot_{k,l} g_2$, satisfies the formula (i) of Proposition 7.3 up to conjugacy without restriction on signs.

For a given 3- or 4-regular plane graph, denote $\mathscr{CP}(G_0)$ the set of equivalence classes of $\mathscr{P}(G_0)$ under the action of $D(Mov(G_0))$. Also, denote by $Stab(G_0)$ the stabilizer of the pair $(L, R) \in \mathscr{CP}(G_0)$ under the action of $SL_2(\mathbb{Z})$ on $\mathscr{CP}(G_0)$.

Proposition 7.5 *If G_0 be a 3- or 4-regular plane graph, then the following hold:*

(i) *$Stab(G_0)$ is a finite index subgroup of $SL_2(\mathbb{Z})$, whose index I is equal to the size of the orbit of $(L, R) \in \mathscr{CP}(G_0)$ under the action of $SL_2(\mathbb{Z})$.*

(ii) *If $(k_1, l_1) = (k_0, l_0)M$ with $M \in Stab(G_0)$, then $GC_{k_0,l_0}(G_0)$ and $GC_{k_1,l_1}(G_0)$ have the same ZC-vector.*

(iii) *There exist a finite set $\{(k_1, l_1), \ldots, (k_I, l_I)\}$ with $gcd(k_i, l_i) = 1$, such that, denoting by P_i the ZC-vector of $GC_{k_i,l_i}(G_0)$, the following hold: for every (k, l) with $gcd(k, l) = 1$, there is an $i_0 \in \{1, \ldots, I\}$ and an $M \in Stab(G_0)$, such that $(k, l)M = (k_{i_0}, l_{i_0})$ and $GC_{k,l}(G_0)$ has ZC-vector P_i.*

Proof (i) The group $Mov(G_0)$ is finite; so, $\mathscr{P}(G_0)$, $\mathscr{CP}(G_0)$ and the orbit of (L, R) are finite also. This implies the finite index property by elementary group theory.

(ii) If $(k_1, l_1) = (k_0, l_0)M$, then $L \odot_{k_0,l_0} R$ and $L \odot_{k_1,l_1} R$ are equal, up to a conjugacy. Since conjugacy does not change the cyclic structure, it does not change the corresponding ZC-vector. So, $GC_{k_0,l_0}(G_0)$ has the same ZC-vector as $GC_{k_1,l_1}(G_0)$.

(iii) Since a partition vector partitions a finite set, there exist a finite number of possibilities for it. Denote by $M_1, \ldots, M_I$ the set of coset representatives of $Stab(G_0)$ in $SL_2(\mathbb{Z})$. The group $SL_2(\mathbb{Z})$ is transitive on the set of pairs $(k, l) \in \mathbb{Z}^2$ with $gcd(k, l) = 1$. So, for any $(k, l) \in \mathbb{Z}^2$ with $gcd(k, l) = 1$, there exists $P \in SL_2(\mathbb{Z})$, such that $(k, l)P = (1, 0)$. Write $P = MM_i$ with $M \in Stab(G_0)$ and one obtains $(k, l)M = (k_i, l_i)$ with $(k_i, l_i) = (1, 0)M_i^{-1}$. $\square$

Remark 7.6 (i) The hexagonal (or square) lattice have a point group of isometry of order 6 (or 4) of rotations of angle $\frac{\pi}{3}$ (or $\frac{\pi}{2}$). So, $GC_{k,l}(G_0)$ is isomorphic to

$GC_{-l,k+l}(G_0)$ (or to $GC_{-l,k}(G_0)$). The $Stab(G_0)$ contains a subgroup, which is isomorphic to this group if G_0 is Dodecahedron and Octahedron, but not Tetrahedron, for which $-I_2 \notin Stab(Tetrahedron)$. Perhaps, one will get this group as subgroup with a less-strict definition of membership in $Stab(G_0)$.

(ii) It seems that there are no constraints on the values of the coefficients of elements of $Stab(G_0)$.

Conjecture 7.2 (i) $Stab(Dodecahedron)$ is generated by

$$\begin{pmatrix} 1 & -1 \\ 1 & 0 \end{pmatrix}, \begin{pmatrix} -4 & -3 \\ 3 & 2 \end{pmatrix} \text{ and } \begin{pmatrix} -4 & -1 \\ 1 & 0 \end{pmatrix};$$

(ii) $Stab(Cube)$ is generated by

$$\begin{pmatrix} -1 & 1 \\ -1 & 0 \end{pmatrix} \text{ and } \begin{pmatrix} 0 & -1 \\ 1 & 2 \end{pmatrix};$$

(iii) $Stab(Octahedron)$ is generated by

$$\begin{pmatrix} 0 & -1 \\ 1 & 0 \end{pmatrix}, \begin{pmatrix} -4 & -3 \\ 3 & 2 \end{pmatrix} \text{ and } \begin{pmatrix} -4 & -1 \\ 1 & 0 \end{pmatrix}.$$

Conjecture 7.3 For any matrix $A = \begin{pmatrix} a & b \\ c & d \end{pmatrix}$, define A' by:

(i) if either $a \neq d$, or $a = d = 0$, then $A' = \begin{pmatrix} a & -c \\ -b & d \end{pmatrix}$;

(ii) otherwise, $A' = \begin{pmatrix} a & c \\ b & d \end{pmatrix}$.

Let G_0 be a 3-regular graph. If $A \in Stab(G_0)$, then $A' \in Stab(G_0)$.

7.5 The GC-Construction on Basic Plane Graphs

Consider the GC-construction $GC_{k,l}(G_0)$ for some bifaced plane graphs of high symmetry. Observe that if $gcd(k, l) = u$, then one can decompose, using Proposition 6.1, the action as two consecutive ones: $GC_{\frac{k}{u}, \frac{l}{u}}(G_0)$ and u-inflation. So, using Proposition 7.1 and 7.2, it suffices to consider only the case $gcd(k, l) = 1$.

We will consider below the following problems:

- what are the possible ZC-vectors of $GC_{k,l}(G_0)$?
- how can those ZC-vectors be expressed in terms of (k, l)?

Given a graph G_0, the first problem can be solved by using Theorem 7.6.

For the second problem, one can prove in some cases (see Theorem 7.11) simple congruence conditions which determine the ZC-vector, by using the normal subgroups of the moving group and Proposition 7.4.

While the moving group allows us to prove most of the results below, in some cases (see Theorem 7.10) the geometric considerations are sufficient. An important case, considered in Theorems 7.8 and 7.9, is the one, when $Rot(G_0)$ is transitive on

$\mathscr{DE}$. Given a group Γ, the enumeration of 3-regular maps M with $Rot(G_0) = \Gamma$ being transitive on $\mathscr{DE}$, is carried on in [Jon85].

Theorem 7.8 *If G_0 is a 3- or 4-regular plane graph, then the following hold:*

   (i) *The actions of $Rot(G_0)$ and $Mov(G_0)$ on $\mathscr{DE}$ commute.*
  (ii) *The action of $Rot(G_0)$ on $\mathscr{DE}$ is free.*
 (iii) *If the action of $Rot(G_0)$ on $\mathscr{DE}$ is transitive, then:*
(iii.1) *the action of $Mov(G_0)$ on $\mathscr{DE}$ is free,*
(iii.2) *every directed edge $\overrightarrow{e} \in \mathscr{DE}$ defines an injective group morphism*

$$\begin{cases} \phi_{\overrightarrow{e}} : Mov(G_0) \rightarrow Rot(G_0) \\ \quad\quad u \mapsto \phi_{\overrightarrow{e}}(u) \end{cases} \text{ with } u^{-1}(\overrightarrow{e}) = \phi_{\overrightarrow{e}}(u)(\overrightarrow{e}),$$

(iii.3) *if $\overrightarrow{e}, \overrightarrow{e}' \in \mathscr{DE}$, then there is a $w \in Rot(G_0)$, such that $\phi_{\overrightarrow{e}'}(u) = w^{-1} \circ \phi_{\overrightarrow{e}}(u) \circ w$,*
(iii.4) *for any $\overrightarrow{e} \in \mathscr{DE}$, $\phi_{\overrightarrow{e}}(Mov(G_0))$ is the normal subgroup of $Rot(G_0)$, formed by all elements preserving the orbit partition of $\mathscr{DE}$ under the action of $Mov(G_0)$.*

   Proof (i) The action of $Mov(G_0)$ is defined, in geometric terms, on Fig. 7.3; so, any rotation of G_0 preserves this picture and two actions commute.

    (ii) The only rotation, preserving a directed edge, is, clearly, identity.

(iii.1) Let $\overrightarrow{e}$ be a directed edge and u be an element stabilizing $\overrightarrow{e}$. It implies the equality $u(\overrightarrow{e}) = \overrightarrow{e}$. If $\overrightarrow{e}'$ is another directed edge, then, by transitivity, there exists a $y \in Rot(G_0)$, such that $\overrightarrow{e} = y(\overrightarrow{e}')$. One gets $y^{-1} \circ u \circ y(\overrightarrow{e}') = \overrightarrow{e}'$ and, by commutativity, $u(\overrightarrow{e}') = \overrightarrow{e}'$. So, u is the identity.

(iii.2) If $\overrightarrow{e}$ is a directed edge of G_0 and $u \in G_0$, then, by transitivity and (ii), there is an unique $y \in Rot(G_0)$, such that $u^{-1}(\overrightarrow{e}) = y(\overrightarrow{e})$. If y denotes $\phi_{\overrightarrow{e}}(u)$, then the following hold:

$$\begin{aligned} \phi_{\overrightarrow{e}}(u) \circ \phi_{\overrightarrow{e}}(u')\overrightarrow{e} &= \phi_{\overrightarrow{e}}(u) \circ u'^{-1}(\overrightarrow{e}) \\ &= u'^{-1} \circ \phi_{\overrightarrow{e}}(u)(\overrightarrow{e}), \text{ by commutativity of } Rot(G_0) \text{ and } Mov(G_0), \\ &= u'^{-1} \circ u^{-1}(\overrightarrow{e}) = (u \circ u')^{-1}(\overrightarrow{e}) \\ &= \phi_{\overrightarrow{e}}(u \circ u')(\overrightarrow{e}), \text{ by the definition of } \phi_{\overrightarrow{e}}. \end{aligned}$$

      Hence, (iii.1) yields equality $\phi_{\overrightarrow{e}}(u \circ u') = \phi_{\overrightarrow{e}}(u) \circ \phi_{\overrightarrow{e}}(u')$ and injectivity of $\phi_{\overrightarrow{e}}$.

(iii.3) If $\overrightarrow{e}'$ is another directed edge, then there is an unique y, such that $\overrightarrow{e} = w(\overrightarrow{e}')$. So, one gets again, by commutativity, $u(\overrightarrow{e}') = y^{-1} \circ v \circ y(\overrightarrow{e}')$, i.e., $\phi_{\overrightarrow{e}'}(u) = y^{-1} \circ \phi_{\overrightarrow{e}}(u) \circ y$.

(iii.4) It can be checked, using the construction of orbit done in Theorem 7.3, that any element of $Rot(G_0)$, which leaves invariant one orbit, say, O_1, will leave invariant other orbits. By construction, any element u of the form $\phi_{\overrightarrow{e}}(u)$ will leave invariant the orbit of $\overrightarrow{e}$ and so, any orbit. Moreover, using freeness of the action, one proves, that if $f \in Rot(G_0)$ preserves the partition of $\mathscr{DE}$

into orbits under the action of $Mov(G_0)$, then there exists an $u \in Mov(G_0)$, such that $\phi_{\vec{e}}(u) = f$. So, $\phi_{\vec{e}}(Mov(G_0))$ is the group of transformations preserving the partition of $\mathscr{DE}$ into orbits and it is normal by (iii.3). $\qquad\square$

Theorem 7.9 *Let G_0 be a 3- or 4-regular v-vertex plane graph, such that $Rot(G_0)$ is transitive on $\mathscr{DE}$. Let (k, l) with $gcd(k, l) = 1$ and let r denote the number of ZC-circuits of $GC_{k,l}(G_0)$. The following hold:*

(i) *$GC_{ku,lu}(G_0)$ is ZC-uniform and it holds:*
(i.1) *if u is even, then there are $\frac{u}{2}$ orbits of ZC-circuits of size $2r$ each,*
(i.2) *if u is odd, then there are $\frac{u-1}{2}$ orbits of ZC-circuits of size $2r$ and one orbit of size r,*
(i.3) *$GC_{ku,lu}(G_0)$ is ZC-transitive if and only if $u = 1$ or 2.*
(ii) *If i_0 denotes the number of faces of nonzero curvature, which are incident to a fixed ZC-circuit ZC of $GC_{k,l}(G_0)$ with $gcd(k, l) = 1$, then:*
(ii.1) *i_0 is even, $r = \frac{|\mathscr{S}(G_0)|}{i_0}$ and the stabilizer of ZC is the point subgroup $D_{i_0/2}$ (or C_2) of $Rot(G_0)$, if $i_0 > 2$ (or $i_0 = 2$, respectively),*
(ii.2) *r is equal to:*

$$\begin{cases} \dfrac{3v}{2\,Ord(IPM_1(G_0,k,l))} & \text{in the 3-regular case,} \\[2mm] \dfrac{2v}{Ord(IPM_1(G_0,k,l))} & \text{in the 4-regular case.} \end{cases}$$

Proof We consider only 3-regular case, since a proof for 4-regular case is similar.

Not all faces are 6-gonal, since we consider finite plane graphs. The transitivity of $Rot(G_0)$ on $\mathscr{DE}$ implies transitivity on the set of faces; so, all faces have the same number q of edges, where $q < 6$. This yields $GC_{k,l}(G_0)$ being tight if $gcd(k, l) = 1$. Since $G_1 = GC_{k,l}(G_0)$ is tight, every zigzag Z is incident on the right to a non 6-gonal face F; this incidence corresponds to a directed edge $\vec{e} \in \mathscr{DE}$. The directed edge $\vec{e}$ belongs to F and comes, in fact, from G_0. The transitivity of $Rot(G_0)$ on $\mathscr{DE}$ yield the transitivity on the set of zigzags of G_1, since $\vec{e}$ defines the zigzag Z.

Now denote $G_2 = GC_{ku,lu}(G_0) = GC_{u,0}(G_1)$. Every zigzag Z of G_1 corresponds to a set of zigzags $Z_1, ..., Z_u$ of G_2. If Z has positions $(\vec{e}, 1)$ and $(\vec{e}', k+l)$, then there exists a transformation $g \in Rot(G_0)$, such that $g(\vec{e}) = \overleftarrow{e}'$ with $\overleftarrow{e}'$ being the reverse of $\vec{e}'$. This transformation reverses the orientation of Z and maps Z_1 to Z_u in G_2 and, more generally, Z_s to Z_{u+1-s}.

(ii.1) Suppose that ZC has the right incidences $\vec{e}_1, ..., \vec{e}_s$ and the left incidences $\vec{e}'_1, ..., \vec{e}'_{s'}$ with $i_0 = s+s'$. By transitivity on $\mathscr{DE}$, there exists an element $g_0 \in Rot(G_0)$, such that $g_0(\vec{e}_1) = \vec{e}'_1$. This yields $s = s'$ and $i_0 = 2s$. Consider now the group $Stab_2$ of transformations preserving the set $\{\vec{e}_1, ..., \vec{e}_{i_0/2}\}$. $Stab_2$ is a normal subgroup of the stabilizer $Stab_1$. The stabilizer $Stab_2$ can do only cyclic shifts on the right incidences $\vec{e}_1, ..., \vec{e}_{i_0/2}$ and so, it is isomorphic to $C_{i_0/2}$ and $Stab_1$ is isomorphic to $D_{i_0/2}$.
(ii.2) Take a zigzag Z of $GC_{k,l}(G_0)$ and define the sequence $\vec{e}_1, ..., \vec{e}_{Ord(Z)}$ by $\vec{e}_{i+1} = IPM_1(G_0, k, l)\vec{e}_i$. By z-transitivity, all zigzags have the same

length; so, $Ord(Z) = Ord(IPM_1(G_0, k, l))$. The length of Z is $Ord(Z)$ $2t(k, l)$. Since $Rot(G)$ is z-transitive, one obtains, by direct enumeration and using that every edge is covered two times, $r \, Ord(Z)2t(k, l) = 3nt(k, l)$ and so, $r = \frac{3v}{2Ord(IPM_1(G_0, k, l))}$. ☐

Remark 7.7 (i) Every element of $Rot(G_0)$ yields a restriction on the possibilities for $Mov(G_0)$ by Theorem 7.8 (i).

(ii) A 3-regular (respectively, 4-regular) plane graph G_0 with v vertices has $3v$ (respectively, $4v$) directed edges. The generators L and R of $Mov(G_0)$ are even permutations of those directed edges by Proposition 7.5. So, $Mov(G_0)$ is isomorphic to a subgroup of $Alt(3v)$ (respectively, of $Alt(4v)$).

(iii) In the extreme case of $Rot(G_0)$ being transitive, the group $Mov(G_0)$ is isomorphic to a subgroup of $Rot(G_0)$; so, it has at most $3v$ (or $4v$) elements.

(iv) The smallest 3-regular plane graphs, for which $Mov(G_0) = Alt(3v)$, are given in the picture below with their symmetry groups.
Does there exist an example of a 4-regular plane graphs with $Mov(G_0) = Alt(4v)$?

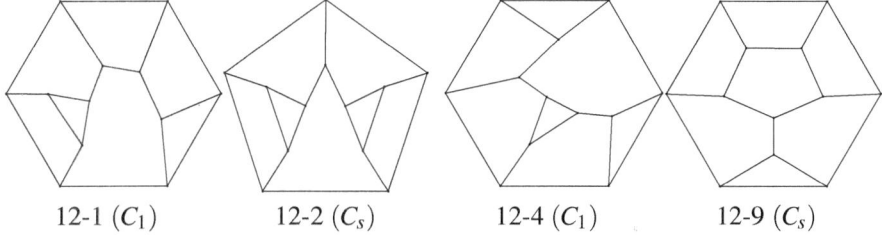

12-1 (C_1) 12-2 (C_s) 12-4 (C_1) 12-9 (C_s)

A face F of a 3- (or 4-regular) plane graph is called 1-*colored* if all its vertices (or, respectively, edges) belong to one zigzag (or, respectively, central circuit).

Lemma 7.3 *If G is a 3- or 4-regular tight plane graph, whose nonflat faces are all 1-colored, then it is ZC-knotted.*

Proof Let G be a 4-regular tight plane graph with all faces of nonflat faces being 1-colored. Let $C_1, \ldots, C_r$ be the central circuits of G.

If two central circuits C_i and C_j have opposite edges of a square, then they define a road, which is a pseudo-road, since G is tight and finish on a q-gonal face with $q \neq 4$. The 1-coloring property yields $C_i = C_j$.

Assume that two central circuits C_i and C_j intersect in one vertex, say, V. If V belongs to a nonsquare face, then one obtains $C_i = C_j$ by 1-coloring property. If not, then one can find a vertex V', which is adjacent to V, such that $\{V, V'\}$ belong to a square. Using the above reasoning, one finds that C_i and C_j intersect in V'. Since G is connected, C_i and C_j intersect in a vertex of a q-gonal face with $q \neq 4$; so, $C_i = C_j$.

The proof in 3-regular case is similar.   ☐

Theorem 7.10 *If $0 \le l \le k$ with $gcd(k, l) = 1$, then in 8 cases below only following ZC-vectors of $GC_{k,l}(G_0)$ occurs; "index" gives the index of $Stab(G_0)$ in $SL_2(\mathbb{Z})$.*

G_0	Possible ZC-vector	Index
Tetrahedron $\{3; 3\}$	2^3	6
Cube $\{4; 3\}$	3^4 or 2^6	8
Dodecahedron $\{5; 3\}$	5^6 or 3^{10}, or 2^{15}	10
Bundle$_3$ $\{2; 3\}$	1^3 or 3	8
Klein map $\{3; 7\}$	3^{28} or 4^{21}	7
Octahedron $\{3; 4\}$	4^3 or 3^4, or 2^6	9
$Foil_2$ $\{2; 4\}$	2^2	6
Trefoil $2 \times K_3$	2^3 or 6	6

Proof The group $Rot(G_0)$ is transitive on $\mathscr{DE}$ for considered cases, except Trefoil 3_1. So, by Theorem 7.9, the partition vector is l^r. By Theorem 7.9, the number r of ZC-circuits is equal to $\frac{3v}{2g}$ or $\frac{2v}{g}$, with g being the order of an element of $Mov(G_0)$, which, using embedding $\phi_{\vec{e}}$ of Theorem 7.8, is an element of $Rot(G_0)$.

If G_0 is Bundle$_3$, then the orders of elements of $Rot(G_0)$ are 1, 2, or 3, which yields $r = 1$ or 3 as the only possibilities; those values of r are attained for $(k, l) = (1, 0)$ and $(1, 1)$, respectively.

If G_0 is Tetrahedron or $Foil_2$ (i.e., Bundle$_4$), then $Mov(G_0) = \mathbb{Z}_2 \times \mathbb{Z}_2$ and Theorem 7.4 (iii) with $K = \{Id\}$ implies $L \odot_{k,l} R \neq Id$. So, $L \odot_{k,l} R$ is of order 2, which proves the required results. Another possibility is to use Theorem 5.2 from [DeDu05] (respectively, Theorem 5 of [DeDuSh03]), which gives that a tight 3_n (respectively, 4-hedrite) has exactly three zigzags (respectively, two central circuits).

In all other cases, $Mov(G_0)$ is noncommutative and so, $l > 1$ by Corollary 7.1.

If G_0 is Dodecahedron, then $r = 6, 10$, or 15, which are attained for $(k, l) = (1, 0)$, $(1, 1)$, and $(2, 1)$, respectively.

If G_0 is Octahedron, then $r = 3, 4$, or 6, which are attained for $(k, l) = (1, 0)$, $(1, 1)$, and $(2, 1)$, respectively.

If G_0 is Cube, then $r = 6, 4$, or 3. Assume that $r = 3$; then, by Theorem 7.9, the stabilizer of any zigzag Z is D_4, with the 4-fold axis going through two squares, say, sq_1 and sq_6. Z cannot be incident to sq_1 or sq_6, since it would yield 1-coloring property and so, G being z-knotted. So, Z is incident exactly once to each of the squares, say, $sq_2, \ldots, sq_5$. One can construct a zigzag Z', which is parallel to Z and incident to both, sq_1 and sq_2. Either $\{sq_2, sq_4\}$, or $\{sq_3, sq_5\}$ form the 4-fold axis of Z; so, either sq_2, or sq_3 are 1-colored. Therefore, $r = 3$ is not possible and the values $r = 4, 6$ are attained for $(k, l) = (1, 0)$ and $(1, 1)$.

If G_0 is Klein map, then the orders of nonzero element of $Rot(G_0)$ are 2, 3, 4, or 7. In order to show the impossibility of 2 and 7, we use Theorem 7.6.

$GC_{k,l}(Trefoil)$ is tight; so, by Theorem 4 in [DeDuSh03], it has at most three central circuits. Assume that $GC_{k,l}(Trefoil)$ has two central circuits, say, C_1 and C_2. Since $GC_{k,l}(Trefoil)$ has a 3-fold rotation axis. By going through triangles, say, T_1 and T_2, one obtains, that those two triangles are 1-colored. Two parallel edges, say, e_1 and e_2, of a square will define a pseudo-road, which finish either on a 2- or 3-gon, giving e_1, e_2 in the same central circuit. The proof goes in the same way, as in Lemma 7.3, and one obtains, that $GC_{k,l}(Trefoil)$ has one central circuit.

cc-transitivity is trivial in the case $r = 1$; in the case $r = 3$, the 3-fold axis of symmetry around T_1, T_2 gives cc-transitivity and so, **cc** $= 2^3$.   □

Remark 7.8 The above proof of Theorem 7.10 uses Corollary 7.1 (iii). Another, more combinatorial, method is possible: the maximum number of ZC-circuits a tight graph a_v or i-hedrite is bounded; see [DeDu05], [DeDuSh03] and Chaps. 2–4.

For the link 7_6^2, the index is 1,764, and all possibilities of ZC-vectors are, with their first appearance (k, l):

14 (4, 1)	1^2, 12 (9, 7)	1^2, 2^2, 8 (5, 1)	1^4, 2, 4^2 (21, 19)
1^4, 3^2, 4 (7, 5)	2, 12	(10, 3) 2, 6^2	(5, 2) 2^2, 10 (2, 1)
2^2, 4, 6 (11, 2)	2^3, 4^2 (5, 3)	2^4, 6 (9, 2)	2^7 (29, 21)
3^2, 8 (1, 1)	4, 10 (1, 0)	4, 5^2 (3, 1)	4^2, 6 (9, 1)
6, 8 (3, 2)			

We expect also, that if $GC_{k,l}(7_6^2)$ has two central circuits, then the closest integer to $\frac{|C_1 \cap C_2|}{t(k,l)}$ is 3.

Theorem 7.11 *The ZC-vectors of $GC_{k,l}(G_0)$ are distributed in the following way (see Table 6.2 above):*

G_0	ZC-vector if I	ZC-vector if II	Index
Bundle$_3$	1^3	3	8
Cube	2^6	3^4	8
Dyck map {3; 8}	4^{12}	3^{16}	8
Trunc. Cube	2^{12}, 3^4 or 2^{18}	6^6 or 9^4	64
Trunc. Dodecahedron	2^{30}, 3^{10} or 2^{30}, 5^6 or 2^{45}	15^6 or 6^{15} or 9^{10}	80
Trunc. Cuboctahedron	2^{12}, 4^{12} or 2^{24}, 3^8 or 2^{36} or 3^8, 4^{12}	6^{12} or 9^8	256
Trefoil 3_1	2^3	6	6
Octahedron	3^4	4^3 or 2^6	9
knot 4_1	1^2, 3^2 or 2, 6	2^2, 4 or 8	72

Proof All those results follow from repeated application of Proposition 7.4. The groups were computed using GAP [GAP] and PlanGraph [Dut02].

The group $Mov(Bundle_3) = \mathbb{Z}_3$ is commutative. Using 7.4 (ii) with $K = \{Id\}$, i.e., the trivial normal subgroup, the result follows:

The group $Mov(Cube)$ is isomorphic to the noncommutative group $Alt(4)$ and has the normal subgroup $K = \langle(1, 2)(3, 4), (1, 3)(2, 4)\rangle$. So, $L \odot_{k,l} R \neq Id$ and $L \odot_{k,l} R \in K$ if and only if $k \equiv l \pmod 3$. The sets of elements of order 3 and of order 2 in $Alt(4)$ are $Alt(4) - K$ and $K - \{Id\}$, respectively. This yields the required result.

The group $Mov(Trefoil)$ has order 36 and has one normal subgroup K_1 of order 9, for which $Mov(Trefoil)/K_1 = \mathbb{Z}_2 \times \mathbb{Z}_2$. So, applying 7.4 (ii), one obtains $L \odot_{k,l} R \notin K_1$. $\{Id, \overline{LR}\}$ is a normal subgroup of $\mathbb{Z}_2 \times \mathbb{Z}_2$, which corresponds to a normal subgroup K_2 of $Mov(G_0)$ of order 18. Using 7.4 (ii), one obtains $L \odot_{k,l} R \in K_2$ if and only if $k \equiv l \pmod 2$. The elements of $Mov(G_0) - K_2$ correspond to $GC_{k,l}(Trefoil)$ having one central circuit, while the elements of $K_2 - K_1$ correspond to $GC_{k,l}(Trefoil)$ having three central circuits. So, the result follows.

The group $Mov(Octahedron)$ is isomorphic to $Sym(4)$, which possess normal subgroups $K_1 = Alt(4)$ and $K_2 = \langle (1, 2)(3, 4), (1, 3)(2, 4) \rangle$. $Mov(Octahedron)/K_2$ is noncommutative; so, $L \odot_{k,l} R \notin K_2$. By 7.4 (ii), it holds $L \odot_{k,l} R \in K_1$ if and only if $k \equiv l \pmod 2$. The elements of $K_1 - K_2$ have order 3, while the elements of $Mov(G_0) - K_1$ have order 2 or 4. So, $r = 4$ if and only if $k \equiv l \pmod 2$ and $r \in \{3, 6\}$ if and only if $k - l \equiv 1 \pmod 2$.

The group $Mov(Dyck\ map)$ has 48 elements and two normal subgroups, K_1 and K_2, of order 4 and 16, respectively. The quotient $Mov(Dyck\ map)/K_1$ is noncommutative; so, $L \odot_{k,l} R \notin K_1$. The quotient $Mov(Dyck\ map)/K_2$ is commutative and $\overline{L} = \overline{R}^{-1}$. So, $L \odot_{k,l} R \in K_2$ if and only if $k \equiv l \pmod 3$. However, elements of $Mov(G_0) - K_1$ correspond to $\mathbf{z} = 3^{16}$, while elements of $K_1 - K_2$ correspond to $\mathbf{z} = 4^{12}$. So, the result follows.

For the remaining cases of knot 4_1, Truncated Cube, Truncated Dodecahedron, and Truncated Cuboctahedron, the technique was always the same:

- first compute the set $\mathscr{S}$ of possibilities for ZC, using Theorem 7.6,
- find a normal subgroup K of index 2 or 3 in $Mov(G_0)$,
- the sets $ZC(K) \cap \mathscr{S}$ and $ZC(Mov(G_0) - K) \cap \mathscr{S}$ are disjoint, which yield the required result.

Those computer computations had to deal with the size of the moving groups; for example $Mov(Trunc.\ Cuboctahedron)$ has 1,327,104 elements. $\square$

Remark 7.9 (i) One can prove easily, that for pairs $(k, l) = (2l - 1, 1), (2l - 7, l)$, $(2l - 17, l)$ the graph $GC_{k,l}(Octahedron)$ has 3 central circuits for every l. Also, $GC_{2l-3,l}(Octahedron)$ has 6 central circuits for every l. We expect that for other values of i the number of central circuits of $GC_{2l-i,l}(Octahedron)$ depends on l.

(ii) $GC_{k,3}(Octahedron)$ with $k \equiv 1, 2 \pmod 3$ has 6 central circuits for any k; we expect, that for other values of l, the number of central circuits depends on k.

Examples of 3-regular z-uniform graphs are Tetrahedron, $Prism_3$, Cube, 10-2, 10-3, $Prism_5$ (see Fig. 1.7). In Tables 7.2 and 7.3, we present the z- and cc-vectors of such graphs for pairs (k, l) with $t(k, l) \leq 200$. We add $*$ to k, l in the first column if $k \equiv l \pmod 3$ or $k \equiv l \pmod 2$ in 3- or 4-regular case, respectively. For Cube, Dodecahedron, Trefoil, and Octahedron we also indicate the intersection vectors.

Table 7.2 z-structure of $GC_{k,l}(G_0)$, $t(k,l) \le 200$, for some 3-regular graphs G_0; see Fig. 1.7

$G_0 =$		Cube		$Prism_3$	$Prism_6$	10-2	10-3	Dodecahedron	
k,l	t(k,l)	z	Int	z	z	z	z	z	Int
1,0	1	3^4	$(0,0); 2^3$	9	9^2	15	15	5^6	$(0,0); 2^5$
1,1*	3	2^6	$(0,0); 2^4, 4$	$2^3; 3$	$2^6, 3^2$	$2^2; 3, 8$	$2^3, 9$	3^{10}	$(0,0); 2^9$
2,1	7	3^4	$(3,0); 12^3$	3^3	3^6	15	$3, 4^3$	2^{15}	$(0,0); 2^{14}$
3,1	13	3^4	$(9,0); 20^3$	9	9^2	6, 9	$2^3, 3^3$	3^{10}	$(0,3); 8^9$
3,2	19	3^4	$(9,0); 32^3$	9	9^2	6, 9	5^3	5^6	$(0,15); 32^5$
4,1*	21	2^6	$(4,0); 14^4, 20$	$1^3, 2^3$	$1^6, 2^6$	$2^2, 3^2, 5$	15	5^6	$(5,10); 36^5$
5,1	31	3^4	$(18,0); 50^3$	9	9^2	15	5^3	5^6	$(15,10); 52^5$
4,3	37	3^4	$(19,0); 62^3$	9	9^2	6, 9	15	5^6	$(0,25); 64^5$
5,2*	39	2^6	$(12,0); 26^4, 28$	$2^3, 3$	$2^6, 3^2$	$1^2, 2^5, 3$	15	5^6	$(0,25); 68^5$
6,1	43	3^4	$(30,0); 66^3$	3^3	3^6	15	$2^3, 9$	3^{10}	$(0,9); 24^3, 28^6$
5,3	49	3^4	$(36,0); 74^3$	9	9^2	15	$2^3, 9$	3^{10}	$(0,9); 28^3, 32^6$
7,1*	57	2^6	$(12,0); 38^4, 52$	$2^3, 3$	$2^6, 3^2$	$1^3, 2^5$	$1^3, 4^3$	2^{15}	$(0,4); 14^4, 18^{10}$
5,4	61	3^4	$(30,0); 102^3$	3^3	3^6	15	$1^3, 4^3$	2^{15}	$(0,4); 14^4, 18^{10}$
7,2	67	3^4	$(45,0); 104^3$	9	9^2	15	$3, 4^3$	2^{15}	$(0,8); 18^{14}$
8,1	73	3^4	$(45,0); 116^3$	9	9^2	15	$2^3, 9$	3^{10}	$(0,18); 42^3, 46^6$
7,3	79	3^4	$(54,0); 122^3$	9	9^2	15	$2^3, 9$	3^{10}	$(0,18); 46^3, 50^6$
6,5	91	3^4	$(45,0); 152^3$	9	9^2	15	15	5^6	$(0,70); 154^5$
9,1	91	3^4	$(63,0); 140^3$	9	9^2	15	5^3	5^6	$(30,40); 154^5$

(continued)

Table 7.2 (continued)

$G_0 =$		Cube		$Prism_3$	$Prism_6$	10-2	10-3	Dodecahedron	
7,4*	93	2^6	$(24,0); 62^4, 76$	$2^3, 3$	$2^6, 3^2$	$2^2, 3^2, 5$	15	5^6	$(0, 70); 158^5$
8,3	97	3^4	$(72,0); 146^3$	9	9^2	15	15	5^6	$(10, 70); 162^5$
9,2	103	3^4	$(63,0); 164^3$	9	9^2	$3, 6^2$	$2^3, 9$	3^{10}	$(12, 12); 58^3, 66^6$
7,5	109	3^4	$(81,0); 164^3$	9	9^2	$3, 6^2$	15	5^6	$(40, 40); 186^5$
10,1*	111	2^6	$(24,0); 74^4, 100$	$2^3, 3$	$2^6, 3^2$	$2^2, 3^2, 5$	15	5^6	$(50, 40); 186^5$
7,6	127	3^4	$(63,0); 212^3$	9	9^2	15	5^3	5^6	$(0, 90); 218^5$
8,5*	129	2^6	$(40,0); 86^4, 92$	$1^3, 2^3$	$1^6, 2^6$	$2, 5, 8$	15	5^6	$(20, 80); 218^5$
9,4	133	3^4	$(99,0); 200^3$	9	9^2	15	5^3	5^6	$(0, 90); 230^5$
11,1	133	3^4	$(84,0); 210^3$	3^3	3^6	$6, 9$	$2^3, 3^3$	3^{10}	$(0, 30); 74^3, 86^6$
10,3	139	3^4	$(102,0); 210^3$	3^3	3^6	15	15	5^6	$(10, 100); 234^5$
11,2*	147	2^6	$(48,0); 98^4, 100$	$2^3, 3$	$2^6, 3^2$	$2^2, 3^2, 5$	$3, 4^3$	2^{15}	$(4, 12); 38^8, 42^6$
9,5	151	3^4	$(99,0); 236^3$	9	9^2	15	$2^3, 3^3$	3^{10}	$(0, 30); 86^3, 98^6$
12,1	157	3^4	$(108,0); 242^3$	9	9^2	15	$3, 4^3$	2^{15}	$(0, 12); 38^8, 50^6$
11,3	163	3^4	$(108,0); 254^3$	9	9^2	15	$2^3, 9$	3^{10}	$(6, 42); 98^9$
8,7	169	3^4	$(84,0); 282^3$	3^3	3^6	15	$3, 4^3$	2^{15}	$(0, 12); 38^4, 50^{10}$
11,4	181	3^4	$(135,0); 272^3$	9	9^2	15	$3, 4^3$	2^{15}	$(0, 20); 46^4, 50^{10}$
13,1*	183	2^6	$(40,0); 122^4, 164$	$1^3, 2^3$	$1^6, 2^6$	$2^2, 3, 8$	$2^3, 9$	3^{10}	$(0, 45); 104^3, 116^6$
9,7	193	3^4	$(144,0); 290^3$	9	9^2	15	$2^3, 9$	3^{10}	$(0, 45); 108^3, 124^6$
13,2	199	3^4	$(135,0); 308^3$	9	9^2	15	15	5^6	$(20, 125); 340^5$

Table 7.3 cc-structure of $GC_{k,l}(G_0)$, $t(k,l) \leq 200$, for some 4-regular graphs G_0

$G_0 =$		$Trefoil\ 3_1$		4_1	7^2_6	Octahedron		$APrism_4$
k,l	t(k,l)	cc	Int	cc	cc	cc	Int	cc
1,0	1	6	(3,0)	8	4;10	4^3	(0, 0); 2^2	16
1, 1*	2	2^3	(0, 0); 2^2	2, 6	3^2; 8	3^4	(0, 0); 2^3	4; 12
2,1	5	6	(15,0)	2^2; 4	2^2; 10	2^6	(0, 0); 2^5	2^4; 8
3, 1*	10	2^3	(2, 0); 8^2	1^2; 3^2	4, 5^2	3^4	(3, 0); 8^3	2^2; 3^4
3,2	13	6	(39,0)	8	6,8	4^3	(4, 4); 18^2	16
4,1	17	6	(51,0)	8	14	4^3	(4, 4); 26^2	16
4,3	25	6	(75,0)	8	14	4^3	(8, 8); 34^2	16
5, 1*	26	2^3	(8, 0); 18^2	$1^2, 3^2$	1^2; 2^2, 8	3^4	(9, 0); 20^3	1^4; 6^2
5,2	29	6	(87,0)	8	2, 6^2	4^3	(8, 8); 42^2	16
5, 3*	34	2^3	(8, 0); 26^2	$1^2, 3^2$	$2^3, 4^2$	3^4	(9, 0); 28^3	1^4; 6^2
6,1	37	6	(111,0)	2^2, 4	14	2^6	(2, 2); 10, 14^4	2^4, 8
5,4	41	6	(123,0)	2^2, 4	14	2^6	(2, 2); 14^4, 18	2^4, 8
7, 1*	50	2^3	(16, 0); 34^2	2, 6	2^2, 10	3^4	(18, 0); 38^3	4, 12
7,2	53	6	(159,0)	2^2, 4	14	2^6	(4, 4); 18^5	2^4, 8
7, 3*	58	2^3	(16, 0); 42^2	2, 6	4, 5^2	3^4	(18, 0); 46^3	4, 12
6,5	61	6	(183,0)	8	2^2, 10	4^3	(20, 20); 82^2	16
7,4	65	6	(195,0)	8	6,8	4^3	(20, 20); 90^2	16
8,1	65	6	(195,0)	8	4,10	4^3	(16, 16); 98^2	16
8,3	73	6	(219,0)	8	4,10	4^3	(24, 24); 98^2	16
7, 5*	74	2^3	(18, 0); 56^2	2, 6	$1^4, 3^2, 4$	3^4	(24, 0); 58^3	4, 12
9, 1*	82	2^3	(26, 0); 56^2	2, 6	$4^2, 6$	3^4	(30, 0); 62^3	4, 12
7,6	85	6	(255,0)	8	4,10	4^3	(28, 28); 114^2	16
9,2	85	6	(255,0)	2^2, 4	$2^4, 6$	2^6	(6, 6); 26, 30^4	2^4, 8
8,5	89	6	(267,0)	8	2^2, 10	4^3	(28, 28); 122^2	16
9,4	97	6	(291,0)	8	4,10	4^3	(28, 28); 138^2	16
10,1	101	6	(303,0)	2^2, 4	4, 10	2^6	(6, 6); 26, 38^4	2^4, 8
9, 5*	106	2^3	(34, 0); 72^2	2, 6	3^2, 8	3^4	(30, 0); 86^3	4, 12
10,3	109	6	(327,0)	8	2, 12	4^3	(36, 36); 146^2	16
8,7	113	6	(339,0)	2^2, 6	14	2^6	(6, 6); 38^4, 50	2^4, 8
11, 1*	122	2^3	(40, 0); 82^2	$1^2, 3^2$	$4^2, 6$	3^4	(45, 0); 92^3	$1^4, 6^2$
11,2	125	6	(375,0)	8	$2^2, 4, 6$	4^3	(40, 40); 170^2	16
9, 7*	130	2^3	(32, 0); 98^2	$1^2, 3^2$	1^2, 12	3^4	(45, 0); 100^3	$1^4, 6^2$
11, 3*	130	2^3	(40, 0); 90^2	2, 6	$2^4, 6$	3^4	(48, 0); 98^3	4, 12
11,4	137	6	(411,0)	2^2, 4	2^2, 10	2^6	(10, 10); 46^4, 50	2^4, 8
9,8	145	6	(435,0)	8	14	4^3	(48, 48); 194^2	16
12,1	145	6	(435,0)	8	4, 10	4^3	(36, 36); 218^2	16
11, 5*	146	2^3	(48, 0); 98^2	$1^2, 3^2$	3^2, 8	3^4	(45, 0); 116^3	$1^4, 6^2$
10,7	149	6	(447,0)	2^2, 4	$4^2, 6$	2^6	(12, 12); 50^5	2^4, 8

(continued)

Table 7.3 (continued)

$G_0 =$		$Trefoil\ 3_1$		4_1	7_6^2	Octahedron		$APrism_4$
11,6	157	6	(471,0)	8	4, 10	4^3	(48, 48); 218^2	16
12,5	169	6	(507,0)	8	4, 10	4^3	(52, 52); 234^2	16
11, 7*	170	2^3	(50, 0); 120^2	$1^2, 3^2$	$3^2, 8$	3^4	(63, 0); 128^3	$2^2, 3^4$
13, 1*	170	2^3	(56, 0); 114^2	$1^2, 3^2$	$2^2, 10$	3^4	(63, 0); 128^3	$2^2, 3^4$
13,2	173	6	(519,0)	8	14	4^3	(52, 52); 242^2	16
13, 3*	178	2^3	(56, 0); 122^2	2, 6	$1^2, 12$	3^4	(66, 0); 134^3	4, 12
10,9	181	6	(543,0)	8	6, 8	4^3	(60, 60); 242^2	16
11,8	185	6	(555,0)	$2^2, 4$	14	2^6	(14, 14); $62^4, 66$	$2^4, 8$
13,4	185	6	(555,0)	8	14	4^3	(60, 60); 250^2	16
12,7	193	6	(579,0)	8	$2^2, 10$	4^3	(64, 64); 258^2	16
13, 5*	194	2^3	(56, 0); 138^2	$1^2, 3^2$	$2^2, 10$	3^4	(69, 0); 148^3	$1^4, 6^2$
14,1	197	6	(591,0)	$2^2, 4$	14	2^6	(12, 12); 50, 74^4	$2^4, 8$

Conjecture 7.4 *(i) For $GC_{k,l}(Icosidodecahedron)$, cc-vector is:*

(2^{30}), $(3^{20})or(5^{12})$ $if\ k \equiv l$ (mod 2),
(10^6), $(4^{15})or(6^{10})$, *otherwise.*

(ii) For $GC_{k,l}(truncated\ Icosidodecahedron)$, z-vector is:

$2^{30}, 3^{40}or2^{30}, 5^{24}or3^{20}, 5^{24}or2^{60}, 3^{20}or2^{60}, 5^{12}or3^{40}, 5^{12}or2^{90}or3^{60}or5^{36}$ *if $k \equiv l$ (mod 2),*
$9^{20}, 6^{30}or15^{12}$, *otherwise.*

Theorem 7.6 yield the list of all possible ZC-vectors. The Proposition 7.4 does not yield the expected partition of cc-vectors for Icosidodecahedron, while Truncated Icosidodecahedron was too complex to be treated.

Now we indicate the properties, which we expect to hold for ZC-structure and moving group of $Foil_m$, $Prism_m$, and $APrism_m$. We extracted those conjectures from extensive computation and expect that the proofs will come from better understanding of the moving group and the (k, l)-product.

Conjecture 7.5 *For $GC_{k,l}(Foil_m)$ with $gcd(k, l) = 1$ holds: cc-vector is 2^m if $k - l$ is odd and, otherwise, it is m or $(\frac{m}{2})^2$ for m odd or even, respectively.*

This conjecture was checked for $m \le 20$; in the computation were used normal subgroups of $Mov(Foil_m)$ and Proposition 7.4.

Conjecture 7.6 *On z-structure of $GC_{k,l}(Prism_m)$ with $gcd(k, l) = 1$, we conjecture:*

(i) $GC_{k,l}(Prism_m)$ is z-balanced and tight.
(ii) All possible z-vectors for $GC_{k,l}(Prism_m)$ are:

(ii.1) *if* $k \equiv l$ *(mod 3): all* 2^m, $(\frac{m}{j})^j$,

       where j is any divisor of m, such that $j \equiv 2$ *(mod 4), if* $m \equiv 0$ *(mod 2).*

(ii.2) *if* $k - l \equiv 1, 2$ *(mod 3): all* $(\frac{3m}{j})^j$,

       where j is any divisor of m, such that $j \equiv m$ *(mod 4), if* $m \equiv 0$ *(mod 2).*

(iii) *Denoting* $m^* = \frac{m}{gcd(m,4)}$, *the following hold:*

(iii.1) $\mathbf{z} = 3^m$ *in the case* $l = k - 1$ *if and only if* $k = 2, 2m^* - 1$ *(mod* $2m^*$*);*

       $\mathbf{z} = 3^m$ *in the case* $l = 1$ *if and only if* $k = 2, 3m^* - 3$ *(mod* $3m^*$*),*

(iii.2) $\mathbf{z} = 2^m$, $(\frac{m}{2})^2$ *in the case* $m \equiv 0$ *(mod 4),* $k \equiv l$ *(mod 3),*

(iii.3) *if* $m \equiv 1, 2, 3$ *(mod 4), then:*

- $\mathbf{z} = 2^m$, 1^m *in the case* $l = k - 3$ *if and only if* $k \equiv 3m^* - 5, 3m^* - 1, 3m^* + 4, 3m^* + 8$ *(mod* $6m^*$*);*
- $\mathbf{z} = 2^m$, 1^m *in the case* $l = 1$ *if and only if* $k \equiv \frac{m^*-1}{2}$ *(mod* m^**),*

(iii.4) *in the case* $(k, l) = (1, 1)$, $\mathbf{z} = 2^m$, $(\frac{m}{2})^2$ *or* 2^m, *m, if m is even or odd, respectively.*

(iv) *The order of* $Mov(Prism_m)$ *is* $12(m^*)^3$ *and its largest normal subgroup has index 3. The orders of all other normal subgroups are exactly the numbers of the form* $2^i q^3$, *where* $0 \le i \le max(3t - 6, 0)$, *t is the exponent of 2 in the factorization of m and q is any odd divisor of m.*

(v) $\begin{pmatrix} 2m + 1 & -2m \\ 2m & 1 - 2m \end{pmatrix} \in Stab(Prism_m).$

(vi) *The index of* $Stab(Prism_m)$ *is* $\frac{64}{9}(m^*)^2$ *if* $m \equiv 0$ *(mod 3) and* $8(m^*)^2$, *otherwise.*

Conjecture 7.7 *For z-structure of* $GC_{k,l}(APrism_m)$ *with* $gcd(k, l) = 1$, *it holds:*

(i) $GC_{k,l}(APrism_m)$ *is z-balanced and tight.*

(ii) *All possible cc-vectors for* $GC_{k,l}(APrism_m)$ *are:*

(ii.1) *if* $k - l \equiv 1$ *(mod 2), then* $\mathbf{cc} = 2^m$, $(\frac{2m}{j})^j$ *and* $(\frac{4m}{j})^j$,

       where j is any odd divisor of m, such that $j \equiv 0$ *(mod 3) if* $m \equiv 0$ *(mod 3).*

(ii.2) *if* $k \equiv l$ *(mod 2), then* $\mathbf{cc} = (\frac{m}{i})^i$, $(\frac{3m}{j})^j$, *where i, j are any divisors of m, such that:*

- $j \equiv 0$ *(mod 3) if* $m \equiv 0$ *(mod 3) and*
- *either i, j are odd and* $gcd(i, j) = 1$, *or* $gcd(i, j) = 2$ *and* $i + j \equiv 2$ *(mod 4).*

(iv) *Denote* $m^* = \frac{m}{gcd(m,3)}$. *The order of* $Mov(APrism_m)$ *is* $24\frac{(m^*)^4}{gcd(m,2)}$.

      Let $m^* = \Pi_{t=1}^T p_t$ *with* $2 \le p_1 \le p_2 \le \cdots \le p_T$ *and all* p_t *are prime.*

(iv.1) *If* m^* *is odd, then the orders of normal subgroups are all numbers of form*

      $\Pi_{t=1}^T p_t^{j_t}$ *with* $j_t \in \{0, 1, 3, 4\}$ *and* $4\Pi_{t=1}^T p_t^{j_t}$ *or* $12\Pi_{t=1}^T p_t^{j_t}$ *with* $j_t \in \{3, 4\}$.

(iv.2) *If* m^* *is even, then the same expressions hold, but* $j_1 \ne 4$.

      In terms of index: the indexes of the above groups are:

      if m^* *is odd: 2g, 6g for any divisor g of* m^* *and any* $24\Pi_{t=1}^T p_t^{j_t}$ *for* $j_t \in \{0, 1, 3, 4\}$;

if m^* is even: $2g$, $6g$ for any divisor g of $\frac{m^*}{2}$ and any $24\Pi_{t=2}^{T} p_t^{j_t}$, $96\Pi_{t=2}^{T} p_t^{j_t}$, $192\Pi_{t=2}^{T} p_t^{j_t}$ for $j_t \in \{0, 1, 3, 4\}$.

(v)   It holds $\begin{pmatrix} 3m+1 & 3m \\ -3m & -3m+1 \end{pmatrix} \in Stab(APrism_m)$ and $Stab(APrism_m)$ is stable by transposition.

(vii)   the index of $Stab(APrism_m)$ in $SL_2(\mathbb{Z})$ is $gcd(m,4)m^2$ if $m \equiv 0$ (mod 3) and $9gcd(m,4)m^2$, otherwise.

7.6   Projections of ZC-Transitive $GC_{k,l}(G_0)$ for Some Graphs G_0

We consider in this section the case, when $GC_{k,l}(G_0)$ is ZC-transitive. Such situation occurs if $Rot(G_0)$ is transitive on the set $\mathcal{DE}$ of directed edges and in some other cases, for example, for G_0 being Trefoil 3_1.

By transitivity of $Aut(G_0)$ on the set of ZC-circuits (apropos, transitivity of $Rot(G_0)$ on $\mathcal{DE}$ implies ZC-transitivity by Theorem 7.9), all ZC-circuits have the same signature, which we denote by (α_1, α_2).

Definition 7.4   Let G_0 be 3- or 4-regular plane graph, such that $GC_{k,l}(G_0)$ is ZC-transitive. Call *projection of G* and denote by $Proj_{k,l}(G_0)$ the plane graph, obtained by the deletion of all but one central circuits of $Med(GC_{k,l}(G_0))$ (or, respectively, $GC_{k,l}(G_0)$). It has $\alpha_1 + \alpha_2$ vertices.

Tables 7.5 and 7.6 represent the projections of $GC_{k,l}(G_0)$ with G_0 being Cube, Dodecahedron and Trefoil, Octahedron. The first column contains (k, l) and mark * if $k \equiv l$ (mod 3) (respectively, $k \equiv l$ (mod 2)). For each graph G_0 and considered pair (k, l), we indicate the ZC-vector, the number **Nr** of its projection, its symmetry group, and p-vector. The Figs. 7.6, 7.7, 7.8 and 7.9, 7.10, 7.11 present pictures of projections given in Tables 7.5 and 7.6, respectively, by their numbers in figures.

Projections **Nr.1, 2, 9, 11, 12, 13, 14** of $GC_{k,l}(Cube)$ coincide with projections **Nr.1, 3, 6, 7, 8, 9, 10** of $GC_{k,l}(Dodecahedron)$. Also, in Table 7.6, for Trefoil 3_1, we omit projections in the cc-knotted case, since it coincides with the graph itself.

The plane graph $Proj_{k,l}(G_0)$ is 4-regular with one central circuit; hence, one can use the notion of type of intersection presented in Fig. 1.2. However, this intersection does not correspond to the self-intersection of the corresponding central circuit in $GC_{k,l}(G_0)$. For instance, central circuits of $GC_{13,3}(Octahedron)$ have self-intersection $(66, 0)$, while their projection have self-intersection $(33, 33)$.

Proposition 7.6   *If G_0 is a 3-regular plane graph, then $Med(G_0)$ appears as a projection of $Med(GC_{k,0}(G_0))$.*

Proof   The zigzags $(Z_i)_{1 \le i \le p}$ of G_0 correspond to the central circuits $(C_i)_{1 \le i \le p}$ in $Med(G_0)$. Let the set of zigzags of $GC_{k,0}(G_0)$ be $(Z_{i,j})_{1 \le i \le p \ 1 \le j \le k}$. Those zigzags become central circuits $C_{i,j}$ in $Med(GC_{k,0}(G_0))$. The central circuit C_i correspond

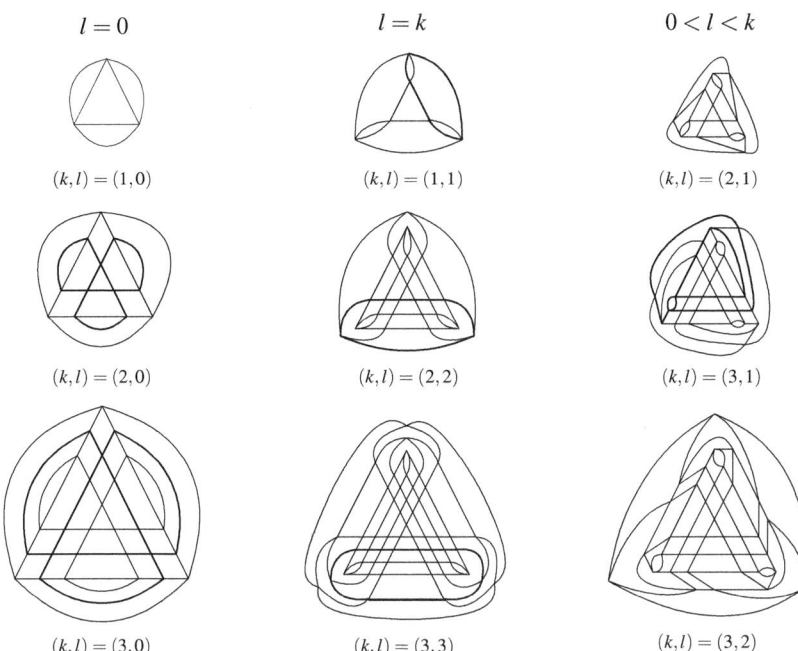

$l=0$ $l=k$ $0<l<k$

$(k,l)=(1,0)$ $(k,l)=(1,1)$ $(k,l)=(2,1)$

$(k,l)=(2,0)$ $(k,l)=(2,2)$ $(k,l)=(3,1)$

$(k,l)=(3,0)$ $(k,l)=(3,3)$ $(k,l)=(3,2)$

Fig. 7.5 Graphs $GC_{k,l}(G_0)$ with G_0 =Trefoil for $0 \le l \le k \le 3$; in nonknotted case a projection is marked by double line

Table 7.4 Conjectured ZC-vector and signature for $GC_{k,l}(G_0)$ with $l = k - 1, 1$ and G_0 being Dodecahedron, Octahedron or Cube

G_0	ZC	$(k, k-1)$			$(k, 1)$		
		k	α_1	α_2	$(k, 1)$	α_1	α_2
Dodec.	2^{15}	2 (mod 3)	0	$4\binom{\lfloor \frac{k}{3}\rfloor+1}{2}$	2 (mod 5)	0	$4\binom{\lfloor \frac{k}{5}\rfloor+1}{2}$
	3^{10}	none	—	—	1, 3 (mod 5)	0	$3\binom{\lceil (k-1)/2\rceil}{2}$
	5^{6}	1, 3 (mod 3)	0	?	0, 4 (mod 5)	$5\binom{\lceil \frac{2}{5}(k+1)\rceil}{2}$	$10\lceil \frac{k}{5}\rceil^2$
Octah.	2^{6}	2 (mod 3)	$\frac{(k-2)(k-1)}{9}$	$\frac{(k-2)(k-1)}{9}$	2 (mod 4)	$2\binom{\frac{k+2}{4}}{2}$	$2\binom{\frac{k+2}{4}}{2}$
	3^{4}	none	—	—	1, 3 (mod 4)	$3\binom{\frac{k+1}{2}}{2}$	0
	4^{3}	0, 1 (mod 3)	$\equiv 0$ (mod 4)	$\equiv 0$ (mod 4)	0 (mod 4)	$\frac{k^2}{4}$	$\frac{k^2}{4}$
Cube	2^{6}	none	—	—	1 (mod 3)	$4\binom{\lceil (k+l)/3\rceil}{2}$	0
	3^{4}	all	$3\binom{k}{2}$	0	0, 2 (mod 3)	$3\binom{k-\lfloor (k-1)/3\rfloor}{2}$	0

Table 7.5 Projections of $GC_{k,l}(G_0)$, $t(k,l) \le 200$, with G_0 being Cube or Dodecahedron

		$GC_{k,l}(Cube)$			$GC_{k,l}(Dodecahedron)$			
k,l	z	Projection	Group	$p_2, \ldots, p_6$	z	Projection	Group	$p_2, \ldots, p_6$
1,0	3^4	0_1	$D_{\infty h}$	0,0,0,0,0	5^6	0_1	$D_{\infty h}$	0,0,0,0,0
1, 1*	2^6	0_1	$D_{\infty h}$	0,0,0,0,0	3^{10}	0_1	$D_{\infty h}$	0,0,0,0,0
2,1	3^4	9 = Trefoil	D_{3h}	3,2,0,0,0	2^{15}	0_1	$D_{\infty h}$	0,0,0,0,0
3,1	3^4	10	D_{3h}	6,2,0,0,3	3^{10}	6 = Trefoil	D_{3h}	3,2,0,0,0
3,2	3^4	11 = (3 × 3)*	D_{3h}	0,8,3,0,0	5^6	13 = (3 × 5)*	D_{5h}	0,10,5,2,0
4, 1*	2^6	1 = (2 × 2)*	D_{2d}	2,4,0,0,0	5^6	13 = (3 × 5)*	D_{5h}	0,10,5,2,0
5,1	3^4	12	D_3	0,8,12,0,0	5^6	14 = (5 × 5)*	D_{5h}	0,10,15,2,0
4,3	3^4	12	D_3	0,8,12,0,0	5^6	14 = (5 × 5)*	D_{5h}	0,10,15,2,0
5, 2*	2^6	5	D_2	2,8,0,4,0	5^6	14 = (5 × 5)*	D_{5h}	0,10,15,2,0
6,1	3^4	25	D_3	6,12,0,12,2	3^{10}	7 = (3 × 3)*	D_{3h}	0,8,3,0,0
5,3	3^4	17	D_3	6,14,6,6,6	3^{10}	7 = (3 × 3)*	D_{3h}	0,8,3,0,0
7, 1*	2^6	2	D_2	2,4,8,0,0	2^{15}	1 = (2 × 2)*	D_{2d}	2,4,0,0,0
5,4	3^4	13	D_3	0,8,24,0,0	2^{15}	1 = (2 × 2)*	D_{2d}	2,4,0,0,0
7,2	3^4	26	D_3	0,24,12,6,5	2^{15}	2 = (2 × 4)*	D_{4d}	0,8,2,0,0
8,1	3^4	18	D_3	0,14,27,6,0	3^{10}	8	D_3	0,8,12,0,0
7,3	3^4	19	D_3	0,20,24,12,0	3^{10}	8	D_3	0,8,12,0,0
6,5	3^4	14	D_3	0,8,39,0,0	5^6	15	D_5	0,10,60,2,0
9,1	3^4	20	D_3	6,26,9,18,6	5^6	18	D_5	0,40,10,12,10
7, 4*	2^6	6	D_2	2,8,12,4,0	5^6	15	D_5	0,10,60,2,0
8,3	3^4	28	D_3	0,36,12,24,2	5^6	19	D_5	0,20,50,12,0
9,2	3^4	27	D_3	0,24,27,12,2	3^{10}	11	D_3	0,14,6,6,0
7,5	3^4	29	D_3	6,50,0,0,27	5^6	20	D_5	0,50,0,22,10
10, 1*	2^6	3	D_2	2,4,20,0,0	5^6	21	D_5	0,40,30,12,10
7,6	3^4	15	D_3	0,8,57,0,0	5^6	16	D_5	0,10,80,2,0
8, 5*	2^6	7	D_2	2,24,0,12,4	5^6	22	D_5	0,30,30,50,22
9,4	3^4	32	D_3	6,48,6,30,11	5^6	16	D_5	0,10,80,2,0
11,1	3^4	30	D_3	0,24,48,12,2	3^{10}	9	D_3	0,8,24,0,0
10,3	3^4	33	D_3	6,54,12,6,26	5^6	17	D_5	0,20,80,12,0
11, 2*	2^6	8	D_2	2,28,4,8,8	2^{15}	4 = (4 × 4)*	D_{4d}	0,8,10,0,0
9,5	3^4	31	D_3	0,42,30,24,5	3^{10}	9	D_3	0,8,24,0,0
12,1	3^4	22	D_3	6,44,24,24,12	2^{15}	3	D_2	2, 4, 8,0,0
11,3	3^4	21	D_3	0,38,48,18,6	3^{10}	12	D_3	0,14,30,6,0
8,7	3^4	16	D_3	0,8,78,0,0	2^{15}	3	D_2	2,4,8,0,0
11,4	3^4	34	D_3	6,72,18,6,35	2^{15}	5	D_2	0,8,14,0,0
13, 1*	2^6	4	D_2	2,4,36,0,0	3^{10}	10	D_3	0,8,39,0,0
9,7	3^4	24	D_3	6,80,12,12,36	3^{10}	10	D_3	0,8,39,0,0
13,2	3^4	23	D_3	0,56,51,12,18	5^6	23	D_5	0,50,65,22,10

Table 7.6 Projections of $GC_{k,l}(G_0)$, $t(k,l) \leq 200$, with G_0 being Trefoil 3_1 or Octahedron

	$GC_{k,l}(Trefoil)$				$GC_{k,l}(Octahedron)$			
k,l	cc	Projection	Group	p_1, p_2, p_3, p_4	cc	Projection	Group	p_2, p_3, p_4
1,0	6		D_{3h}	0,3,2,0	4^3	0_1	$D_{\infty h}$	0,0,0
1, 1*	2^3	0_1	D_∞	0,0,0,0	3^4	0_1	$D_{\infty h}$	0,0,0
2,1	6		D_3	0,3,2,12	2^6	0_1	$D_{\infty h}$	0,0,0
3, 1*	2^3	1	C_{2v}	2,1,0,1	3^4	8	D_{3h}	3,2,0
3,2	6		D_3	0,3,2,36	4^3	20	D_{4d}	0,8,2
4,1	6		D_3	0,3,2,48	4^3	20	D_{4d}	0,8,2
4,3	6		D_3	0,3,2,72	4^3	21	D_{4d}	0,8,10
5, 1*	2^3	2	C_2	0,3,2,5	3^4	9	D_{3h}	0,8,3
5,2	6		D_3	0,3,2,84	4^3	21	D_{4d}	0,8,10
5, 3*	2^3	2	C_2	0,3,2,5	3^4	9	D_{3h}	0,8,3
6,1	6		D_3	0,3,2,108	2^6	1	D_{2d}	2,4,0
5,4	6		D_3	0,3,2,120	2^6	1	D_{2d}	2,4,0
7, 1*	2^3	3	C_2	0,3,2,13	3^4	10	D_3	0,8,12
7,2	6		D_3	0,3,2,156	2^6	20	D_{4d}	0,8,2
7, 3*	2^3	3	C_2	0,3,2,13	3^4	10	D_3	0,8,12
6,5	6		D_3	0,3,2,180	4^3	22	D_4	0,8,34
7,4	6		D_3	0,3,2,192	4^3	22	D_4	0,8,34
8,1	6		D_3	0,3,2,192	4^3	24	D_4	0,8,26
8,3	6		D_3	0,3,2,216	4^3	25	D_4	0,8,42
7, 5*	2^3	4	C_2	0,3,2,15	3^4	11	D_3	0,8,18
9, 1*	2^3	5	C_2	0,3,2,23	3^4	12	D_3	0,8,24
7,6	6		D_3	0,3,2,252	4^3	23	D_4	0,8,50
9,2	6		D_3	0,3,2,252	2^6	4	D_2	0,8,6
8,5	6		D_3	0,3,2,264	4^3	26	D_4	0,8,50
9,4	6		D_3	0,3,2,288	4^3	23	D_4	0,8,50
10,1	6		D_3	0,3,2,300	2^6	2	D_2	2,4,8
9, 5*	2^3	6	C_2	0,3,2,31	3^4	12	D_3	0,8,24
10,3	6		D_3	0,3,2,324	4^3	28	D_4	0,8,66
8,7	6		D_3	0,3,2,336	2^6	2	D_2	2,4,8
11, 1*	2^3	8	C_2	0,3,2,37	3^4	14	D_3	0,8,39
11,2	6		D_3	0,3,2,372	4^3	29	D_4	0,8,74
9, 7*	2^3	7	C_2	0,3,2,29	3^4	13	D_3	0,8,39
11, 3*	2^3	9	C_2	0,3,2,37	3^4	15	D_3	0,8,42
11,4	6		D_3	0,3,2,408	2^6	5	D_2	0,8,14
9,8	6		D_3	0,3,2,432	4^3	27	D_4	0,8,90
12,1	6		D_3	0,3,2,432	4^3	30	D_4	0,8,66
11, 5*	2^3	10	C_2	0,3,2,45	3^4	14	D_3	0,8,39

(continued)

Table 7.6 (continued)

	$GC_{k,l}(Trefoil)$				$GC_{k,l}(Octahedron)$			
10,7	6		D_3	0,3,2,444	2^6	6	D_2	0,8,18
11,6	6		D_3	0,3,2,468	4^3	27	D_4	0,8,90
12,5	6		D_3	0,3,2,504	4^3	31	D_4	0,8,98
11, 7*	2^3	11	C_2	0,3,2,47	3^4	16	D_3	0,8,57
13, 1*	2^3	12	C_2	0,3,2,53	3^4	17	D_3	0,8,57
13,2	6		D_3	0,3,2,516	4^3	32	D_4	0,8,98
13, 3*	2^3	13	C_2	0,3,2,53	3^4	19	D_3	0,8,60
10,9	6		D_3	0,3,2,540	4^3	33	D_4	0,8,114
11,8	6		D_3	0,3,2,552	2^6	7	D_2	0,8,22
13,4	6		D_3	0,3,2,552	4^3	34	D_4	0,8,114
12,7	6		D_3	0,3,2,576	4^3	35	D_4	0,8,122
13, 5*	2^3	14	C_2	0,3,2,53	3^4	18	D_3	0,8,63
14,1	6		D_3	0,3,2,588	2^6	3	D_2	2,4,20

to the set of central circuits $(C_{i,j})_{1 \le j \le k}$ forming a parallel class. So, after removing the central circuits $C_{i,j}$ with $1 \le i \le p$ and $2 \le j \le k$, one obtains $Med(G_0)$. □

The Proposition 7.6 means that one can consider projection only for $GC_{k,l}(G_0)$ with $gcd(k, l) = 1$. Every symmetry preserving a ZC-circuit in $GC_{k,l}(G_0)$ yields a symmetry of the projection graph. This symmetry group is denoted by $Rot_{k,l}(G_0)$. Note, that the group of all symmetries of $Proj_{k,l}(G_0)$ can be larger than $Rot_{k,l}(G_0)$. We expect equality $Rot_{k,l}(G_0) = Aut(Proj_{k,l}(G_0))$ in all, but a finite number of, cases.

If G_0 is a Cube, Dodecahedron, or Octahedron, then one can apply Theorem 7.9 and get that $Rot_{k,l}(G_0) = D_m$. The group $Rot(Trefoil) = D_3$ is not transitive on directed edges. If the graph $GC_{k,l}(Trefoil)$ has 3 central circuits, then the stabilizer of a central circuit has order 2 and the group itself is C_2.

See on Fig. 7.5 a list of first 5-hedrites of symmetry D_{3h} and D_3 with their projections marked by double lines.

The following proposition is to compare with Theorem 7.3.

Proposition 7.7 *If G_0 is a 3- or 4-regular plane graph, whose faces have gonality divisible by 3 or 2, respectively, then all ZC-circuits of $GC_{k,l}(G_0)$ are simple.*

Proof If G_0 satisfy this property, then $GC_{k,l}(G_0)$ satisfy it too. The 3-regular case was proved in [Mo64]. Let us consider the 4-regular case.

If a central circuit of G_0 self-intersects, then one gets an 1-*gonal regular patch* P, i.e., a patch with an angle $\frac{\pi}{2}$ (see [DeDuSh03] for details). The local Euler formula ([DeSt03]) gives $3 = 4 - t = \sum_i (4 - i)p'_i$ with p'_i being the number of i-gonal faces in P. We got a contradiction, since the right-hand side is even. □

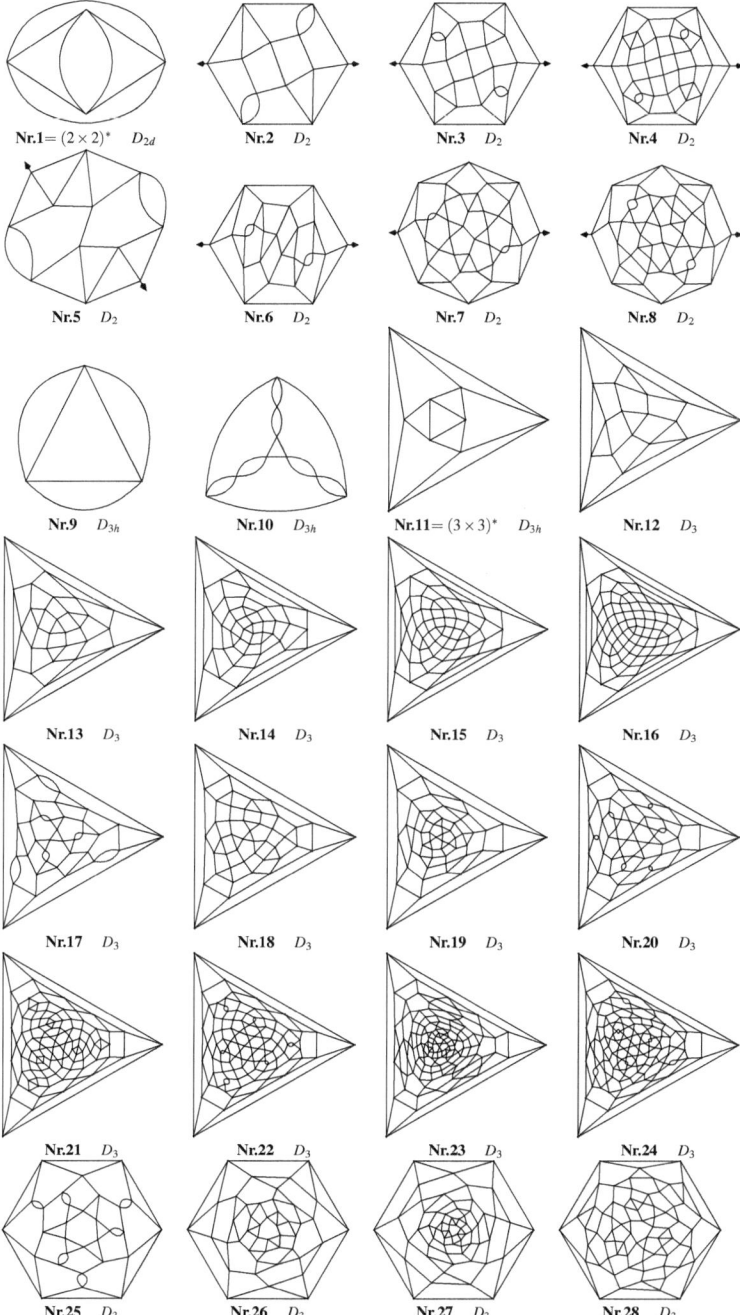

Fig. 7.6 Projections of $GC_{k,l}(Cube)$ from Table 7.5 (part 1)

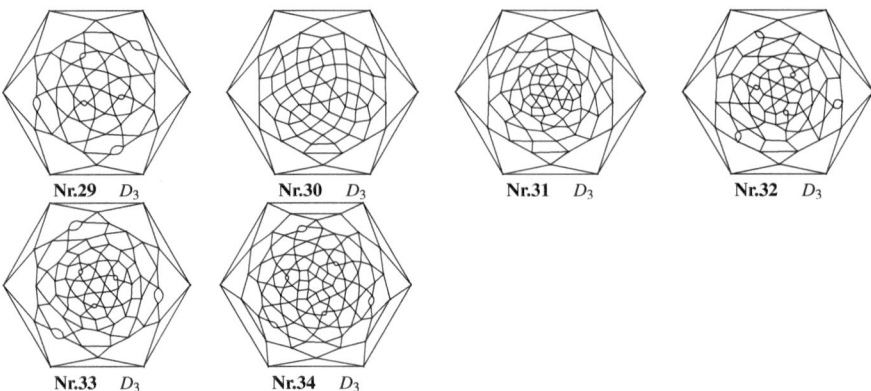

Nr.29 D_3 Nr.30 D_3 Nr.31 D_3 Nr.32 D_3

Nr.33 D_3 Nr.34 D_3

Fig. 7.7 Projections of $GC_{k,l}(Cube)$ from Table 7.5

Proposition 7.8 *For $GC_{k,l}(Dodecahedron)$ with $r = 6$, for $GC_{k,l}(Cube)$ with $r = 4$ and for $GC_{k,l}(Octahedron)$ with $r = 3$ or 4, the symmetry group is transitive on pairs of ZC-circuits and their pairwise intersection has the same size for every two different ZC-circuits.*

Proof The stabilizers of ZC-circuits are point groups D_m by Theorem 7.9.

If $GC_{k,l}(Dodecahedron)$ has 6 zigzags $Z_1,...,Z_6$, then $Stab(Z_1) = D_5$. The conjugacy class of D_5 in $Rot(Dodecahedron) = Alt(5)$ has 6 elements. The pairwise intersection of those subgroups has size 2. So, the action of $Stab(Z_1)$ on Z_2 yields five zigzags $Z_2, ..., Z_6$, i.e., G is transitive on pairs of zigzags.

If $GC_{k,l}(Cube)$ has 4 zigzags $Z_1,...,Z_4$, then $Stab(Z_1) = D_3$. The conjugacy class of D_3 in $Rot(Cube) = Sym(4)$ has 4 elements. The pairwise intersection of those subgroups has size 2 and the proof is as above.

If $GC_{k,l}(Octahedron)$ has 3 central circuits C_1, C_2, and C_3, then pairs of central circuits correspond to central circuits and so, we get again transitivity. If it has 4 central circuits, then the proof is the same as for $GC_{k,l}(Cube)$. $\square$

Conjecture 7.8 (i) *If G_0, G_1 are two 4-regular plane graphs, then the set of pairs (k, l) with $gcd(k, l) = 1$, such that $G_0 = Proj_{k,l}(G_1)$, is finite.*
(ii) *3-regular plane graph, then the set of pairs (k, l) with $gcd(k, l) = 1$, such that $G_0 = Proj_{k,l}(G_1)$, is finite.*

A 4-regular plane graph can have central circuits of the same length, but with different number of self-intersections. For example, $GC_{5,3}(G_0 = 7_6^2)$ (see Table 7.3) has one central circuit of length 68 with self-intersection 2, while any of two other central circuits of length 68 have self-intersection 4.

Conjecture 7.9 (i) *Each central circuit of $GC_{k,l}(Trefoil)$ has self-intersection of the form $(x, 0)$.*
(ii) *If $gcd(k, l) = 1$, then $Proj_{k,l}(Trefoil)$ is a 5-hedrite, except of the cases $(k, l) = (1, 1)$ or $(3, 1)$.*

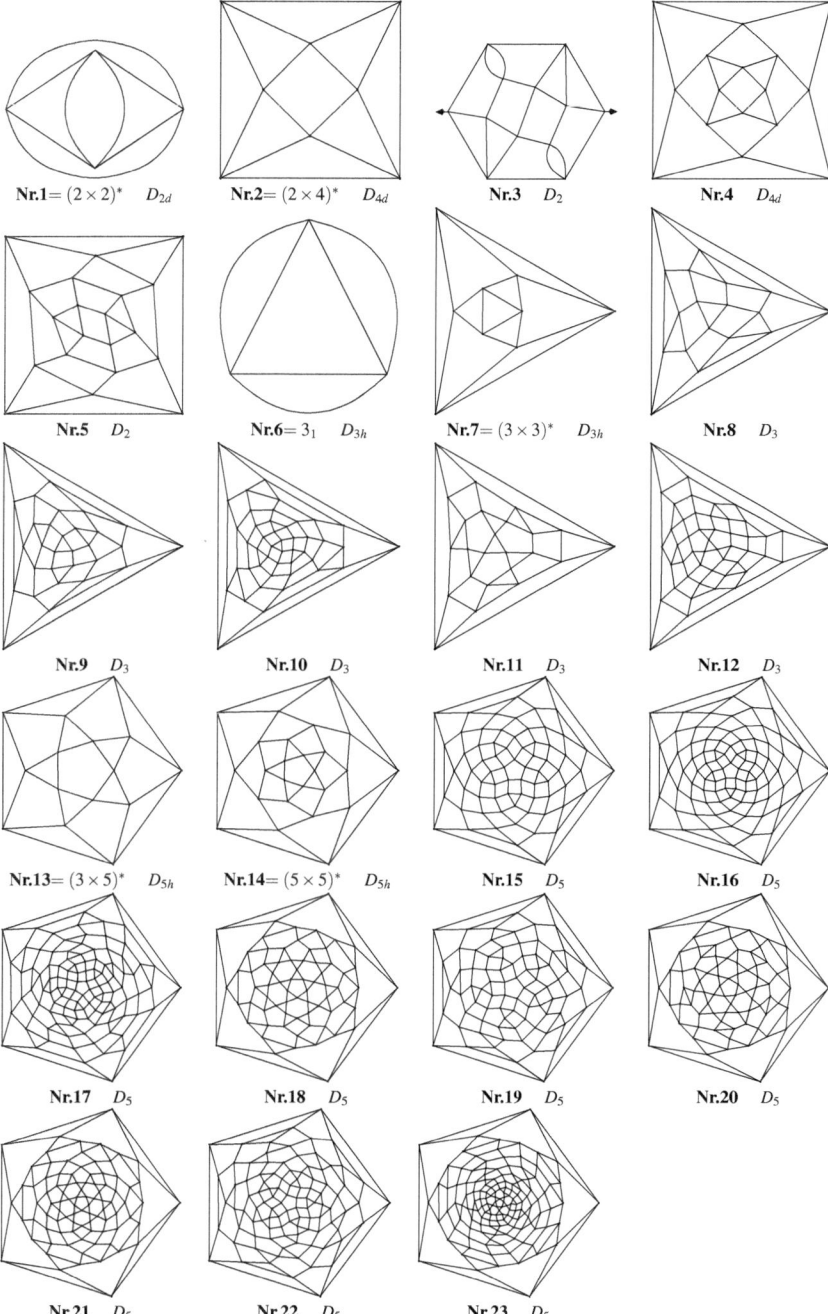

Fig. 7.8 Projections of $GC_{k,l}(Dodecahedron)$ from Table 7.5

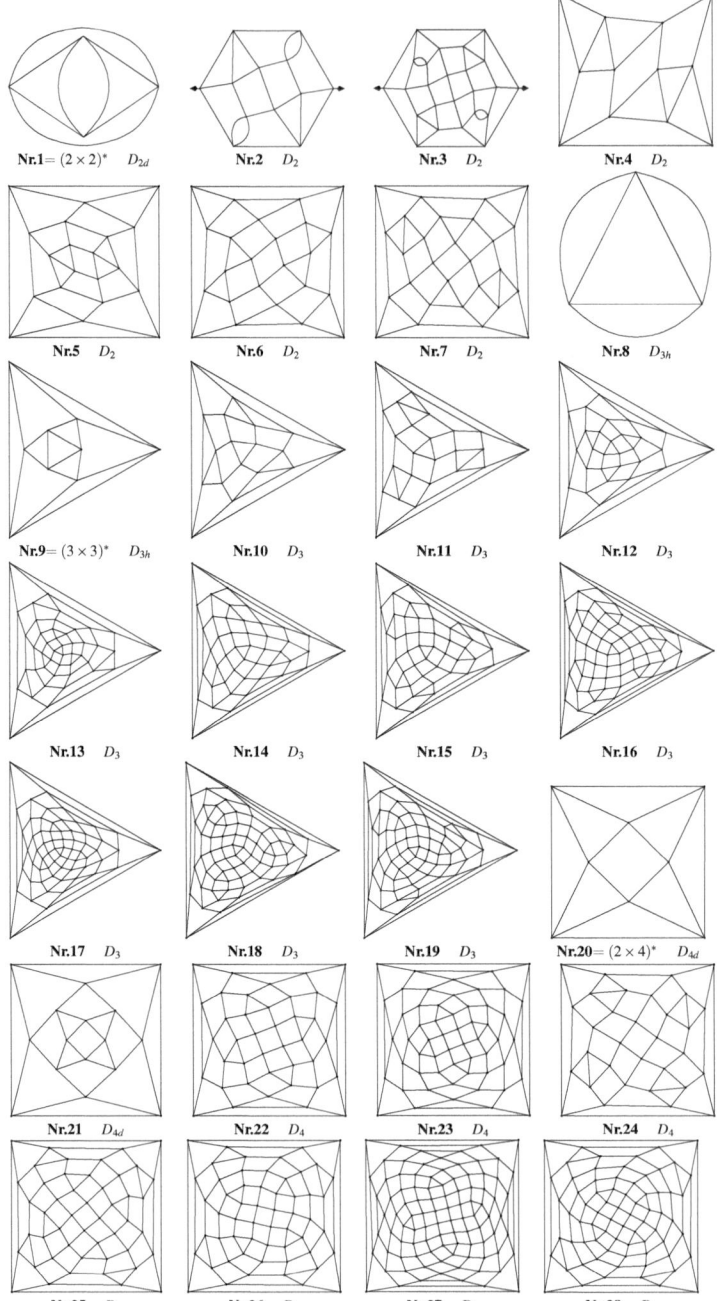

Fig. 7.9 Projections of $GC_{k,l}(Octahedron)$ from Table 7.6 (part 1)

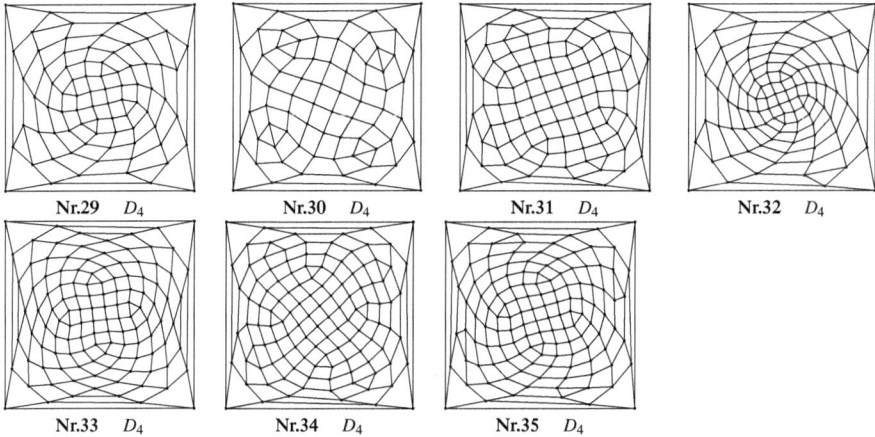

Fig. 7.10 Projections of $GC_{k,l}(Octahedron)$ from Table 7.6 (part 2)

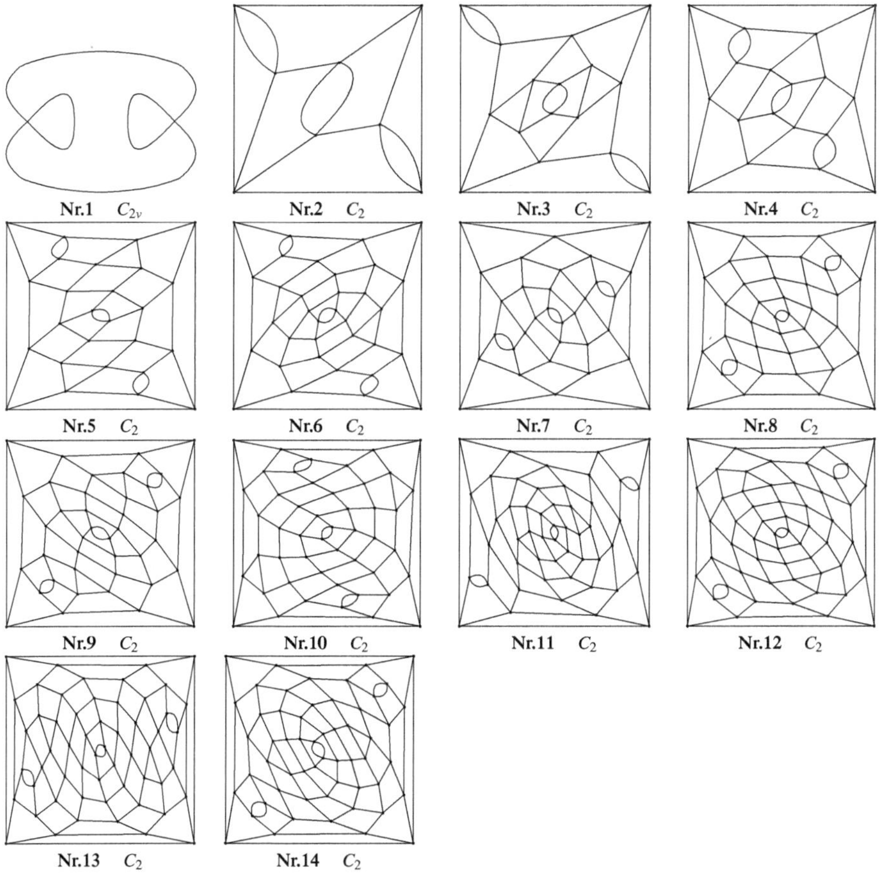

Fig. 7.11 Projections of $GC_{k,l}(Trefoil)$ from Table 7.6

Remark, that for $GC_{k,l}(4_1)$, all central circuits satisfy to $\alpha_2 = 0$ if $k \equiv l$ (mod 2) and $\alpha_1 = \alpha_2$, otherwise.

Conjecture 7.10 (i) *The 2-fold axis of the point group $Rot_{k,l}(G_0)$ do not go through vertices of $Proj_{k,l}(Cube)$ or $Proj_{k,l}(Dodecahedron)$, if the rotation group is D_2.*

(ii) *$Proj_{k,l}(Cube)$ and $Proj_{k,l}(Dodecahedron)$ do not have q-gonal, $q > 6$, faces.*

(iii) *Denote by p_2 the number of 2-gons; for a projection of $GC_{k,l}(Cube)$ it holds:*

(iii.1) *if $r = 6$, then $p_2 = 0$ or 2,*

(iii.2) *if $r = 4$, then $p_2 = 0$ or 6, except of $Proj_{2,1}(Cube)$, for which $p_2 = 3$.*

(iv) *For a projection of $GC_{k,l}(Dodecahedron)$, one can have $p_2 > 0$ only in case $\mathbf{z} = 2^{15}$, for which $p_2 = 2$; in this case, α_1 and α_2 are divisible by 4.*

The projections, considered in this section, are often one of the following forms:

(i) The Conway graph $(k \times m)^*$, defined in Sect. 2.4.
(ii) The D_m-*spiral* alternating knot is a 4-regular plane graph with symmetry D_m having p-vector ($p_m = 2$, $p_3 = 2m$, p_4, other $p_i = 0$) and only one central circuit.

Conjecture 7.11 *If (k, l) has the form $(3b + 4, 1)$, then $G = Proj_{k,l}(Cube)$ is a D_2-spiral alternating knot. Moreover, we expect the following:*

(i) *four triangles of G occur in two pairs of adjacent ones,*
(ii) *there are four pseudo-roads, linking each 2-gon to triangles, and having the same length b,*
(iii) *G has $4\binom{b+2}{2}$ vertices.*

See the cases $b = 0$, 1, and 2 on the picture below.

 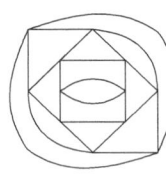 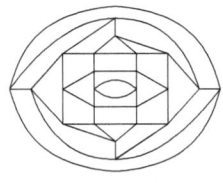

$$b = 0 \qquad\qquad b = 1 \qquad\qquad b = 2$$

Conjecture 7.12 *If d is the length of each central circuit, r the number of central circuits, and (α_1, α_2) the signature of each of them in $GC_{k,l}(Octahedron)$, then:*

(i) *$\alpha_1 \equiv 0$ (mod 3) and $\alpha_2 = 0$ if $k \equiv l$ (mod 2); otherwise, $\alpha_1 = \alpha_2$.*
(ii) *$\alpha_1 = \alpha_2 \geq \frac{d-4}{16}$ if $r = 3$ with equality if and only if $(k, l) = (4p, 1)$; $\alpha_1 \leq \frac{d-6}{8}$ if $r = 4$ with equality if $l = 1$.*

Research in Chaps. 6, 7 leaves many open questions, for example:

• to extend Thurston's idea to classes of plane graphs, defined by > 1 parameter,

- to consider the self-intersection number of ZC-circuits in $GC_{k,l}(G_0)$,
- to prove the conjectures of expression of ZC for $Foil_m$, using another idea than the moving group formalism,
- to prove that one can have $\mathbf{ZC} = 1^p$ only for Bundle$_3$,
- to extend the Goldberg–Coxeter construction to higher dimension and, more precisely, to simplicial and cubical complexes.

7.7 Zigzags of Other Parameter Constructions

All zigzags of $(\{v_3 = 4, v_6\}, 3)$-spheres are simple and the z-vector is:

$$(4s_1)^{m_1}, (4s_2)^{m_2}, (4s_3)^{m_3} \quad \text{with} \quad s_i, m_i \in \mathbb{N} \text{ and } s_i m_i = \frac{v}{4}.$$

This was first established in [GrünMo63] but there is another way to establish it: Any $(\{v_3 = 4, v_6\}, 3)$-sphere is obtained as a quotient of a $(\{v_6\}, 3)$-torus by a group of order 2 formed by inversion. The four vertices of degree 3 come from the four invariant vertices of the torus. All zigzags of a $(\{v_6\}, 3)$-torus are partitioned into three parallel classes that cover the vertex-set. All the zigzags in a parallel class are of the same length and when passing to the quotient, the parallel classes are preserved; hence, the above result is proved. The same argument applies for $(\{v_2 = 4, v_4\}, 4)$-spheres and their central circuits [DeDuSh03].

For other classes of maps, the structure is more complicated and it seems very difficult to obtain simple description of the zigzags of fullerenes. But for $(\{v_4 = 6, v_6\}, 3)$-spheres, we have a simple conjecture for the ones with simple zigzags:

Conjecture 7.13 *All $(\{v_4 = 6, v_6\}, 3)$-spheres with only simple zigzags are:*

- $GC_{k,k}(Octahedron)$ *and*
- *the family of graphs with parameters (m, k) with $v = 4h(2h - 3k)$ triangles; see Fig. 7.12. Their symmetry is O_h if $k = 0$, D_{6h} if $h = 3k$ and D_{3d}, otherwise. The z-vector is $(6h - 6k)^{3h-3k}, (6h)^{h-2k}, (12h - 18k)^k$.*

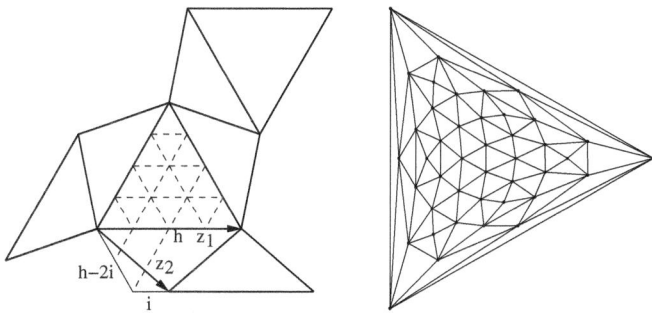

Fig. 7.12 The parameter description of a family of $(\{v_4 = 6, v_6\}, 3)$-spheres with simple zigzags, $z_1 = h\omega$ and $z_2 = h - 2k + k\omega$ and the case $(h, k) = (4, 1)$

References

[Dut02] Dutour, M.: PlanGraph, a GAP package for Planar Graph, http://www.liga.ens.fr/dutour/PlanGraph

[DeSt03] Deza, M., Shtogrin, M.: Octahedrites, special issue "Polyhedra in Science and Art", Symmetry: Culture and Science. The Quarterly of the International Society for the Interdisciplinary Study of Symmetry **11**(1–4), 27–64 (2003)

[DeDu05] Deza, M., Dutour, M.: Zigzag structure of simple two-faced Polyhedra. Comb. Probab. Comput. **14**, 31–57 (2005)

[DeDuSh03] Deza, M., Dutour, M., Shtogrin, M.: 4-valent plane graphs with 2-, 3- and 4-gonal faces, "Advances in Algebra and Related Topics" (in memory of B.H.Neumann; Proceedings of ICM Satellite Conference on Algebra and Combinatorics, Hong Kong 2002). World Scientific publishing Co. 73–97 (2003)

[GrünMo63] Grünbaum, B., Motzkin, T.S.: The number of hexagons and the simplicity of geodesics on certain polyhedra. Can. J. Math. **15**, 744–751 (1963)

[GrZa74] Grünbaum, B., Zaks, J.: The existence of certain planar maps. Discrete Math. **10**, 93–115 (1974)

[Jon85] Jones, G.A.: Enumerating regular maps and normal subgroups of the modular group, Séminaire Lotharingien de Combinatoire (1985)

[Mo64] Motzkin, T.S.: The evenness of the number of edges of a convex polyhedron. In: Proceedings of the National Academy of Sciences, vol. 52, pp. 44–45. U.S.A. (1964)

[Ne72] Newman, M.: Integral Matrices. Academic Press, New York (1972)

[No87] Norton, S.: Generalized Moonshine (Appendix to the paper Finite Groups and Modular Functions by G. Mason in the same volume, pp. 181–207). Proc. of the Symposium in Pure Mathematics, (The Arcata Conference on Representation of Finite Groups), ed. by P. Fong, AMS 47–1, 208–209 (1987)

[Rol76] Rolfsen, D.: Knots and Links, Mathematics Lecture Series 7, Publish or Perish, Berkeley (1976); second corrected printing: Publish or Perish. Houston (1990)

[GAP] The GAP Group, GAP - Groups, Algorithms, and Programming, Version 4.3; 2002. http://www.gap-system.org

Chapter 8
Zigzags of Polytopes and Complexes

In this chapter, based mainly on [DeDu04], we focus on generalization of zigzags for higher dimension. Inspired by Coxeter's notion of Petrie polygon for d-polytopes (see [Cox73]), we generalize the notion of zigzag circuits on complexes and compute the zigzag structure for several interesting families of d-polytopes, including semiregular, regular-faced, Wythoff Archimedean ones, Conway's 4-polytopes, half-cubes, and folded cubes.

Also considered are regular maps and Wilson–Lins triality relations on maps.

Zigzags can also be defined for maps on orientable surface; moreover, this notion, being local, is defined even for non-oriented maps. See Sect. 8.4 on maps. Also, the notion of zigzag extends naturally on infinite plane graphs.

We use for polytopes notations and terminology of [Cox73]; for example, α_d, β_d, γ_d and $\frac{1}{2}\gamma_d$ denote d-dimensional simplex, cross-polytope, cube and half-cube, respectively. Their skeleton graphs are denoted by K_{d+1}, $K_{d\times 2}$, H_d and $\frac{1}{2}H_d$, respectively. We use also Schläfli notation from [Cox73] in Tables 8.1 and 8.4.

The *medial of a polytope* P, denoted by $Med(P)$, is the convex hull of all edge-midpoints of P. It is also defined combinatorially (see Sect. 1.3) on maps (V, E, F) (vertices, edges, faces) on surfaces as the map $(V' = E, E', F' = V \cup F)$ with $e_1, e_2 \in E$ forming an edge $(e_1, e_2) \in E'$ if e_1, e_2 share incident ones vertex and face. This notion of medial can also be also be generalized on d-dimensional complexes; see Sect. 8.3.

8.1 Zigzags for d-Dimensional Complexes

We extend here the definitions of zigzags to any complex. A *chain of length* k in a partially ordered set is a sequence $(x_0, \ldots, x_k)$, such that $x_i < x_{i+1}$. A chain C is a *subchain* of another chain C' if it is obtained by removing some elements in C'.

A chain is *maximal chain* if it is not a subchain of another chain. The *rank*(x) of an element x is the maximal length of chains, beginning at the smallest element 0

© Springer India 2015   191
M. Deza et al., *Geometric Structure of Chemistry-Relevant Graphs*,
Forum for Interdisciplinary Mathematics 1, DOI 10.1007/978-81-322-2449-5_8

and terminating at x. A partially ordered set is called *ranked* if there is a smallest element 0 and a greatest element 1 and if, given two elements $x < y$ with no elements z satisfying to $x < z < y$, one has $rank(y) = 1 + rank(x)$.

A partially ordered set is called an *order lattice* if for any two elements x and y, there is a unique smallest element s and a unique greatest element t, such that $x \leq s$, $y \leq s$ and $x \geq t$, $y \geq t$.

The *dimension of an element* is defined as $dim(x) = rank(x) - 1$.

A *d-dimension of an element* $\mathcal{K}$ is a finite partially ordered set, such that it holds:

(i) $\mathcal{K}$ has a smallest 0 and greatest element 1,
(ii) $\mathcal{K}$ is ranked and all maximal chains have length $d + 2$,
(iii) given two elements x and y with $x \leq y$ and $dim(y) = 2 + dim(x)$, there are exactly two elements u, u', such that $x \leq u \leq y$ and $x \leq u' \leq y$.

In a d-dimensional complex, a maximal chain is called a *flag*; it necessarily begins at 0 and terminates at 1.

A d-dimensional complex is called *simplicial complex* if for every element x of dimension d, there is exactly $d + 1$ elements of dimension 0 contained in it.

Using (iii), one can define the following permutation operator on flags. For $1 \leq i \leq d + 1$, denote by σ_i the operator transforming $(0, x_1, \ldots, x_i, \ldots, x_{d+1}, 1)$ into the flag $(0, x_1, \ldots, x_i', \ldots, x_{d+1}, 1)$ with x_i' being the unique element satisfying to $x_i' \neq x_i$ and $x_{i-1} \leq x_i' \leq x_{i+1}$. One has $\sigma_i^2 = 1$ and $\sigma_i \sigma_j = \sigma_j \sigma_i$ if $i < j - 1$.

Definition 8.1 Let $\mathcal{K}$ be a d-dimensional complex, then:

(i)   denote by $\mathcal{F}(\mathcal{K})$ the set of flags of $\mathcal{K}$,
(ii) denote by $\mathcal{G}(\mathcal{K})$ the graph having, as vertex-set, $\mathcal{F}(\mathcal{K})$, with two flags being adjacent if they are obtained one from the other by a permutation σ_i,
(iii) the complex $\mathcal{K}$ is said to be *orientable* if $\mathcal{G}(\mathcal{K})$ is bipartite; an *orientation* of $\mathcal{K}$ consists in selecting one of the two connected components.

In the case $d = 2$, i.e., if $\mathcal{K}$ is a map, the elements of dimension 0, 1 and 2 are called *vertices*, *edges* and *faces*, respectively.

The definition of orientability, given above, corresponds to the fact that, given an orientation on a cell complex and a maximal chain $(f_1, \ldots, f_{d-1})$ of faces, one can find the last face f_d that makes it a flag.

A $(d + 1)$ *-polytope* P is defined as the convex hull of a set of points in $\mathbb{R}^{d+1}$. The set of faces of P defines an order lattice and so, a d-dimensional complex, which is an order lattice, since the boundary of a d-polytope is homeomorphic to S^{d-1}.

Call a d-dimensional complex a *regular complex* if its symmetry group is transitive on the set of flags.

Theorem 8.1 *If $\mathcal{K}$ is a d-dimensional complex and $x = (0, x_1, \ldots, x_{d+1}, 1)$ is its flag, then there exists a unique sequence of faces $(x_{i,j})_{1 \leq i \leq d+1, 1 \leq j \leq d+2-i}$, namely:*

$$x_{1,1}, \quad \ldots\ldots, \quad x_{1,j}, \quad \ldots\ldots, \quad x_{1,d+1}$$
$$x_{2,1}, \quad \ldots\ldots, \quad x_{2,j}, \quad \ldots\ldots, \quad x_{2,d}$$
$$\vdots$$
$$x_{d-1,1}, x_{d-1,2}, x_{d-1,3}$$
$$x_{d,1}, x_{d,2}$$
$$x_{d+1,1}$$

such that it holds:

(i) $x_{i,1} = x_i$,
(ii) $dim(x_{i,j}) = i - 1$,
(iii) $x_{i,j} \leq x_{i+1,j}$ *for* $1 \leq i \leq d$ *and* $1 \leq j \leq d + 1 - i$,
(iv) $x_{i,j} \leq x_{i+1,j-1}$ *for* $1 \leq i \leq d$ *and* $2 \leq j \leq d + 2 - i$.
     Moreover, if $\mathcal{K}$ *is an order lattice, then the elements* $(x_{i,j})$ *are uniquely defined by the vertex sequence* $(x_{1,j})_{1 \leq j \leq d+1}$.

Proof Using property (iii), one can find $x_{1,2}, \ldots, x_{d,2}$, then $x_{1,3}$ and so on.

If $\mathcal{K}$ is an order lattice, then $x_{i,j}$ can be characterized as the smallest element greater than $x_{i-1,j}$ and $x_{i-1,j+1}$. □

Definition 8.2 Let $\mathcal{K}$ be a d-dimensional complex.

(i)   Denote by $T = \sigma_{d+1}\sigma_d \ldots \sigma_1$ the *translation operator* of $\mathcal{K}$.
(ii) A *zigzag* in $\mathcal{K}$ is a circuit $(f_1, \ldots, f_l)$ of flags, such that $f_{j+1} = T(f_j)$ and l denote the length of the zigzag.
(iii) Given a flag f, the *reverse flag* f^t of f is defined as $(0, x_{1,d+1}, x_{2,d}, \ldots, x_{d,2}, x_{d+1,1}, 1)$ with $(x_{i,j})$ as in Theorem 8.1.
(iv) The *reverse zigzag* of a zigzag $(f_1, \ldots, f_l)$ is the zigzag $(f_l^t, f_{l-1}^t, \ldots, f_1^t)$.

The above notion (central in this chapter), for the special case of an d-polytope, essentially coincides with the following notion on page 223 of [Cox73]: "A Petrie polygon of an d-dimensional polytope or of an $(d - 1)$-dimensional honeycomb, is a skew polygon, such that any $(d - 1)$ consecutive sides but no d, belong to a Petrie polygon of a cell."

The choice of a zigzag $(f_1, \ldots, f_l)$ over its reverse $(f_1^t, \ldots, f_l^t)$ amounts to choosing an orientation on the zigzag. In the sequel a zigzag is identified with its reverse.

Note that if $\mathcal{K}$ is a d-dimensional simplicial complex with f facets, then one has $|\mathcal{F}(\mathcal{K})| = (d + 1)!f$. Note also that the stabilizer of a flag is trivial and so, if $\mathcal{K}$ has p orbits of flags, then $|\mathcal{F}(\mathcal{K})| = p|Sym(\mathcal{K})|$.

Proposition 8.1 *If the complex* $\mathcal{K}$ *is oriented and of even dimension, then the length of any zigzag is even.*

Proof Since $\mathcal{K}$ is oriented, the set $\mathcal{F}(\mathcal{K})$ is split into two parts, $\mathcal{F}_1$ and $\mathcal{F}_2$. Since d is even, the translation $T = \sigma_{d+1}\sigma_d \ldots \sigma_1$ of all its flags has an odd number of components; so it interchanges $\mathcal{F}_1$ and $\mathcal{F}_2$. □

In Sect. 5.91 of [Cox73] the evenness of the length h of zigzags was obtained for complexes arising from Coxeter groups of dimension 3; there was given the formula $g = h(h + 2)$ with g being the size of the group.

Definition 8.3 Take a zigzag $Z = (f_1, \ldots, f_l)$ and its reverse $Z^t = (f_l^t, \ldots, f_1^t)$.

 (i) Given a flag f_j, if $\sigma_1(f_j)$ belongs to Z, then self-intersection is called *type I*, while if $\sigma_1(f_j)$ belongs to Z^t, then it is called *type II*.
 (ii) The *signature of the zigzag* Z (see Sect. 1.3) is the pair (n_I, n_{II}) with n_I being the number of self-intersections of type I and n_{II} the number of self-intersections of type II. The signature does not change if one interchanges Z and Z^t.
 (iii) Take two zigzags Z_1 and Z_2 with associated circuits $(f_{1,1}, \ldots, f_{1,l})$, $(f_{1,l}^t, \ldots, f_{1,1}^t)$ and $(f_{2,1}, \ldots, f_{2,l})$, $(f_{2,l}^t, \ldots, f_{2,1}^t)$. If $f_{1,j}^t$ belongs to Z_2, then it is called an intersection of *type I*, while if it belongs to Z_2^t, it is called an intersection of *type II*.
 (iv) The *signature* (n_I, n_{II}) is the pair enumerating such intersections. If Z_2 and Z_2^t are interchanged, then the types of intersections are also interchanged.

The *z-vector of a complex* $\mathcal{K}$ is the vector enumerating the lengths of all its zigzags with their signature as subscript. The simple zigzags are put in the beginning, in increasing order of length, without their signature $(0, 0)$, and separated by a semicolon from others. Self-intersecting zigzags are also ordered by increasing lengths. If there are $m > 1$ zigzags of the same length l and the same signature $(\alpha_1, \alpha_2) \neq (0, 0)$, then we write l_{α_1, α_2}^m. It turns out that Snub Cube, Snub Dodecahedron, $Pyr(\beta_{d-1})$, and $BPyr(\alpha_{d-1})$ are the only polytopes in the tables of this chapter, having self-intersecting zigzags.

Given two zigzags Z and Z', their *normalized signature* is the pair (n_I, n_{II}) enumerating intersection of type I and II with orientation chosen so that $n_I \leq n_{II}$. For a zigzag Z, its *intersection vector* $\mathbf{Int}(Z) = \ldots, (c_{k,I}, c_{k,II})^{m_k}, \ldots$ is such that $(\ldots, (c_{k,I}, c_{k,II}), \ldots)$ is a sequence $(c_{k,I}, c_{k,II})$ of its nonzero normalized signature with all others zigzags, and m_k denote respective multiplicities. If the zigzag has signature (n_1, n_{II}), then its length l satisfies

$$l = 2(n_I + n_{II}) + \sum_k m_k(c_{k,I} + c_{k,II}) .$$

The *dual of a complex* $\mathcal{K}$ is the complex $\mathcal{K}^*$ with the same elements as $\mathcal{K}$, but with $x \leq y$ in $\mathcal{K}^*$ being equivalent to $y \leq x$ in $\mathcal{K}$.

Theorem 8.2 *Every zigzag Z in $\mathcal{K}$ corresponds to a unique zigzag Z^* in $\mathcal{K}^*$ with the same length.*

Proof Given a flag $f = (0, x_1, \ldots, x_{d+1}, 1)$ of $\mathcal{K}$, one can associate to it a flag $f' = (1, x_{d+1}, \ldots, x_1, 0)$ of $\mathcal{K}^*$. Denote by σ_i' the operator on $\mathcal{K}^*$, which acts by changing the i-th element. It is easy to see that its action on f' corresponds to the action of σ_{d+2-i} on f. So, one has $T'(f') = (T^{-1}f)'$ and every zigzag $(f_1, \ldots, f_l)$ of $\mathcal{K}$ corresponds to a zigzag $(f_l', \ldots, f_1')$ of $\mathcal{K}^*$.   □

In the case of maps (i.e., for $d = 2$), every intersection in $\mathcal{K}$ corresponds to an intersection in $\mathcal{K}^*$ with type I or II interchanged. This is not, a priori, the case of complexes of dimension $d > 2$.

A d-dimensional complex $\mathcal{K}$ is said to be *z-transitive* if its symmetry group $Sym(\mathcal{K})$ is transitive on zigzags. It is said to be *z-knotted*, if it has only one zigzag, and *z-pure*, if none of its zigzags self-intersects. Note that the stabilizer of a flag is necessarily the trivial group, i.e., every orbit of flags has the size $|Sym(\mathcal{K})|$.

Denote by $Z(\mathcal{K})$ the graph formed by the set of zigzags of a complex $\mathcal{K}$ with two zigzags being adjacent if the signature of their intersection is different from $(0, 0)$. For $d = 2$, we prove in Sect. 8.4 that $Z(\mathcal{K})$ is connected. But there is no reason to think that connectivity will still hold for complexes of dimension $d > 2$.

Proposition 8.2 *If a d-dimensional complex $\mathcal{K}$ is regular, then:*

(i) $\mathcal{K}$ is z-transitive,
(ii) if $Z(\mathcal{K})$ is connected, then $\mathcal{K}$ is either z-pure, or z-knotted.

Proof The transitivity on zigzags is obvious. If a zigzag has a self-intersection, then, by transitivity, all flags correspond to a self-intersection of zigzags. Since $Z(\mathcal{K})$ is connected, it means that there is only one zigzag. □

Conjecture 8.1 *Any odd-dimensional complex is z-pure.*

We do not see why it would be true but we could not find a counterexample.

8.2 Z-Structure of Some Generalizations of Regular d-Polytopes

A *regular d-polytope* is one whose symmetry group is transitive on flags.

A *regular-faced d-polytope* is one having only regular facets. A *semiregular d-polytope* is a regular-faced d-polytope whose symmetry group is transitive on vertices. All semiregular, but not Platonic, 3-polytopes (i.e., 13 Archimedean 3-polytopes and *Prism$_m$*, *APrism$_m$* for any $m \geq 3$) were discovered by Kepler [Ke1619]. The list of all 7 semiregular, but not regular, d-polytopes with $d \geq 4$ was given by Gosset in 1897 [Gos00], but his proofs were never published; see also [BlBl91]. This list consists of 5 polytopes, denoted by n_{21} (where $n \in \{0, 1, 2, 3, 4\}$) of dimension $n + 4$, and two exceptional ones (both 4-dimensional): *snub 24 -cell* $s(3; 4; 3)$ and *octicosahedric polytope*. 0_{21}, 24-cell, $s(3; 4; 3)$ and the octicosahedric polytope are the medials of α_4, β_4, 24-cell and 600-cell, respectively; see also Sect. 8.3 and Table 8.5 for the notion of *Wythoff Archimedean*.

The $s(3; 4; 3)$ is obtained also by eliminating some 24 vertices of 600-cell (see [Cox73]). 1_{21} is $\frac{1}{2}\gamma_5$; 2_{21} and 3_{21} are *Delaunay polytopes* of the root lattices E_6 and E_7. The skeleton of 4_{21} is the *root graph* of all 240 roots of the root system E_8.

The *pyramid operation* $Pyr(\mathcal{K})$ (respectively, *bipyramid operation* $BPyr(\mathcal{K})$) on a d-dimensional complex $\mathcal{K}$ is the $(d+1)$-dimensional complex with one (respectively, two) new vertices, connected to all vertices of the original complex.

All 92 *Johnson solids*, i.e., regular-faced 3-polytopes were found in [Jo66]. All regular-faced, but not semiregular, d-polytopes, $d \geq 4$ are also known [BlBl80]. This list consists of two infinite families of d-polytopes $(Pyr(\beta_{d-1})$ and $BPyr(\alpha_{d-1}))$, three particular 4-polytopes $(Pyr(Ico)$, $BPyr(Ico)$ and the union of $0_{21} + Pyr(\beta_3)$, where β_3 is a facet of 0_{21}) and, finally, any 4-polytope (except of snub 24-cell), arising from 600-cell by the following special cut of vertices. If E is a subset of the 120 vertices of 600-cell, such that any two vertices in E are not adjacent, then this polytope is the convex hull of all vertices of 600-cell, except those in E.

Conway [Con67] enumerated all *Archimedean 4-polytopes*, i.e., those having a vertex-transitive group of symmetry and whose cells are regular or Archimedean polyhedra and prisms or antiprisms with regular faces. The list consists of:

1. 45 polytopes obtained by Wythoff's kaleidoscope construction from regular 4-polytopes (see Table 8.5 and, more generally, Sect. 8.3);
2. 17 prisms on Platonic, other than Cube, and Archimedean solids (see Table 8.2);
3. prisms on $APrism_m$ for any $m > 3$ (see Conjecture 8.4);
4. a doubly infinite set of 4-polytopes, which are direct products $C_p \times C_q$ of two regular polygons (if one of polygons is a square, then one gets prisms on $Prism_m$) (see Conjecture 8.3);
5. the snub 24-cell $s(3; 4; 3)$ (see Table 8.1);
6. a 4-polytope, called in [Con67] *Grand Antiprism*; it has 100 vertices (all from 600-cell), 300 cells α_3 and 20 cells $APrism_5$ forming two interlocking tubes.

Remark 8.1 The Grand Antiprism has z-vector $(30^{20}, 50^{40}, 90^{20})$. The corresponding intersection vectors are $(1^{10}, 2^{10})$ and $(1^{10}, 2^{20})$ and $(1^{10}, 2^{10}, 4^5, (4, 4)^5)$.

Remark 8.2 Complete information on z-structure of 92-Johnson polyhedra is available from [Dut04]. 25 among them are z-uniform.

Remark 8.3 The number of polytopes, obtained by special cuts is large but finite. The number of polytopes obtained by cutting k vertices is given in Table 8.5 for $1 \leq k \leq 24$. See [DuMy08] for more details.

k	Nr. polytopes						
1	1	7	334380	13	74619659	19	25265
2	7	8	1826415	14	54482049	20	1683
3	39	9	7355498	15	26749384	21	86
4	436	10	21671527	16	8690111	22	9
5	4776	11	46176020	17	1856685	23	1
6	45775	12	70145269	18	263268	24	1

For 24 vertices, there is only one possible special cut, which yields semiregular snub 24-cell. For more than 25 vertices, there is no special cut possible (i.e., the

Table 8.1 z-structure of regular, semiregular, regular-faced d-polytopes, and Conway's 4-polytopes

Dimension	Complex	z-vector	Int. vectors
$d - 1$	d-simplex $\alpha_d = \{3^{d-1}\}$	$(d+1)^{d!/2}$	1^{d+1} if $d \geq 4$
			$(1, 1)^2$ if $d = 3$
$d - 1$	Cross-d-polytope $\beta_d = \{3^{d-2}; 4\}$	$(2d)^{2^{d-2}(d-1)!}$	2^d
2	Dodecahedron = $\{5; 3\}$	10^6	2^5
2	Great Dodecahedron = $\{5; \frac{5}{2}\}$	6^{10}	2^3
2	Petersen graph on P^2	5^6	1^5
3	600-cell = $\{3; 3; 5\}$	30^{240}	2^{15}
3	24-cell = $\{3; 4; 3\}$	12^{48}	2^6
3	Snub 24-cell = $s(3; 4; 3)$	20^{144}	$(1, 1)^4, 2^4, 4$
3	Octicosahedric polytope	45^{480}	$1^{15}, 2^{15}$
3	$0_{21} = Med(\alpha_4)$	15^{12}	$(1, 2)^5$
4	$1_{21} = \frac{1}{2}\gamma_5 = Med(\beta_5)$	12^{240}	$1^8, 2^2$
5	$2_{21} = $ Schläfli polytope (in E_6)	18^{4320}	$1^6, 2^6$
6	$3_{21} = $ Gosset polytope (in E_7)	90^{48384}	$2^{15}, 4^{15}$
7	4_{21} (240 roots of E_8)	$36^{29030400}$	$1^{24}, 4^3$
2	92 Johnson solids	See Remark 8.2	
3	$Pyr(Icosahedron)$	25^{12}	$10, 3^5$
3	$BPyr(Icosahedron)$	40^{12}	$20, 4^5$
3	$0_{21} + Pyr(\beta_3)$	42^6	$(1, 1), (8, 8), (12, 12)$
3	Special cuts of 600-cell	See Remark 8.3	
$d - 1$	$Pyr(\beta_{d-1})$	See Conjecture 8.2	
$d - 1$	$BPyr(\alpha_{d-1})$	See Conjecture 8.2	
3	45 Wythoff Archimedean 4-polytopes	See Table 8.5	
3	17 prisms on Platonic and Archimedean solids	See Table 8.2	
3	Grand Antiprism	See Remark 8.1	
3	$C_p \times C_q$	See Conjecture 8.3	
3	Prisms on $APrism_m$	See Conjecture 8.4	

skeleton of 600-cell has independence number 24, see page 82 of [Mar94]). Due to difficulty of computation and very large size of data, we computed the z-structure of special cuts of 600-cell only up to 3 vertices; see [Dut04].

Remark 8.4 In Table 8.3 note that:

(i) Among those eight polytopes only $\{5; \frac{5}{2}; 5\}$ and $\{\frac{5}{2}; 5; \frac{5}{2}\}$ are self-dual.

(ii) In the case of Great Stellated Dodecahedron $\{\frac{5}{2}; 3\}$, the item h in Table 1 on page 292 of [Cox73] (corresponding to the length of a zigzag) was $\frac{10}{3}$, while

Table 8.2 z-structure of prisms on Platonic and Archimedean solids

Polyhedron P	P		$Prism(P)$	
	z	Int. vectors	z	Int. vectors
Tetrahedron α_3	4^3	$(1,1)^2$	16^6	$(3,3)^2, 4$
Octahedron β_3	6^4	2^3	8^{24}	2^4
Dodecahedron $\{5;3\}$	10^6	2^5	40^{12}	$6^5, 10$
Icosahedron $\{3;5\}$	10^6	2^5	40^{12}	$6^5, 10$
Cuboctahedron	8^6	$0, 2^4$	32^{12}	$6^4, 8$
Icosidodecahedron	10^{12}	$0^6, 2^5$	40^{24}	$6^5, 10$
Truncated Tetrahedron	12^3	$(3,3)^2$	16^{18}	$3^4, 4$ or $(3,3)^2, 4$
Truncated Octahedron	12^6	$2^4, 4$	16^{36}	$2^4, 4^2$
Truncated Cube	18^4	$(2,4)^3$	24^{24}	$2^3, 4^3, 6$ or $(2,4)^3, 6$
Truncated Icosahedron	18^{10}	2^9	24^{60}	$2^9, 6$
Truncated Dodecahedron	30^6	$(2,4)^5$	40^{36}	$2^5, 4^5, 10$ or $(2,4)^5, 10$
Rhombicuboctahedron	12^8	$0, 2^6$	16^{48}	$2^6, 4$
Rhombicosidodecahedron	20^{12}	$0, 2^{10}$	80^{24}	$6^{10}, 20$
Truncated Cuboctahedron	18^8	$2^6, 6$	24^{48}	$2^6, 6^2$
Truncated Icosidodecahedron	30^{12}	$2^{10}, 10$	40^{72}	$2^{10}, 10^2$
Snub Cube	$30^4_{3,0}$	$(4,4)^3$	40^{24}	$2^4, (2,2)^4, 16$
Snub Dodecahedron	$50^6_{5,0}$	$(4,4)^5$	200^{12}	$(12,12)^5, 80$

in Table 8.3, we put the value 10. In fact, our notion is combinatorial, while Coxeters define Petrie polygon as a skew polygon (see Fig. 6.1a on p. 93 of [Cox73]).

Proposition 8.3 *For infinite series of regular polytopes we have:*

(i) $\mathbf{z}(\alpha_d) = (d+1)^{d!/2}$ *with* **Int** $= 1^{d+1}$ *for* $d \geq 4$ *and* $(1,1)^2$ *for* $d = 3$.
(ii) $\mathbf{z}(\beta_d) = (2d)^{2^{d-2}(d-1)!}$ *with* **Int** $= 2^d$.

Proof Both polytopes are z-uniform since they are regular polytopes. In order to know the length of a zigzag, one needs to compute the successive images of a flag under $T = \sigma_d \ldots \sigma_2 \sigma_1$.

Denote by $\{0, \ldots, d\}$ the vertices of α_d. Clearly, the image of the flag $f = (\{0\}, \{0,1\},\ldots,\{0,\ldots,d-1\})$ is $(\{1\},\{1,2\},\ldots,\{1,\ldots,d\})$, i.e., it is the image of f under a cycle of length $d+1$. So, its length is $d+1$ and there is no self-intersection; hence, the z-vector is as in (i). Also, the intersection vector is 1^{d+1} since two different zigzags intersect at most once if $d \geq 4$. The case $d = 3$ is trivial.

Denote by $\pm e_i$ with $1 \leq i \leq d$ the vertices of β_d. Clearly, the image of the flag $f = (\{e_1\},\{e_1, e_2\},\ldots, \{e_1, \ldots e_d\})$ is the flag.

Table 8.3 z-structure of half-d-cubes for $d \leq 13$

Dimension	Half-d-cube	z-vector	Int. vectors
2	$\frac{1}{2}\gamma_3 = \alpha_3$	4^3	$(1,1)^2$
3	$\frac{1}{2}\gamma_4 = \beta_4$	8^{24}	2^4
4	$\frac{1}{2}\gamma_5 = Med(\beta_5)$	12^{240}	$1^8, 2^2$
5	$\frac{1}{2}\gamma_6$	32^{1440}	$2^4, 3^8$
6	$\frac{1}{2}\gamma_7$	120^{6720}	$3^{24}, 12^4$
7	$\frac{1}{2}\gamma_8$	36^{430080}	$2^{12}, 4^3$
8	$\frac{1}{2}\gamma_9$	$84^{3870720}$	$4^6, 5^{12}$
9	$\frac{1}{2}\gamma_{10}$	$192^{38707200}$	$5^{24}, 12^6$
10	$\frac{1}{2}\gamma_{11}$	$216^{851558400}$	$3^{48}, 18^4$
11	$\frac{1}{2}\gamma_{12}$	$160^{30656102400}$	$6^8, 7^{16}$
12	$\frac{1}{2}\gamma_{13}$	$880^{159411732480}$	$7^{80}, 40^8$

Table 8.4 z-structure of nonconvex regular 3- and 4-polytopes (adapted from [Cox73])

Schläfli symbol of P	$|Aut(P)|$	z-vector
$\{\frac{5}{2}; 5\}$	120	6^{10}
$\{\frac{5}{2}; 3\}$	120	10^6
$\{\frac{5}{2}; 5; 3\}$	14400	20^{360}
$\{5; \frac{5}{2}; 5\}$	14400	15^{480}
$\{\frac{5}{2}; 3; 5\}$	14400	12^{600}
$\{\frac{5}{2}; 5; \frac{5}{2}\}$	14400	15^{480}
$\{3; \frac{5}{2}; 5\}$	14400	20^{360}
$\{\frac{5}{2}; 3; 3\}$	14400	30^{240}

$f' = (\{e_2\}, \{e_2, e_3\}, \ldots, \{e_2, \ldots, e_d\}, \{-e_1, e_2, \ldots, e_d\})$. Denote by ϕ the composition of the cycle $(1, \ldots, d)$ on the coordinates with the symmetry $(x_1, \ldots, x_d) \mapsto (-x_1, x_2, \ldots, x_d)$. The order of ϕ is $2d$ and $\phi(f) = f'$. So, all zigzags have length $2d$ and there is no self-intersection. If two zigzags are intersecting, then they, moreover, intersect twice, since $\phi^d = -Id$ and one gets $\mathbf{Int} = 2^d$. □

In Table 8.4 is given the z-structure of half-d-cubes for $d \leq 13$; note that the length of any zigzag there divides $2(d-2)$.

Proposition 8.4 *For half-d-cube it holds that*

(i) *There are $d!2^{d-1}(d-2)$ flags, forming one orbit for $d = 3, 4$ and $d - 2$ orbits for $d \geq 5$.*
(ii) *It is z-uniform.*

Proof Let us write the set of vertices of $\frac{1}{2}\gamma_d$ as $\{S \subset \{1, \ldots, d\}$ with $|S|$ even$\}$. One has $\frac{1}{2}\gamma_3 = \alpha_3$ and $\frac{1}{2}\gamma_4 = \beta_4$, which are regular polytopes and whose structure is known. Therefore, one can assume $d \geq 5$. The list of facets of $\frac{1}{2}\gamma_d$ consists of:

1. $2d$ facets $x_i = 0$ and $x_i = 1$ (those facets are incident to 2^{d-2} vertices of $\frac{1}{2}\gamma_d$, which form a polytope $\frac{1}{2}\gamma_{d-1}$).
2. 2^{d-1} simplex facets generated by vertices $\{S_1, \ldots, S_d\}$ with $|S_i \Delta S_j| = 2$ if $i \neq j$.

From the above list of facets, one can easily deduces the list of i-faces of $\frac{1}{2}\gamma_d$:

1. all $\frac{1}{2}\gamma_i$ with $4 \leq i \leq d - 1$ and
2. all k-sets $\{S_1, \ldots, S_k\}$ with $|S_i \Delta S_j| = 2$ if $i \neq j$.

The first kind of faces is obtained by intersecting hyperplanes $x_l = 0, 1$, while the second is obtained by taking any subset of a simplex face of $\frac{1}{2}\gamma_d$. The symmetry group of $\frac{1}{2}\gamma_d$ has size $2^{d-1}d!$. It is generated by permutations of d coordinates and operation $S \mapsto S_0 \Delta S$ for a fixed $S_0 \in \frac{1}{2}\gamma_d$. There is one orbit of k-dimensional faces if $k \leq 2$ and two orbits, otherwise.

Take a flag $F_0 \subset F_1 \subset \cdots \subset F_{d-1}$. If F_i is a simplex face, then all faces contained in it are also simplexes. Therefore, the orbit, to which a flag belongs, is determined by the greatest index i, for which it is still a simplex. Since $2 \leq i \leq d-1$, this makes $d - 2$ orbits. This yields (ii), since the stabilizer of a flag is trivial.

Let us denote by O_i with $2 \leq i \leq d - 1$, the orbit formed by all flags, whose greatest index is i. One has $\sigma_4(O_2) \subset O_3$ and $\sigma_k(O_2) \subset O_2$ for $k \neq 2$. If $i = d - 1$, then $\sigma_d(O_{d-1}) \subset O_{d-2}$, while $\sigma_k(O_{d-1}) \subset O_{d-1}$ if $k \neq d$. If $2 < i < d$, then one has $\sigma_{i+2}(O_i) \subset O_{i+1}$ and $\sigma_{i+1}(O_i) \subset O_{i-1}$; for other k, one has $\sigma_k(O_i) \subset O_i$.

Recalling $T = \sigma_d\sigma_{d-1}\ldots\sigma_1$, one gets $T(O_i) \subset O_{i-1}$ if $i > 2$ and $T(O_2) \subset O_{d-1}$. So, all orbits of flags are touched by any zigzag of $\frac{1}{2}\gamma_d$. It proves z-uniformity. $\square$

Proposition 8.5 *For $Pyr(\beta_{d-1})$, it holds that*

(i) *there are $(d + 1)(d - 1)!2^{d-2}$ flags partitioned into $d + 1$ orbits;*
(ii) *it is z-uniform.*

Proof Denote by v the vertex, on which we do the pyramid construction. Take a flag $(F_1, \ldots, F_d)$ of $Pyr(\beta_{d-1})$. The sequence of faces $(F_1 \cap \beta_{d-1}, \ldots, F_d \cap \beta_{d-1})$ cannot be a flag for three possible reasons:

1. $F_1 \cap \beta_{d-1} = \emptyset$, it means that $F_1 = \{v\}$.
2. $F_i \cap \beta_{d-1} = F_{i+1} \cap \beta_{d-1}$, it means that $F_{i+1} = conv(F_i, v)$.

3. $F_d \cap \beta_{d-1} = \beta_{d-1}$, it means that $F_d = \beta_{d-1}$.

This implies, since β_{d-1} is regular, $Pyr(\beta_{d-1})$ has the following orbits of flags:

1. O_i, with $1 \leq i \leq d$, being the orbit of flags of $Pyr(\beta_{d-1})$, whose first face containing v is in position i;
2. the orbit O_{d+1} of flags obtained by adding β_{d-1} to a flag of β_{d-1}.

The operator σ_i with $1 \leq i \leq d$, which acts on the flag $(F_1, \ldots, F_d)$ by exchanging the term F_i, acts on the orbit by permuting the orbits O_i and O_{i+1} and leaving the others preserved. Hence, the product T acts on the set of orbits O_i as the cycle $(1, 2, \ldots, d+1)$. So, $Pyr(\beta_{d-1})$ is z-uniform. □

Conjecture 8.2 (checked for $d \leq 10$)

(i) For z-structure of $Pyr(\beta_{d-1})$ it holds:

(i.1) z-vector is:

$$
\begin{cases}
(d^2 - 1)^{(d-2)!2^{d-2}} & \text{for } d \text{ even,} \\
2(d^2 - 1)_{2d-2,0}^{(d-2)!2^{d-3}} & \text{for } d \text{ odd and } d > 3, \\
16_{8,8} & \text{for } d = 3.
\end{cases}
$$

(i.2) Intersection vectors are:

$$
\begin{cases}
(d-1)^{d-1}, 2d - 2 & \text{for } d \text{ even and } d \geq 4, \\
(2d - 2)^{d-1} & \text{for } d \text{ odd.}
\end{cases}
$$

(ii) For z-structure of $BPyr(\alpha_{d-1})$, it holds that

(ii.1) z-vector is:

$$
\begin{cases}
(d^2)^{(d-1)!} & \text{for } d \text{ even and } d \geq 4, \\
(2d^2{}_{2d,0})^{\frac{(d-1)!}{2}} & \text{for } d \text{ odd and } d > 3, \\
18_{6,3} & \text{for } d = 3.
\end{cases}
$$

(ii.2) Intersection vectors are:

$$
\begin{cases}
2d, (d-2)^d & \text{for } d \text{ even and } d > 4, \\
(2d - 4)^d & \text{for } d \text{ odd and } d > 3, \\
8, (2, 2)^2 & \text{for } d = 4.
\end{cases}
$$

Clearly, $Pyr(\beta_2)$ and $BPyr(\alpha_2)$ are square pyramid and dual $Prism_3$, respectively.

Conjecture 8.3 (checked for $p, q \leq 15$) Let t denote $\gcd(p, q)$ and s denote $\frac{pq}{t^2}$. Then for z-structure of $C_p \times C_q$, it holds that

(i) If p, q are both even, then
$z = (2ts)^{6t}$ with $\mathbf{Int} = (0, 2s)^t$ for all zigzags.
(ii) If exactly one of p, q is odd, then
$z = (2ts)^{6t}$ with $\mathbf{Int} = s^{2t}$ for $4t$ zigzags and $\mathbf{Int} = (s, s)^t$ for other $2t$ zigzags.

(iii) If p, q are both odd, then

$\mathbf{z} = (2ts)^{2t}, (4ts)^{2t}$ with **Int** $= (s, s)^t$ *for zigzags of length 2ts and* **Int** $=$ $(2s, 2s)^t$ *for zigzags of length 4ts.*

For any zigzag of *Prism(P)* with $\mathbf{z}(P) = a^b$ and P being Platonic or Archimedean 3-polytope, one has $\mathbf{z} = (\frac{4a}{gcd(a,3)})^{2gcd(a,3)b}$. In general, $\mathbf{z} = (\frac{da}{gcd(a,d-1)})^{2gcd(a,d-1)b}$. Cube is not included in Table 8.2, because the prism on it is just γ_4. The above relation works also for prisms on antiprisms.

Conjecture 8.4 *For z-structure of prism on APrism$_m$ it holds:*

$\mathbf{z} = (\frac{8m}{gcd(m,3)})^{8gcd(m,3)}$ *with* **Int** $= (\frac{2m}{3})^4$ *if* $gcd(m, 3) = 3$ *and, otherwise, two zigzags have* **Int** $= 2m^4$, *two zigzags have* **Int** $= 2m, (2m, 4m)$ *and four zigzags have* **Int** $= 2m^2, 4m.$

Denote by $I(Z_1, Z_2) = (n_I, n_{II})$ the pair of intersection numbers between two zigzags, Z_1 and Z_2, corresponding to intersections of types I and II. Given a map f acting on a complex $\mathcal{K}$ without any fixed face, the *folded complex* $\tilde{\mathcal{K}}$ is defined as the quotient space of $\mathcal{K}$ under f; it is not always an order lattice.

Proposition 8.6 *Let $\mathcal{K}$ be a complex and f a fixed-point free involution on $\mathcal{K}$; then one has:*

(i) *For any zigzag Z of $\mathcal{K}$, such that $f(Z) = Z$, the length and the signature of its image $\tilde{Z}$ in $\tilde{K}$ are the half of the length and the signature, respectively, of Z.*

(ii) *If $Z_2 = f(Z_1)$ with $Z_2 \neq Z_1$, then we put compatible orientation on Z_1 and Z_2. The zigzags Z_1 and Z_2 are mapped to a zigzag $\tilde{Z}$ of $\mathcal{K}$ with its signature being equal to the signature of Z_1 plus $\frac{1}{2}I(Z_1, Z_2)$.*

Concerning intersection vectors, one has:

(i) *Two zigzags of $\mathcal{K}$, which are invariant under f, are mapped to zigzags of $\tilde{\mathcal{K}}$ with halved intersection.*

(ii) *Take an invariant zigzag Z of $\mathcal{K}$ and $Z_2 = f(Z_1)$ two equivalent zigzags of $\mathcal{K}$. They are mapped to $\tilde{Z}$ and $\tilde{Z}'$ and one has $I(\tilde{Z}, \tilde{Z}') = I(Z, Z_1)$.*

(iii) *Take two pairs (Z_1, Z_1') and (Z_2, Z_2') with $Z_i' = f(Z_i)$. They are mapped to $\tilde{Z}_1$ and $\tilde{Z}_2$ and their intersection $I(\tilde{Z}_1, \tilde{Z}_2)$ is equal to $I(Z_1, Z_2) + I(Z_1, Z_2')$.*

For example, Petersen graph, embedded on the projective plane $\mathbb{P}^2$, is a folding of Dodecahedron by central inversion. Another example is a map on the torus $\mathbb{T}^2$, which is folded onto the Klein bottle $\mathbb{K}^2$.

The *folded cube* $\square_d$ is obtained from γ_d by *folding*, i.e., by identifying its opposite faces. Obtained complex is $(d - 1)$-dimensional, like γ_d, but it is not an order lattice; so this complex does not admit a realization as polyhedral complex.

Proposition 8.7 *For $\square_d$ one has* $\mathbf{z} = d^{2^{d-2}(d-1)!}$ *with* **Int** $= 1^d.$

Proof Every zigzag of β_d corresponds to a zigzag of $(\beta_d)^* = \gamma_d$; hence, by the proof of Proposition 8.3, the zigzags of γ_d are centrally symmetric. By applying Theorem 8.6, one obtains $\mathbf{z} = d^{2^{d-2}(d-1)!}$. Furthermore, one can prove easily that zigzags of γ_d have **Int** $= 2^d$; hence, the intersection vector of $\square_d$ is 1^d. $\square$

A $(d-1)$-dimensional complex $\mathcal{K}$ is said to be *of type* $\{3, 4\}$ if every $(d-2)$-dimensional face is contained in 3 or 4 faces of dimension $d-1$. These simplicial complexes are classified in terms of partitions: given such a simplicial complex, there exist a partition $(P_1, \ldots, P_t)$ of $\{1, \ldots, d\}$, such that $\mathcal{K}^*$ is isomorphic to $\Delta_1 \times \Delta_2 \times \cdots \times \Delta_t$ with Δ_i being the simplex of dimension $|P_i|$; see [DDS04] for details.

Conjecture 8.5 (checked up to $d = 8$)

 (i) *A simplicial complex of type* $\{3, 4\}$ *is not z-uniform if and only if the sizes of parts in the corresponding partition are either $(\frac{d}{2}, \frac{d}{2})$, or all even (except simplex).*
    In non-z-uniform case, $\gcd(l_1, l_2) = \min(l_1, l_2)$ for any two lengths of zigzags.
 (ii) *In special case* $\{1, \ldots, \frac{d}{2}\}, \{\frac{d}{2} + 1, \ldots, d\}$ *one has* $\max(l_i) = \frac{d(d+2)}{2}$ *and* $\min(l_i) = d + 2$. *In the other extreme case* $\{1, 2\}, \ldots, \{d-1, d\}$ *one has* $\max(l_i) = 3d$ *and* $\min(l_i) = \frac{3d}{2}$.
 (iii) *For partition* $\{1\}, \{2, \ldots, d\}$ *the simplicial complex of type* $\{3, 4\}$ *is, in fact,* $BPyr(\alpha_{d-1})$.
 (iv) *For partition* $\{1\}, \{2\}, \ldots, \{d-2\}, \{d-1, d\}$ *the simplicial complex of type* $\{3, 4\}$ *has the following z-structure:*
 (iv.1) $\lfloor \frac{d}{2} \rfloor$ *orbits, each zigzag has length $6d$ and intersection vector $(d, 0)^6$ (intersection vectors are $(12, 6)$ for $d = 3$ and $8, (2, 2)^2$ for $d = 4$).*
 (iv.2) *For odd d, all orbits have $2^{d-3}(d-2)!$ zigzags. For even d, one orbit has size* $2^{d-4}(d-2)!$ *and* $\frac{d-2}{2}$ *orbits have size* $2^{d-3}(d-2)!$.

8.3 Wythoff Kaleidoscope Construction

Wythoff construction is defined for any d-dimensional complex $\mathcal{K}$ and nonempty subset V of $\{0, \ldots, d\}$. It was introduced in [Wy07, Cox35].

The set of all *partial flags* $(f_{i_0}, \ldots, f_{i_m})$, with $f_{i_j} \subset f_{i_{j+1}}$ and $i_j \in V$, is the vertex-set of a complex, which we denote by $\mathcal{K}(V)$ and call *Wythoff construction with respect to the complex $\mathcal{K}$ and the set V.*

In general, one has $\mathcal{K}(V) = \mathcal{K}^*(d - V)$ with $d - V$ denoting the set of all $d - i$, $i \in V$. If a complex $\mathcal{K}$ is self-dual, then one has $\mathcal{K}(V) = \mathcal{K}(d - V)$. One has, in general, $\mathcal{K}(\{0\}) = \mathcal{K}$, $\mathcal{K}(\{d\}) = \mathcal{K}^*$ and $\mathcal{K}(\{1\}) = Med(\mathcal{K})$. Dual $\mathcal{K}(\{0, \ldots, n\})$ is a simplicial $(d - 1)$-complex called *order-complex* [Sta97].

It is easy to see that a general d-dimensional complex admits at most $2^{d+1} - 1$ nonisomorph Wythoff constructions, while a self-dual d-dimensional complex admits at most $2^d + 2^{\lceil \frac{d-1}{2} \rceil} - 1$ such nonisomorph constructions. Curiously, in the regular complexes considered, we obtain exactly $2^{d+1} - 1$ and $2^d + 2^{\lceil \frac{d-1}{2} \rceil} - 1$ nonisomorph complexes.

If $\mathcal{K}$ is a 2-dimensional complex, then it is easy to see that $\mathcal{K}(V)$ with $V = \{0\}$, $\{0, 1\}, \{0, 1, 2\}, \{0, 2\}, \{1, 2\}, \{1\}$ and $\{2\}$ correspond, respectively, to following maps:

original map $\mathcal{M}$, Truncated $\mathcal{M}$, Truncated $Med(\mathcal{M})$, $Med(Med(\mathcal{M}))$, Truncated $\mathcal{M}^*$, $Med(\mathcal{M})$ and $\mathcal{M}^*$, i.e., dual $\mathcal{M}$.

Call *Wythoff Archimedean* any Wythoff construction with respect to some regular d-polytope. By applying the Wythoff construction to the three 3-regular Platonic solids (Tetrahedron, Cube and Dodecahedron) one obtains all Archimedean 3-polytopes, except Snub Cube and Snub Dodecahedron; their z-structure is indicated in columns 2 and 3 of Table 8.2. See in Table 8.5 the z-structure of Wythoff Archimedean 4-polytopes.

8.4 Wilson–Lins Triality

In the case of *embedded maps*, i.e., maps on surfaces, flags are triples (v, e, f) with $v \in e \subset f$, where v, e and f are incident vertex, edge, and face, respectively. Denote by a, b and c the three maps σ_1, σ_2 and σ_3. Vertex, edge, and face are identified with the set of flags containing them; so with orbits on flags of the groups $\langle b, c \rangle$, $\langle a, c \rangle$ and $\langle a, b \rangle$. Zigzags were defined in Sect. 8.1 above as circuits of flags $(f_i)_{1 \le i \le l}$ with $f_{i+1} = cbaf_i$. So, they correspond to orbits of the group $\langle ac, b \rangle$.

Given a map $\mathcal{M}$, let $(v_i)_{i \ge 1}$, $(p_j)_{j \ge 1}$ and $(z_k)_{k \ge 1}$ be its v-, p- and z-vectors and let V, F, Z and E be its sets of vertices, faces, zigzags, and edges, respectively. Then $\sum_{i \ge 1} iv_i = \sum_{j \ge 1} jp_j = \sum_{k \ge 1} kz_k = 2|E|$.

One can reconstruct the map from the flag-set and the triple (a, b, c) of operations, such that $a^2 = b^2 = c^2 = (ac)^2 = 1$, acting on it using the representation of vertices, edges, and faces as orbits. In fact, given a set X and fixed-point-free involutions A, B, C on X with $AB = BA$ and $\langle A, B, C \rangle$ transitive on X, the quadruple $(X; A, B, C)$ defines a map M with sets $V(M), E(M), F(M), Z(M)$ of vertices, edges, faces, zigzags being orbit-sets of (acting on X) groups $\langle A, C \rangle$, $\langle A, B \rangle$, $\langle C, B \rangle$, $\langle C, AB \rangle$, respectively. $M = (X; A, B, C)$ is *orientable* if $[\langle A, B, C \rangle : \langle CA, CB \rangle] = 2$.

Usual (or *Poincaré, geometrical*) duality $dual(M)$ interchanges roles of A and B; so, vertices and faces leaving edges, zigzags. The maps (a, b, c) to (a, b, ac) or (ac, b, c) produce the maps called in [Li82] *phial*$(\mathcal{M})$ and *skew*$(\mathcal{M})$. The *skew*(M) (or *Petri dual, Petrial*) interchanges B and AB; so, faces and zigzags. The *phial*(M) (or *Wilson dual, Wilsonial*) interchanges vertices and zigzags. The operations *dual*, *skew*, and *phial* are reflexions. It was found in [Wil79] that the group $\langle dual, skew \rangle$ is $\simeq S_3 \simeq Sym_3$. It was proved in [JoTh87] that there is no other "good" notions of dualities for maps on surfaces than the six ones given in Table 8.6. See also [Fran15].

As the simplest example, the *phial* and *skew* of Tetrahedron are obtained by identifying opposite points of Octahedron and, respectively, of its dual; see Fig. 8.1.

It was proved in [NSZ01] that *skew*(M) of orientable map M is orientable if and only if the skeleton of M is bipartite. We conjecture that, for a bipartite graph embedded in oriented surface, the skew operation preserves the *skeleton* (a graph formed by the vertices and edges) of the map and only reverses orientation of one of the parts of the bipartition. In fact, the orientation of surface induces, for each vertex x of $\mathcal{M}$, a cyclic clockwise order on vertices, to which x is adjacent.

Table 8.5 z-structure of Wythoff Archimedean 4-polytopes

Wythoff 4-polytope	z-vector	Intersection vectors
$\alpha_4 = \alpha_4(\{0\}) = \alpha_4(\{3\})$	5^{12}	$(0, 1)^5$
$\alpha_4(\{0, 1\}) = \alpha_4(\{2, 3\})$	20^{12}	$(0, 4)^5$
$\alpha_4(\{0, 1, 2\}) = \alpha_4(\{1, 2, 3\})$	20^{36}	$(0, 1)^5, (0, 3)^5$ or $(1, 3)^5$
$\alpha_4(\{0, 1, 2, 3\})$	20^{72}	$(0, 2)^{10}$
$\alpha_4(\{0, 1, 3\}) = \alpha_4(\{0, 2, 3\})$	48^{20}	$(0, 5)^6, (3, 15)$
$\alpha_4(\{0, 2\}) = \alpha_4(\{1, 3\})$	45^{12}	$(4, 5)^5$
$\alpha_4(\{0, 3\})$	$10^{12}, 30^{12}$	$(0, 2)^5$ or $(0, 6)^5$
$0_{21} = \alpha_4(\{1\}) = \alpha_4(\{2\})$	15^{12}	$(1, 2)^5$
$\alpha_4(\{1, 2\})$	$10^{12}, 20^{12}$	$(0, 2)^5$ or $(0, 4)^5$
$\beta_4 = \beta_4(\{0\})$	8^{24}	$(0, 2)^4$
$\beta_4(\{0, 1\})$	16^{48}	$(0, 1)^8, (0, 4)^2$
$\beta_4(\{0, 1, 2\}) = 24\text{-cell}(\{0, 1\}) = 24\text{-cell}(\{2, 3\})$	24^{96}	$(0, 2)^9, (0, 6)$
$\beta_4(\{0, 1, 2, 3\})$	32^{144}	$(0, 2)^{16}$
$\beta_4(\{0, 1, 3\})$	64^{48}	$(0, 2)^2, (0, 4)^4, (0, 6)^4, (4, 6)^2$
$\beta_4(\{0, 2\}) = 24\text{-cell}(\{1\}) = 24\text{-cell}(\{2\})$	18^{96}	$(0, 2)^9$
$\beta_4(\{0, 2, 3\})$	64^{48}	$(0, 2)^2, (0, 6)^{10}$
$\beta_4(\{0, 3\})$	$16^{24}, 48^{24}$	$(0, 2)^8$ or $(0, 6)^8$
$24\text{-cell} = \beta_4(\{1\}) = 24\text{-cell}(\{0\}) = 24\text{-cell}(\{3\})$	12^{48}	$(0, 2)^6$
$\beta_4(\{1, 2\})$	$16^{24}, 24^{32}$	$(0, 2)^8$ or $(0, 2)^6, (0, 4)^3$
$\beta_4(\{1, 2, 3\})$	32^{72}	$(0, 2)^4, (2, 4)^4$ or $(0, 2)^8, (0, 4)^4$
$\beta_4(\{1, 3\})$	36^{48}	$(0, 2)^8, (0, 4)^3, (0, 8)$
$\beta_4(\{2\})$	24^{24}	$(0, 2)^4, (0, 4)^4$
$\beta_4(\{2, 3\})$	32^{24}	$(0, 2)^4, (0, 6)^4$
$\gamma_4 = \beta_4(\{3\})$	8^{24}	$(0, 2)^4$
$24\text{-cell}(\{0, 1, 2\}) = 24\text{-cell}(\{1, 2, 3\})$	48^{144}	$(0, 2)^{12}, (0, 4)^6$ or $(0, 2)^6, (2, 4)^6$
$24\text{-cell}(\{0, 1, 2, 3\})$	48^{288}	$(0, 2)^{24}$
$24\text{-cell}(\{0, 1, 3\}) = 24\text{-cell}(\{0, 2, 3\})$	96^{96}	$(0, 4)^6, (0, 6)^7, (4, 6)^3$
$24\text{-cell}(\{0, 2\}) = 24\text{-cell}(\{1, 3\})$	54^{96}	$(0, 12), (0, 2)^3, (0, 4)^6, (0, 6)^2$
$24\text{-cell}(\{0, 3\})$	24^{192}	$(0, 2)^{12}$
$24\text{-cell}(\{1, 2\})$	$24^{48}, 48^{48}$	$(0, 2)^{12}$ or $(0, 4)^{12}$
$600\text{-cell} = 600\text{-cell}(\{0\})$	30^{240}	$(0, 2)^{15}$
$600\text{-cell}(\{0, 1\})$	48^{600}	$(0, 2)^{24}$
$600\text{-cell}(\{0, 1, 2\})$	80^{1080}	$(0, 2)^{40}$
$600\text{-cell}(\{0, 1, 2, 3\})$	120^{1440}	$(0, 2)^{60}$
$600\text{-cell}(\{0, 1, 3\})$	320^{360}	$(0, 4)^{20}, (2, 4)^{10}, (6, 12)^{10}$
$600\text{-cell}(\{0, 2\})$	135^{480}	$(0, 2)^{15}, (0, 3)^{15}, (0, 4)^{15}$
$600\text{-cell}(\{0, 2, 3\})$	192^{600}	$(0, 2)^{30}, (0, 4)^{12}, (2, 12)^6$

(continued)

Table 8.5 (continued)

Wythoff 4-polytope	z-vector	Intersection vectors
600-cell($\{0, 3\}$)	60^{960}	$(0, 2)^{30}$
Octicosahedric polytope = 600-cell($\{1\}$)	45^{480}	$(0, 1)^{15}, (0, 2)^{15}$
600-cell($\{1, 2\}$)	$60^{240}, 80^{360}$	$(0, 2)^{30}$ or $(0, 2)^{20}, (0, 4)^{10}$
600-cell($\{1, 2, 3\}$)	120^{720}	$(0, 2)^{15}, (2, 4)^{15}$
		or $(0, 2)^{30}, (0, 4)^{15}$
600-cell($\{1, 3\}$)	108^{600}	$(0, 2)^{12}, (0, 4)^{6}, (2, 8)^{6}$
600-cell($\{2\}$)	90^{240}	$(0, 2)^{15}, (0, 4)^{15}$
600-cell($\{2, 3\}$)	120^{240}	$(0, 2)^{15}, (0, 6)^{15}$
120-cell = 600-cell($\{3\}$)	30^{240}	$(0, 2)^{15}$

Table 8.6 Wilson–Lins triality

(V, F, Z) $\downarrow$	(a, b, c) $\downarrow$	Notation in [Li82]	Euler characteristic of maps
(V, F, Z)	(a, b, c)	map $\mathcal{M}$	$\chi(\mathcal{M}) = |V| - |E| + |F|$
(F, V, Z)	(c, b, a)	$dual(\mathcal{M}) = \mathcal{M}^*$	$\chi(\mathcal{M})$
(V, Z, F)	(ac, b, c)	$skew(\mathcal{M})$	$\chi_s(\mathcal{M}) = \chi(\mathcal{M}) - |F| + |Z|$
(Z, V, F)	(c, b, ac)	$(skew(\mathcal{M}))^* = phial(\mathcal{M}^*)$	$\chi_s(\mathcal{M})$
(Z, F, V)	(a, b, ac)	$phial(\mathcal{M})$	$\chi_p(\mathcal{M}) = \chi(\mathcal{M}) - |V| + |Z|$
(F, Z, V)	(ac, b, a)	$(phial(\mathcal{M}))^* = skew(\mathcal{M}^*)$	$\chi_p(\mathcal{M})$

Fig. 8.1 Folded (antipodal quotients of) Octahedron and Cube on the projective plane $\mathbb{P}^2$

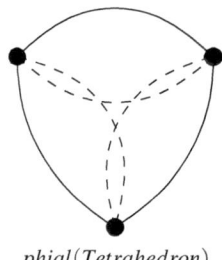 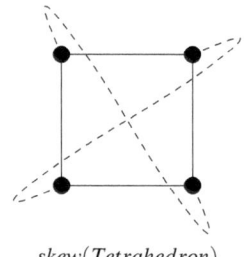

$phial(Tetrahedron)$ $skew(Tetrahedron)$

For example, the complex $skew(Cube)$ is a 3-regular map on the torus $\mathbb{T}^2$ with 8 vertices and 4 hexagonal faces; see Fig. 8.2. The $phial(Cube)$ is a maps on a non-oriented surface of genus 4, i.e., with $\chi = -2$.

As an application of the above, one can count that $\chi_s(Prism_m) = gcd(m, 4) - m$ and $skew(Prism_m)$ is oriented if and only if m is even;

$\chi_p(Prism_m) = 2 + gcd(m, 4) - 2m$ and $phial(Prism_m)$ is non-oriented;
$\chi_s(APrism_m) = 1 + gcd(m, 3) - 2m$ and $skew(APrism_m)$ is non-oriented;
$\chi_p(APrism_m) = 3 + gcd(m, 3) - 2m$ and $phial(APrism_m)$ is oriented.

Fig. 8.2 Two representations of *skew(Cube)*: on the torus and as a Cube with cyclic order of encircled vertices being reversed

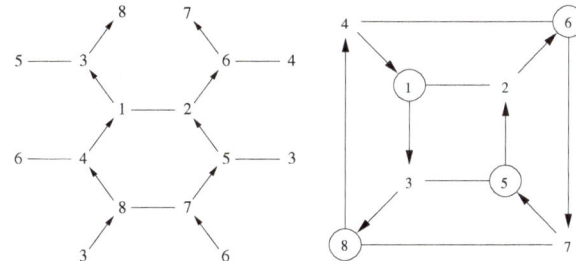

In general, the skew and phial of any map $\mathcal{M}$ cannot be both oriented, because the dual of a bipartite plane graph is not bipartite. Also, $\chi_s(\mathcal{M}) - \chi(\mathcal{M}) = |V| - |F| = 2|V| - 2 - |E|$ if $\mathcal{M}$ is a plane map. So, $\chi_s(\mathcal{M}) - \chi(\mathcal{M}) \geq 2$ if, moreover, $skew(\mathcal{M})$ is oriented, i.e., the skeleton of $\mathcal{M}$ is bipartite.

In [GrWa97], all 14 types of edge-transitive maps were found, using that their *dual* and *skew* maps are edge-transitive also and have the same group (with at most two orbits on V, F, Z).

The *Petri–Coxeter labirinths* are three infinite *skew polyhedra* (i.e., with regular polygonal faces and nonplanar vertex figures) filling 3-space: Petri–Coxeter $\{4; 6|4\}$, $\{6; 4|4\}$, $\{6; 6|3\}$. One can consider them (see below) as projections of the hyperbolic regular tilings $\{4; 6\}$, $\{6; 4\}$, $\{6; 6\}$ on Schwarz minimal P- and D-surfaces of constant negative curvature of genus 3. The petrial (skew) of [4;6]D is [6;6]P and [6;4]D is self-Petrial.

In Table 8.7 are presented several *regular maps* with all vertices of degree 3; see Fig. 1.8. The group of dual Dyck map is denoted by ^4O, because O is a subgroup of index 4 of it; by the same reason, this group (of order 96) is called *tetrakisoctahedral*; it is generated by two elements R, S subject to the relations $R^3 = S^8 = (RS)^2 =$

Table 8.7 z-structure of some regular maps and Goldberg–Coxeter construction $GC_{k,l}$ on them

Regular 3-valent map	Genus	Nr. of vertices	Rotation group	z-vector	$\mathbf{z}(GC_{k,l})/(k^2 + kl + l^2)$
Dodecahedron $= \{5; 3\}$	0	20	$A_5 \simeq PSL(2, 5) = {}^5T$	10^6	10^6 or 6^{10} or 4^{15}
dual Klein map $\{7; 3\}$	3	56	$PSL(2, 7) = {}^7O$	8^{21}	6^{28} or 8^{21}
dual Dyck map $\{8; 3\}$	3	32	^4O	6^{16}	6^{16} or 8^{12}
$\{11; 3\}$	26	220	$PSL(2, 11) = {}^{11}I$	10^{66}	6^{110} or 10^{66} or 12^{55}

$(S^2R^{-1})^3 = 1$. Now, $PSL(2, p)$ for $p = 5, 7, 11$ are denoted by ^5T, ^7O, ^{11}I and called, respectively, *pentakistetrahedral*, *heptakioctohedral*, *undecakisicosahedral*, respectively. A well-known result of Galois is that they are the only ones among all $PSL(2, p)$, which act transitively on less than $p + 1$ elements.

References

[BlBl80]   Blind, G., Blind, R.: Die Konvexen Polytope im $\mathbb{R}^4$, bei denen alle Facetten reguläre Tetraeder sind. Monatshefte für Mathematik **89**, 87–93 (1980)

[BlBl91]   Blind, G., Blind, R.: The semiregular polytopes. Comment. Math. Helv. **66**, 150–154 (1991)

[Con67]   Conway, J.H.: Archimedean, Four-dimensional, polytopes. In: Proceeding of the Colloquium on Convexity, Copenhagen 1965, pp. 38–39. Kobenhavns University Mathematics Institute (1967)

[Cox35]   Coxeter, H.S.M.: Wythoff's construction for uniform polytopes. Proc. London Math. Soc. (2), **38**, 327–339 (1935); reprinted in H.S.M. Coxeter, Twelve geometrical essays, Carbondale, Illinois (1968)

[Cox73]   Coxeter, H.S.M.: Regular Polytopes. Dover Publications, New York (1973)

[Dut04]   Dutour, M.: http://www.liga.ens.fr/dutour/Regular (2004)

[Fran15]   del Río-Francos, M.: Chamfering operation on k-orbit maps. Ars Math. Contemp. **8**, 507–524 (2015)

[DeDu04]   Deza, M., Dutour, M.: Zigzag structure of complexes. Special Issue of SEAMS Math. Bull. **29–2**, 301–320 (2005) (papers/math.CO/0405279 of LANL archive (2004))

[DuMy08]   Dutour Sikirić M., Myrvold, W.: The special cuts of the 600-cell. Beiträge zur Algebra und Geometrie **49**, 269–275 (2008)

[DDS04]   Deza, M., Dutour, M., Shtogrin, M.: On simplicial and cubical complexes with short links. Israel J. Math. **144**, 109–124 (2004)

[Gos00]   Gosset, T.: On the regular and semiregular figures in spaces of n dimensions. Messenger Math. **29**, 43–48 (1900)

[GrWa97]   Graver, J.E., Watkins, M.E.: Locally finite, planar, edge-transitive graphs. Mem. Am. Math. Soc. **126** (1997) (no. 601)

[Jo66]   Johnson, N.W.: Convex polyhedra with regular faces. Can. J. Math. **18**, 169–200 (1966)

[JoTh87]   Jones, G.A., Thornton, G.S.: Operations on maps and outer automorphisms. J. Comb. Theory Ser. B **35**, 93–103 (1987)

[Ke1619]   Kepler, J.: Harmonice Mundi, Opera Omnia, vol. 5. Frankfurt (1864)

[Li82]   Lins, S.: Graph-encoded maps. J. Comb. Theory Ser. B **32**, 171–181 (1982)

[Mar94]   Martini, H.: A hierarchical classification of Euclidean polytopes with regularity properties. In: Bisztriczky, T., McMullen, P., Schneider, R., Ivic Weiss, A. (eds.) Polytopes: Abstract, Convex and Computational, pp. 71–96. Kluwer Academic Publishers, Dordrecht (1994)

[NSZ01]   Nedela, R., Škoviera, M., Zlatoš, A.: Bipartite maps, Petrie duality and exponent groups. Atti Sem. Mat. Fis. Univ. Modena **49**, 109–133 (2001)

[Sta97]   Stanley, R.: Enumerative Combinatorics. Cambridge University Press, Cambridge (1997)

[Wy07]   Wythoff, W.A.: A relation between the polytopes of the C_{600}-family. Koninklijke Akademie van Wetenschappen te Amsterdam, Proceedings of the section of Sciences **20**, 966–970 (1918)

[Wil79]   Wilson, S.E.: Operators over regular maps. Pac. J. Math. **81–2**, 559–568 (1979)

Index

© Springer India 2015
M. Deza et al., *Geometric Structure of Chemistry-Relevant Graphs*,
Forum for Interdisciplinary Mathematics 1, DOI 10.1007/978-81-322-2449-5